Liverpool Marine Biology Committee

Report upon the fauna of Liverpool Bay and the neighboring seas

Liverpool Marine Biology Committee

Report upon the fauna of Liverpool Bay and the neighboring seas

ISBN/EAN: 9783337269081

Printed in Europe, USA, Canada, Australia, Japan

Cover: Foto ©berggeist007 / pixelio.de

More available books at **www.hansebooks.com**

THE FIFTH REPORT

UPON THE

FAUNA OF LIVERPOOL BAY

AND THE

NEIGHBOURING SEAS.

WRITTEN BY THE MEMBERS OF THE

LIVERPOOL MARINE BIOLOGY COMMITTEE

AND OTHER NATURALISTS.

AND EDITED BY

W. A. HERDMAN, D.Sc., F.R.S.,

PROFESSOR OF NATURAL HISTORY IN UNIVERSITY COLLEGE, LIVERPOOL.

WITH TWELVE PLATES.

PRINTED FOR THE
LIVERPOOL MARINE BIOLOGY COMMITTEE,
UNIVERSITY COLLEGE, LIVERPOOL;
BY
C. TINLING & CO.
LIVERPOOL.
1900.

CONTENTS.

iv.

INTRODUCTION.

The present volume, like those already issued, consists of reprints of the Annual Reports (from the ninth to the thirteenth inclusive) on the L.M.B.C. work at Port Erin and elsewhere, and of such papers communicated to the Biological Society of Liverpool as deal with the marine biology of the neighbourhood. It includes four papers on Copepoda by Mr. Isaac Thompson and Mr. Andrew Scott, one on Hydromedusæ by Mr. E. T. Browne, on Turbellaria by Mr. H. L. Jameson, on Actinia by Mr. Clubb, and a note on an abnormal Echinus by Mr. Chadwick.

We have been permitted to reprint in this volume the long list of the marine fauna and flora of the Irish Sea which was prepared by a Committee of the British Association for the Liverpool meeting in 1896.* This list brought together and summed up the results of all the previous marine biological work in the district; and moreover, as references are given after each species to the report or paper in which the occurrence of that species was first recorded, this "B.A. list" forms a useful classified index to the faunistic work of the L.M.B.C. up to the end of 1896. A few mistakes and omissions which were discovered soon after the list was printed will be found noted in the Tenth Annual Report, at p. 102.

A somewhat longer interval than usual has been allowed to elapse before the issue of this volume. That is mainly in consequence of the energies of some of the Committee having been largely diverted from faunistic work during the last couple of years into a new channel, which has

---

* See B.A. Report for 1896, p. 417.

resulted in the production of a new series of publications
(not included in the present volume), "the L.M.B.C.
Memoirs." The following extract from the preface to
the "Memoirs" explains the circumstances and objects
of this new departure : —

"In these twelve years' experience of a Biological
Station (five years at Puffin Island and seven at Port
Erin), where College students and young amateurs formed
a large proportion of the workers, the want has been constantly felt of a series of detailed descriptions of the
structure of certain common typical animals and plants,
chosen as representatives of their groups, and dealt with
by specialists. The same want has probably been felt in
other similar institutions and many College laboratories.

"The objects of the Committee and of the workers at
the Biological Station have hitherto been chiefly faunistic
and speciographic. The work must necessarily be so at
first when opening up a new district. Some of the
workers have published papers on morphological points,
or on embryology and observations on life-histories and
habits: but the majority of the papers in the volumes on
the "Fauna and Flora of Liverpool Bay" have been, as
was intended from the first, occupied with the names and
characteristics and distribution of the many different
kinds of marine plants and animals in our district. And
this faunistic work will still go on. It is far from
finished, and the Committee hope in the future to add
greatly to the records of the Fauna and Flora. But the
papers in the present series are quite distinct from these
previous publications in name, in treatment, and in
purpose. They will be called the "L.M.B.C. Memoirs,"
each will treat of one type, and they will be issued
separately as they are ready, and will be obtainable
Memoir by Memoir as they appear, or later bound up in

convenient volumes. It is hoped that such a series of special studies, written by those who are thoroughly familiar with the forms of which they treat, will be found of value by students of Biology in laboratories and in Marine Stations, and will be welcomed by many others working privately at Marine Natural History.

"It is proposed that the forms selected should, as far as possible, be common L.M.B.C. (Irish Sea) animals and plants of which no adequate account already exists in the text-books. Probably most of the specialists who have taken part in the L.M.B.C. work in the past will prepare accounts of one or more representatives of their groups. The following have already promised their services, and in many cases the Memoir is already far advanced. The first Memoir appeared in October and the second in December, 1899, the third in February and the fourth in April, 1900. Others will follow in rapid succession.

Memoir I. ASCIDIA, W. A. Herdman, 60 pp., 5 Pls.

   II. CARDIUM, J. Johnstone, 92 pp., 7 Pls.

  III. ECHINUS, H. C. Chadwick, 36 pp., 5 Pls

  IV. CODIUM, R. J. H. Gibson and Helen Auld.

     DENDRONOTUS, J. A. Clubb.

     ALCYONIUM, S. J. Hickson.

     PERIDINIANS, G. Murray and F. G. Whitting.

     ZOSTERA, R. J. Harvey Gibson.

     HIMANTHALIA, C. E. Jones.

     DIATOMS, F. E. Weiss.

     FUCUS, J. B. Farmer.

     GIGARTINA, O. V. Darbishire.

     PLAICE, F. J. Cole and J. Johnstone.

     BOTRYLLOIDES, W. A. Herdman.

     CUTTLE-FISH (ELEDONE), W. E. Hoyle.

     OSTRACOD (CYTHERE), Andrew Scott.

PATELLA, J. R. A. Davis and H. J. Fleure.

CALANUS, I. C. Thompson.

ACTINIA, J. A. Clubb.

BUGULA, Laura R. Thornely.

HYDROID, E. T. Browne.

MYXINE, G. B. Howes.

BUCCINUM, M. F. Woodward.

LERNÆA, Andrew Scott.

CALCAREOUS SPONGE, R. Hanitsch.

ARENICOLA, J. H. Ashworth.

ANTEDON, H. C. Chadwick.

OYSTER, W. A. Herdman and J. T. Jenkins.

PORPOISE, A. M. Paterson.

In addition to these, other Memoirs will be arranged for on suitable types, such as *Sagitta* (by Mr. Cole), *Carcinus*, an Amphipod and a Pycnogonid (probably by Dr. A. R. Jackson).

Four Memoirs have now been published,* three Zoological and one Botanical, as follows:—

Memoir I. Ascidia—published in October, 1899, with 60 pp. and five plates.

  ,, II. Cardium—published in December, 1899, with 92 pp., six plates and a map.

  ,, III. Echinus—published in February, 1900, with 36 pp. and five plates.

  ,, IV. Codium—published in April, 1900, with 26 pp. and three plates.

The next two Memoirs, No. V., ALCYONIUM, by Professor Hickson, and No. VI., on the Fish Parasites LERNÆA and LEPEOPHTHEIRUS, by Mr. Andrew Scott, are now in the printer's hands, and will be ready for distribution about the same time as the present volume. LINEUS, by Mr. R.

---

* These can now be obtained from Messrs. Williams & Norgate, 14, Henrietta Street, Covent Garden, London.

C. Punnett, will probably be out early in 1901. Others, such as the OYSTER. SAGITTA, and the PLAICE, are in active preparation.

I would say in conclusion, in issuing this fifth volume of our Reports which brings the record of the L.M.B.C. work down to the end of its sixteenth year, that although our Honorary Treasurer, Mr. Isaac Thompson, wants more money for many purposes, such as a larger laboratory at Port Erin, a fish hatchery, a gas engine and pumps, and a larger boat (and, of course, I agree with him that these needs exist and are pressing), still what I even more earnestly desire to see is *more workers*. The subject of Marine Biology is as wide and as varied as the sea that environs it, and it bristles with problems of every description. The collector and classifier, the observer of habits, the investigator of life-histories, the morphologist studying structure and the physiologist function, the bacteriologist and the chemico-biologist, the most transcendental evolutionist, and even the humble but necessary speciographer, whom it is the fashion now, in some quarters, to despise and deride, will all find in our local Oceanography an ample field for their special researches. Here is work for many minds and many hands for many a year to come.

W. A. HERDMAN.

University College,
Liverpool, November, 1900.

[From Trans. Biol. Soc., L'pool. Vol. X.]

NINTH ANNUAL REPORT of the LIVERPOOL MARINE BIOLOGY COMMITTEE and their BIOLOGICAL STATION at PORT ERIN.

By Professor W. A. HERDMAN, D.Sc., F.R.S.,

DERBY PROFESSOR OF NATURAL HISTORY IN UNIVERSITY COLLEGE, LIVERPOOL;
CHAIRMAN OF THE LIVERPOOL MARINE BIOLOGY COMMITTEE,
AND DIRECTOR OF THE PORT ERIN STATION.

[Read 8th November, 1895.]

THE close of a third triennial period has witnessed the publication (October, 1895) of a Fourth Volume of Collected Reports by our Committee upon the Fauna of Liverpool Bay and the Irish Sea. This volume practically brings the account of the work of the Committee up to the end of the tenth year; the Committee was formed in 1885, the first volume of the "Fauna" was issued in 1886, vol. II. in 1889, vol. III. in 1892, and this fourth volume has now appeared in the autumn of 1895—giving an account of the opening of the Port Erin Station by His Excellency Dr. Spencer Walpole in 1892, and of the investigations conducted in the laboratory and at sea up to the date of our last annual report. The present (ninth) annual report brings on the record to the conclusion of the season 1895.

The Committee have carried on their usual exploring work by means of dredging expeditions and otherwise during the past year. The specimens obtained have been worked up by specialists, and some of the most noteworthy additions to our lists are given below. I am specially indebted to my colleagues on the Committee Mr. Isaac Thompson and Mr. Alfred Walker, to my Assistant Mr. Andrew Scott, and to the various other naturalists who have worked at Port Erin during the year for kind help given me in the preparation of this report.

STATION RECORD.

The following naturalists have worked at the Port Erin Laboratory during the past year :—

DATE.	NAME.	WORK.
February.	I. C. Thompson ...	Copepoda.
—	W. A. Herdman ...	Collecting.
March.	I. C. Thompson	Collecting.
—	W. A. Herdman ...	Collecting.
—	J. C. Sumner	Collecting.
—	R. Boyce ...	Collecting.
—	A. Scott ...	Collecting.
April.	F. G. Baily ⎫	Electric organ of Skate.
—	H. O. Forbes ⎭	
—	W. A. Herdman ...	Tunicata.
—	J. D. F. Gilchrist...	Opisthobranchiata.
—	J. C. Sumner	Collecting.
—	P. M. C. Kermode	Collecting.
—	A. O. Walker ...	Amphipoda.
May.	W. A. Herdman ⎫	Oyster experiments
—	R. Boyce ⎭	
—	P. M. C. Kermode	General.
—	J. C. Sumner ...	Collecting.
June.	I. C. Thompson ..	Copepoda
—	R. Boyce ...	Oysters.
—	A. Leicester ...	Mollusca.
—	W. A. Herdman ...	Tunicata.
—	A. M. Paterson ...	General.
—	W. I. Beaumont ...	Nemertines.
—	T. S. Lea ...	Photographing Algæ.
—	F. W. Gamble	Turbellaria.
—	J. C. Sumner	Collecting.
—	H. O. Forbes	Preserving Animals.
—	A. Scott ...	Copepoda, &c.
July.	W. I. Beaumont ...	Nemertines.
—	J. C. Sumner ...	Collecting.
—	T. S. Lea ...	Photographing Algæ.
August.	H. Meyer Delius ...	Studying fauna

—	J. C. Sumner ...		Collecting.
—	J. D. F. Gilchrist...		Mollusca.
September.	W. A. Herdman ...		General.
—	H. Meyer Delius ...		Studying fauna.
—	J. C. Sumner ...		Collecting.
—	F. W. Gamble		Turbellaria.
—	R. J. Harvey Gibson ...		Marine Algæ.
October.	I. C. Thompson		Copepoda.
—	R. Boyce		General.
—	W. A. Herdman		Collecting.
—	J. C. Sumner		Collecting.
November.	I. C. Thompson		Copepoda.
—	W. A. Herdman		Oyster experiments.
—	J. C. Sumner		General.

The list compares satisfactorily with those of the last
few years. It shows only a slight increase in the number
of workers, but some stayed for long periods, *e.g.*, Mr.
Beaumont from 31st May to July 12th, Mr. Lea from
June 10th to July 4th, and Mr. Delius for the two months
August and September. The work done by the various
naturalists at the station will be referred to further on.

THE AQUARIUM.

There is no new feature to note in connection with
this part of the establishment. About 200 visitors paid
for admission during the season (July and August) when
it was on exhibition, while many other visitors were
taken round the tanks and dishes at other times of the
year when the aquarium was not formally open.

Amongst the animals which have lived in our tanks,
during 1895, may be noted the angler fish (*Lophius
piscatorius*), the top knot (*Zeugopterus punctatus*), the
plumose anemone (*Actinoloba dianthus*) for over six
months, the starfish *Solaster endeca* for over two months,
the wrasse (*Labrus mixtus*), young cod and pollack, and
various other fishes. Amongst other Invertebrates the

Mollusca *Doris tuberculata, Acanthodoris pilosa* and *Aplysia punctata* (the sea hare) spawned freely.

The basement floor of the aquarium was made use of by Professor Boyce and Professor Herdman, during a part of the summer, for some of their investigations on the life conditions and health of the oyster, and the effects of certain diseased conditions. Some further experiments on the same subject are being made in these lower tanks this winter; and the place, from its constant coolness and shade and its proximity to beach and sea, is proving admirably suited for such a purpose.

THE CURATOR.

Mr. J. C. Sumner, from the Royal College of Science, South Kensington, acted as Curator of the Biological Station from March to November, and besides his ordinary routine duties devoted much attention to improving the stock of chemicals and fixing and preserving re-agents in the laboratory. In his report to the Committee he states " I made an inventory of everything in the laboratory, all the apparatus, books, &c.; and then made a list of all the things I thought were wanted. These have been brought or sent to the station from Liverpool during the summer, so that now the place is really very well equipped......the shelves contain all the necessary fixing and killing re-agents, together with some of the commoner stains, &c." (For some faunistic notes from the Curator's diary, see p. 46). The laboratory assistant, William Bridson, is still in the employ of the Committee, at a weekly wage, and continues to give satisfaction.

TEMPERATURE OF THE SEA.

The temperatures of sea and air have not been taken with regularity through the season, but so far as the observations go they entirely corroborate those of the year before last which were printed in full in the Seventh Annual Report. On the whole the sea off Port Erin seems to be of a more equable temperature—slightly warmer in winter and slightly colder in summer—than that of the shallow waters off the Lancashire and Cheshire coasts.

THE PROPOSED SEA-FISH HATCHERY.

It was hoped that before now some arrangement would have been made with the Lancashire Sea-Fisheries Committee or with the Manx Government, or with both these bodies, whereby a Sea-Fish Hatchery for the Irish Sea should be established at Port Erin alongside the Biological Station. We have now advocated that scheme for some years, our Committee has disinterestedly offered to assist by lending tanks for preliminary experiments, by giving the services of their Assistant and in other ways, and successive reports by individuals and committees have shown that the Port Erin site is superior in natural advantages to any of those proposed in Lancashire, Cheshire or North Wales. The water is pure and cool and salt, and the configuration of shore and cliffs is such as to lend itself readily to the formation of a large spawning pond on the beach, while an adjacent creek could easily be converted into a deep vivarium for lobster culture. Our own Committee has no funds to apply to such a purpose, but if any of the powerful bodies interested in promoting the fisheries of the Irish Sea, or in the technical instruction of the fishermen, will provide the money to erect a small experimental hatchery and spawning pond at Port Erin, the Committee is willing to superintend the work for the first few years, and to give time and trouble so as to show what can be done in this locality in the artificial cultivation of food fishes.

DREDGING EXPEDITIONS.

During 1895 the following dredging expeditions in steamers have been carried out, partly with the help, as before, of a Committee of the British Association. This B. A. Committee was re-appointed, for one year, at the Ipswich Meeting, but must bring its labours to a conclusion with a final report to the Liverpool Meeting of the Association in September 1896. With that fuller report in view for next year, the Committee do not propose now to give details* of the separate expeditions, but content themselves with the following brief summary of the occasions and localities :—

I. April 15th, 1895.—Hired steam-trawler "Lady Loch." Localities dredged, to the west and north-west of Port Erin, at depths of 20 to 40 fathoms.

II. April 25th, 1895.—Hired steam-trawler "Lady Loch." Localities dredged, to the west and south of Port Erin, at depths of 30 to 40 fathoms.

At one spot, 6 miles S.E. of Calf Island, 34 fathoms, bottom sand, gravel and shells, such a rich haul was obtained that the trawl-net tore away, and only a small part of the contents was recovered. This contained, however, a number of specimens of a rare shrimp *Pontophilus spinosus*, Leach, along with *Munida rugosa*, *Ebalia tumefacta* and *E. tuberosa*, *Xantho tuberculatus*, *Pandalus brevirostrus*, *Anapagurus hyndmanni*, *Campylaspis* sp., and *Melphidippella macera* amongst Crustaceans, and the following Echinoderma :—*Palmipes membranaceus*, *Porania pulvillus*, *Stichaster roseus*, *Luidea savignii*, *Synapta inhærens*, and other Holothurians. There were also, of course, many Mollusca, Worms, &c., and an unfamiliar Actinian, which

---

* The course of procedure on these expeditions was very fully described last year (Eighth Annual Report, p. 16) and need not be further referred to now.

Professor Haddon considers to be probably his new species *Paraphellia expansa*, previously only known from deep water off the south-west coast of Ireland.

III. June 1st, 1895.—Hired steam-trawler "Lady Loch." Localities dredged, Calf Sound and off S.E. of Isle of Man, at depths of 15 to 20 fathoms.

IV. June 23rd, 1895.—Hired steam-trawler "Rose Ann." Localities dredged, to the W. and N.W. of Peel and Ballaugh, on the "North Bank," at depths about 20 fathoms.

V. August 3rd, 1895.—Lancashire Sea-Fisheries steamer "John Fell." Localities dredged and trawled, Red Wharf Bay and off Point Lynas, on north coast of Anglesey, at depths of 6 to 17 fathoms.

VI. August 19th, 1895.—Steamer "John Fell." Localities dredged, Carnarvon Bay, on south coast of Anglesey, depths 15 to 18 fathoms.

VII. October 27th, 1895.—Hired steamer "Rose Ann." Localities dredged and trawled, off Port Erin and along S.E. side of Island from the Calf Sound to Langness, at depths of 15 to 20 fathoms.

ADDITIONS TO THE FAUNA.

In addition to these "steamer" expeditions there has been frequent dredging and tow-netting from small boats, and a good deal of "shore collecting."

Amongst the more noteworthy animals collected in the district during the year are the following :—

CŒLENTERATA.

Mr. Edward T. Browne has drawn up a list of thirty-four species of MEDUSÆ which are found in the district, and of these the following are specially noteworthy :—

Amphicodon fritillaria (carrying young hydroids in the

umbrellar cavity), *Dysmorphosa minima, Cytæandra areolata* (?), *Lizzia blondina, Laodice calcarata* (new to European seas), and *Eutima insignis.** Mr. Browne writes, in regard to his work at the Biological Station, "Port Erin is a good place for Medusæ. The tide sweeps clean into the bay and I have found very little difference between the pelagic fauna inside the breakwater and that a mile or two off shore. At Plymouth one has to go about two miles outside the Sound before meeting the Channel tide."

Miss L. R. Thornely reports the addition of *Perigonimus repens* and *Tubiclava cornucopia* to the list of Hydroids.

VERMES.

Mr. Beaumont in his recently published report makes the following additions to the list of NEMERTIDA :— *Amphiporus pulcher, A. dissimulans, Tetrastemma flavidum, Prosorhochmus claparédii, Micrura purpurea, M. fasciolata, M. candida,* and *Cerebratulus fuscus.*

During this summer we have dredged from a gravelly bottom, at 10 to 15 fathoms, in two localities near Port Erin, a species of *Polygordius,* either *P. apogon,* M'Intosh, or a new species. It seems to differ from M'Intosh's species in having no eyes. It differs also from all the three species described by Fraipont which have no eyes.

Amongst POLYCHÆTA Mr. Sumner records *Arenicola ecaudata* and *Amphitrite johnstoni* ; Mr. Arnold Watson *Autolytus alexandri* (with egg-sac), and many larval *Pectinaria,* in membranous tubes $\frac{1}{15}$ inch long.

Amongst POLYZOA Miss Thornely reports the rare *Triticella boeckii,* found attached to the prawn *Calocaris macandreæ,* from the deep mud off Port Erin ; also

---

* For Mr. Browne's observations on these and other species see his report in "Fauna of Liverpool Bay," Vol. IV., 1895.

Schizotheca divisa, Mastigophora dutertrei, Schizoporella vulgaris, and *S. cristata, Membranipora solidula, M. nodulosa,* and *M. discreta, Cribrilina gattyæ, Porella minuta, Stomatopora incurvata,* and *Lagenipora socialis* all from the shelly deposit, at 16 to 20 fathoms, to the east of the Calf Sound.

MOLLUSCA.

The following Opisthobranchiata may be mentioned :— *Scaphander lignarius, Pleurobranchus plumula, Oscanius membranaceus, Elysia viridis, Runcina hancocki, Lamellidoris aspera, Jorunna johnstoni, Ægirus punctilucens, Polycera lessoni, Favorinus albus, Cuthona aurantiaca* and *C. nana, Coryphella gracilis, C. lineata* and *C. landsburgi, Facelina drummondi, Eolis arenicola, Cratena concinna, C. amoena* and *C. olivacea, Galvina farrani, G. tricolor* and *G. picta, Embletonia pulchra, Actæonia corrugata, Limapontia nigra, Lomanotus genei,* and a curious little *Doris,* which has been dredged several times in the neighbourhood of Port Erin, and is still unidentified. It may possibly be an unknown species. The Nudibranchs have been chiefly collected and identified by Mr. Beaumont and Mr. Sumner.

CRUSTACEA.

This section is contributed by Mr. I. C. Thompson and Mr. A. O. Walker, Mr. Thompson taking the Copepoda and Mr. Walker the higher forms. The following additional records of Copepoda have, however, been supplied by Mr. Andrew Scott independently of Mr. Thompson's report, viz. :—*Sunaristes paguri,* Hesse ; *Stenhelia reflexa,* T. Scott ; *Laophonte intermedia,* T. S. ; *L. propinqua,* T. and A. S. ; *Cletodes similis,* T. S. ; *Nannopus palustris,* Brady ; *Modiolicola insignis,* Aur. ; and *Dermatomyzon gibberum,* T. and A. S. ; all new to our fauna.

COPEPODA.

In the last report mention was made of a new copepod found by Mr. I. C. Thompson in dredged material taken outside Port Erin at 15 fathoms. This has since been described by Mr. Thompson ("Trans. Liverpool Biol. Soc.," Vol. IX., p. 26, Pls. VI. and VII.) as *Pseudocyclopia stephoides*.

It was by no means easy to decide in which genus to place this well-marked species, as it has strong points of resemblance in common with the three genera, *Pseudocalanus*, *Stephos*, and *Pseudocyclopia*. With *Pseudocyclopia* it agrees in all points excepting in the number of joints in the anterior antennæ, and the primary branch of the posterior antennæ, and as in general appearance and in the first four pairs of swimming feet it strongly resembles *Pseudocyclopia* it was decided provisionally to place it in that genus. Its fifth pair of feet, however, are more like those of *Stephos*. In the "Twelfth Annual Report of the Fishery Board for Scotland" Mr. Thomas Scott added a new species belonging to this genus recently found by him in the Forth area. As the genus *Pseudocyclopia* forms a sort of missing link between the families Calanidæ and Misophriidæ, Mr. Scott has wisely constituted a new family, the Pseudocyclopiidæ, for its reception. The species of *Pseudocyclopia* described by him having respectively sixteen and seventeen joints in the anterior antennæ, he has made that number a family character. The species here described has, however, twenty joints in the anterior antennæ, and as it otherwise agrees in all respects with the family characters of Pseudocyclopiidæ Mr. Thompson suggested that the words "sixteen to seventeen jointed" be altered to "sixteen to twenty jointed" as a character of this new family, with which Mr. Scott at once concurred.

One specimen of *Modiolicola insignis*, Aurivillius, new to the district, was found in the washings of dredged material taken some miles off Peel in June, 1895. This species is known as a messmate within the shell of the "horse mussel" (*Mytilus modiolus*), and has been recorded by Canu ("Les Copep. du Boulon.," p. 238, pl. xxx., fig. 14—20), and more recently by Mr. T. Scott from the Firth of Forth. It had also been found previously by Mr. A. Scott in the "Hole" to the east of the Isle of Man.

The expedition of October 27th in the steamer "Rose Ann" was exceedingly prolific, large numbers of Copepoda being found on the bottom in shallow water (15 to 20 fathoms) although there was very little in the surface tow-nets. From some of the dredged stuff (broken shells, &c.) Mr. Scott obtained 35 species three of which, *Ameira reflexa*, *Idya gracilis* and *Tetragoniceps consimilis*, are new to the district, and eight others seem undescribed forms. Mr. Thompson has obtained already, after only a partial examination of the material, 21 species, of which *Dyspontius brevifurcatus* is new to the British fauna, and a *Cyclopicera* seems new to science. Other rich hauls still remain to be examined by Mr. Thompson.

Mr. A. O. Walker reports the following additions to our lists of the HIGHER CRUSTACEA :—

PODOPHTHALMATA.

Crangon (*Pontophilus*) *spinosus*, Leach.—Several, April 25th, 1895, station 3. Colour: whitish, freckled with reddish-brown on the antennal scales and legs; sparsely on the front and hind margins of thorax and first three abdominal segments, and densely on the last three abdominal segments, hind margin of third and generally front margin of fourth abdominal segments and proximal half of telson and lateral appendages white. Length, 2¼ in.

CUMACEA.

Hemilamprops assimilis, Sars.—Off Galley Head, Co. Cork, November 24th, 1894.

Iphinoe tenella, Sars.—Off North Bank, Peel, June 23rd, 1895. This is new to the British fauna.

Diastylis rugosoides, n.sp.—Galley Head, six males. Very near *D. rugosa* (Sars), from which it differs in the absence of the vertical plica on the carapace, and in the strong dorso-lateral teeth on the first three pleon segments.

ISOPODA.

Cirolana borealis, Lilljeborg.—Galley Head; off Port Erin, April 25th, 1895, station 2.

AMPHIPODA.

A small collection has been made by Mr. R. L. Ascroft, of Lytham, from trawl refuse and a tow-net attached to the trawl beam when working in the southern part of the Irish Sea off Galley Head. The most interesting feature of it is that nearly all the specimens are adult males, in which condition amphipods are less often taken than any other. This may perhaps be attributed to their having been taken late in November, a season at which collectors do not generally dredge.

Parathemisto oblivia, Kröyer.—Galley Head.

Callisoma crenata, Bate.—Galley Head; off Port Erin, April 25th, 1895, station 1.

Hippomedon denticulatus, Bate.—Galley Head.

Orchomenella ciliata, Sars.—Galley Head.

Tryphosites longipes, Bate.—Galley Head.

Lepidepecreum carinatum, Bate.—Galley Head.

Paraphoxus oculatus, Sars.—Off Port Erin, April 25th, 1895, stations 1 and 2.

* Those species marked with a star are now to the British fauna.

Epimeria cornigera, Febr.—Galley Head.

Syrrhoë fimbriata, Stebbing and Robertson.—Off Port April 25th, 1895, station 1.

Leptocheirus hirsutimanus (Bate) = *L. pilosus*, Sars, not Zaddach.—Two miles south-east of Kitterland, 17 fathoms, May 27th, 1894.

Photis longicaudatus, Bate.—Off Port Erin, April 25th, 1895, stations 2 and 3.

*Photis pollex, n.sp.—Colwyn Bay, shore ; Little Orme ; Menai Straits, 5 to 10 fathoms. This species is inter-mediate between *Photis reinhardi* (Kröyer) and *P. tenuicornis* (Sars). The hind margin of the propodos of the second gnathopod in the male is distally produced into a thumb-like process which has its origin much nearer the carpus than in *P. reinhardi*.

Podocerus ocius, Bate.—Sponge *débris*, Port Erin, 1894.

PYCNOGONIDA.

The following rare species found during the year at Port Erin have been named by Mr. G. H. Carpenter, of Dublin :—*Anoplodactylus petiolatus*, Kr., *Ammothea echin-ata*, Hodge, *Nymphon gracile*, Leach, *N. gallicum*, Hoek, *Chætonymphon hirtum*, Kr., and *Pallene producta*, Sars, the last apparently new to Britain.

SOME STATISTICS OF DREDGING RESULTS.

During this year's work we have been paying some attention to the actual numbers of individuals, species, and genera brought up in particular hauls of the dredge or trawl. Our attention has recently been directed to the matter by some statements in Dr. Murray's summary volumes of the "Challenger" Expedition Report which seemed not to be quite in accord with our own experience. Dr. Murray quotes the statistics of the Scottish Sea-

Fisheries Board to show that only 7·3 species of invertebrates and 8·3 species of fishes are captured on the
average by the Fisheries steamer "Garland's" beam
trawl; and he cites as an example of a large and varied
haul from deep water one taken by the "Challenger" at
station 146 in the Southern Ocean, at a depth of 1,375
fathoms, with a 10-foot trawl dragged for at most 2 miles
during at most two hours, when 200 specimens were
captured belonging to fifty-nine genera and seventy-eight
species. Murray then goes on to say: "In depths less
than 50 fathoms, on the other hand, I cannot find in all
my experiments any record of such a variety of organisms
in any single haul, even when using much larger trawls
and dragging over much greater distances." Now our
experience of dredging in the Irish Sea is that quite
ordinary hauls of the dredge or very small trawl (only 4-foot
beam) contain often more specimens, species, and genera
than the special case cited from the "Challenger" results.
On the first of our expeditions after the appearance of Dr.
Murray's volumes we counted the contents of the first
haul of the trawl. The particulars are as follows:—June
23rd, 7 miles W. of Peel, on North Bank, bottom sand
and shells, depth 21 fathoms, trawl 4 feet beam, down for
20 minutes; 232 specimens were counted, but there may
well have been another 100; they belonged to at least 112*
species and 103 genera, a larger number in every respect—
specimens, species, and genera—than in the "Challenger"
haul quoted. The list of these species is here given, and
marine zoologists will see at a glance that it is nothing
out of the way, but a fairly ordinary assemblage of not
uncommon animals such as is frequently met when
dredging in from 15 to 30 fathoms.

* Really an under estimate, several other species have been identified
since from the same haul.

SPONGES:
Reniera, sp.
Halichondria, sp.
Cliona celata
Suberites domuncula
Chalina oculata

CŒLENTERATA:
Dicoryne conferta
Halecium halecinum
Sertularia abietina
Coppinia arcta
Hydrallmania falcata
Campanularia verticillata
Lafoëa dumosa
Antennularia ramosa
Alcyonium digitatum
Virgularia mirabilis
Sarcodictyon catenata
Sagartia, sp.
Adamsia palliata

ECHINODERMATA:
Cucumaria, sp.
Thyone fusus
Asterias rubens
Solaster papposus
Stichaster roseus
Porania pulvillus
Palmipes placenta
Ophiocoma nigra
Ophiothrix fragilis
Amphiura chiajii
Ophioglypha ciliata
O. albida
Echinus sphæra

Spatangus purpureus
Echinocardium cordatum
Brissopsis lyrifera
Echinocyamus pusillus

VERMES:
Nemertes neesii
Chætopterus, sp.
Spirorbis, sp.
Serpula, sp.
Sabella, sp.
Owenia filiformis
Aphrodite aculeata
Polynoe, sp.

CRUSTACEA:
Scalpellum vulgare
Balanus, sp.
Cyclopicera nigripes
Acontiophorus elongatus
Artotrogus magniceps
Dyspontius striatus
Zaus goodsiri
Laophonte thoracica
Stenhelia reflexa
Lichomolgus forficula
Anonyx, sp.
Galathea intermedia
Munida bamffica
Crangon spinosus
Stenorhynchus rostratus
Inachus dorsettensis
Hyas coarctatus
Xantho tuberculatus
Portunus pusillus
Eupagurus bernhardus

E. prideauxii
E. cuanensis
Eurynome aspera
Ebalia tuberosa
POLYZOA:
Pedicellina cernua
Tubulipora, sp.
Crisia cornuta
Cellepora pumicosa and three
  or four undetermined spec-
  ies of Lepralids
Flustra securifrons
Scrupocellaria reptans
Cellularia fistulosa
MOLLUSCA:
Anomia ephippium
Ostrea edulis
Pecten maximus
P. opercularis
P. tigrinus
P. pusio
Mytilus modiolus
Nucula nucleus
Cardium echinatum
Lissocardium norvegicum

Solen pellucidus
Venus gallina
Lyonsia norvegica
Scrobicularia prismatica
Astarte sulcata
Modiolaria marmorata
Saxicava rugosa
Cyprina islandica
Chiton, sp.
Dentalium entale
Emarginula fissura
Velutina lævigata
Turritella terebra
Natica alderi
Fusus antiquus
Aporrhais pes-pelicani
Oscanius membranaceus
Doris, sp.
Coryphella landsburgi
Tritonia plebeia
TUNICATA:
Ascidiella virginea
Styelopsis grossularia
Eugyra glutinans
Botryllus, sp.
B., sp.

The following are two other similar hauls taken with different instruments (dredge and trawl), but both in less than 20 fathoms. On October 27th, 1895, in the steam-trawler "Rose Ann" we counted the first haul of the dredge (2 feet of scraping edge) and the first haul of the small trawl (4 foot beam) with the following results:—

First haul of dredge, across mouth of Port Erin Bay, from Bradda Head towards the Calf Sound, depth 17 fathoms, bottom dead shells, 93 species in 81 genera.

Ascetta primordialis
Cliona celata
Halecium halecinum
Sertularella polyzonias
Hydrallmania falcata
Antennularia antennina
Lafoea dumosa
Obelia, sp.
Asterias rubens
Henricia sanguinolenta
Solaster papposus
Ophiothrix fragilis
Echinus sphæra
Polynoe, sp.
Serpula, sp.
Pomatoceros triqueter
Spirorbis, sp.
Terebella nebulosa
Mucronella peachii
M. ventricosa
Smittia reticulata
Membranipora craticula
M. flemingii
M. dumerilii
M. imbellis
Microporella malusii
M. ciliata
Lichenopora hispida
Schizoporella linearis
S. hyalina
Idmonea serpens
Scrupocellaria reptans
Tubulipora flabellaris
Crisia, sp.

Diastopora suborbicularis
D. patina
Porella concinna
Chorizopora brongniartii
Cellepora costazii
Balanus balanoides
Chthamalus stellatus
Cyclopina gracilis
Misophria pallida
Thalestris clausii
Ectinosoma spinipes
Cyclopicera lata
C. nigripes
Lichomolgus maximus
Dermatomyzon gibberum
Artotrogus magniceps
Zaus goodsiri
Iphimedia obesa
Melita obtusata
Lilljeborgia kinahani
Aora gracilis
Erichthonius abditis
Phtisica marina
Gnathia (Anceus), sp.
Hyas araneus
H. coarctatus
Hippolyte varians
Spirontocaris spinus
Eupagurus bernhardus
Galathea intermedia
Ebalia tuberosa
Portunus, sp.
Achelia echinata
Anomia ephippium

Nucula nucleus
Mytilus modiolus
Pecten opercularis
P. maximus
P. pusio
Saxicava rugosa
Venus lincta
Tapes, sp.
Cyprina islandica
Chiton, sp.
Emarginula fissura
Velutina lævigata
Capulus hungaricus

Buccinum undatum
Fusus antiquus
Trochus cinerarius
Eolis viridis
Polycera quadrilineata
Perophora listeri
Ciona intestinalis
Ascidiella virginea
Ascidia mentula
A. scabra
Styelopsis grossularia
Cynthia morus

The first haul of the small trawl, on the same occasion, off the Halfway Rock, in 18 fathoms yielded 111 species in 93 genera, as follows :—

Leucosolenia coriacea
Suberites domuncula
Cliona celata
Coppinia arcta
Sertularia abietina
Antennularia ramosa
Plumularia, sp.
Sagartia nivea
Sarcodictyon catenata
Palmipes membranaceus
Solaster endeca
Asterias rubens
Henricia sanguinolenta
Porania pulvillus
Echinus sphæra
Echinocyamus pusillus
Lineus marinus
Amphiporus pulcher

Micrura fasciolata
Filograna implexa
Serpula, sp.
Pomatoceros triqueter
Polynoe, sp.
Ætea recta
Scrupocellaria scrupea
S. reptans
Idmonea serpens
Schizotheca fissa
Crisia ramulosa
C. cornuta
Cellepora pumicosa
C. dichotoma
Alcyonidium gelatinosum
A. mytili
Cellaria fistulosa
Membranipora pilosa

M. craticula
M. flemingii
M. imbellis
Chorizopora brongniartii
Smittia trispinosa
S. reticulata
Schizoporella linearis
Mucronella peachii
M. ventricosa
M. coccinea
Porella concinna
Diastopora obelia
Microporella malusii
Hippothoa divaricata
H. distans
Stomatopora johnstoni
Balanus balanoides
Thalestris peltata
Dactylopus flavus
Laophonte spinosa (?)
Ectinosoma atlanticum
Cyclopicera gracilicauda
Lichomolgus liber
Dyspontius striatus
Acontiophorus scutatus
Artotrogus orbicularis
Stenothoe marina
Leucothoe spinicarpa
Amphilochus manudens
Cyproidea brevirostris
Tritæta gibbosa
Cressa dubia
Podocerus cumbrensis
Spirontocaris spinus

Stenorhynchus, sp.
Portunus, sp.
Eupagurus bernhardus
E. cuanensis
Galathea intermedia
G. dispersa
Pandalus annulicornis
Crangon allmani
Xantho tuberculatus
Pycnogonum littorale
Anomia ephippium
Ostrea edulis
Mytilus modiolus
Pecten maximus
P. tigrinus
P. pusio
P. opercularis
Astarte, sp.
Venus casina
Tapes, sp.
Nucula nucleus
Saxicava rugosa
Pectunculus glycimeris
Chiton, sp.
Cyprina islandica
Tectura virginea
Emarginula fissura
Pleurotoma, sp.
Trochus millegranus
T. zizyphinus
Goniodoris nodosa
Amaroucium, sp.
Didemnum, sp.
Leptoclinum maculatum

Botryllus, sp.
Ascidiella virginea
Ascidia mentula
A. plebeia

Corella parallelogramma
Styelopsis grossularia
Cynthia morus

A third haul, on this same occasion (October 27th) gave us, from 16 fathoms, 156 species (see below, p. 34).

In order to get another case, on entirely different ground, not of our own choosing, on the first occasion after the publication of Dr. Murray's volumes when we were out witnessing the trawling observations of the Lancashire Sea-Fisheries steamer "John Fell," I counted, with the help of Mr. Andrew Scott and the men on board, the results of the first haul of the shrimp trawl. It was taken on July 23rd at the mouth of the Mersey estuary, inside the Liverpool Bar, on very unfavourable ground: bottom muddy sand, depth 6 fathoms. The shrimp trawl (1½-inch mesh) was down for 1 hour, and it brought up over seventeen thousand specimens referable to the following thirty-nine species belonging to thirty-four genera:—

Solea vulgaris
Pleuronectes platessa
P.   limanda
Gadus morrhua
G.   æglefinus
G.   merlangus
Clupea spratta
C.   harengus
Trachinus vipera
Agonus cataphractus
Gobius minutus
Raia clavata
R.   maculata
Mytilus edulis

Tellina tenuis
Mactra stultorum
Fusus antiquus
Carcinus mænas
Portunus, sp.
Pagurus bernhardus
Crangon vulgaris
Sacculina, sp.
Amphipoda (undetermined)
Longipedia coronata
Ectinosoma spinipes
Sunaristes paguri
Dactylopus rostratus
Cletodes limicola

Caligus, sp.   *Hydractinia echinata*
Flustra foliacea   *Sertularia abietina*
Aphrodite aculeata   *Hydrallmania falcata*
Pectinaria belgica   *Aurelia aurita*
Nereis, sp.   *Cyanæa*, sp.
Asterias rubens

These numbers have been exceeded on many other hauls in the ordinary course of work by the Fisheries steamer in Liverpool Bay. For example, on this occasion the fish numbered 5,943, and I have records of hauls in which the fish numbered over 20,000. The shrimps probably number as many again, and if the starfishes and other abundant invertebrates are added the total must sometimes reach such enormous numbers as from 45,000 to 50,000 specimens in a single haul of the trawl in shallow water, not including microscopic forms. Hauls such as this are doubtless as prolific of *individuals* as any of those hauls sometimes quoted containing large numbers of specimens (*of a very few species*) of Copepoda and Schizopoda from waters deeper than 50 fathoms,* and are certainly far more prolific in species and genera; while hauls such as the three quoted above under dates June 23rd and October 27th compare favourably as to variety of life, *i.e.*, as to number of *species* and *genera*, with the deep water hauls of the "Challenger" expedition made with a far larger trawl.

On the next occasion, when on board the "John Fell," on our own expedition of August 3rd, two members of this Committee (A. O. Walker and W. A. Herdman) identified the species brought up in the first haul of the

---

* Such as those referred to by Mr. Turbyne in "Nature" for October 24th, 1895, which illustrate an interesting case of distribution of a very few species but do not affect the argument given above for the relative richness, haul for haul, of the shallow as compared with the deeper waters.

trawl (5-inch mesh), taken in Red Wharf Bay, Anglesey, at a depth of 4 to 7 fathoms. They were 78 species, belonging to 67 genera, as follows :—

Solea vulgaris
S. lutea
Pleuronectes platessa
P. limanda
P. flesus
Gadus morrhua
G. æglefinus
G. merlangus
Callionymus lyra
Raia maculata
Fusus antiquus
Buccinum undatum
Natica alderi
Pleurotoma, sp.
Philine, sp.
Eolis, sp.
Polycera quadrilineata
Corbula gibba
Mactra stultorum
Scrobicularia alba
Portunus depurator
Corystes cassivelaunus
Hyas coarctatus
Stenorhynchus phalangium
Eupagurus bernhardus
Crangon vulgaris
Pseudocuma cercaria
Diastylis rathkei
D. spinosa
Balanus balanoides
Paratylus swammerdammii

Harpinia neglecta
Ampelisca lævigata
Monoculodes longimanus
Amphilochus melanops
Pariambus typicus
Achelia echinata
Aphrodite aculeata
Nereis, sp.
Terebella, sp.
(?) *Syllis*, sp.
Serpula, sp.
Spirorbis, sp.
Cellaria fistulosa
Flustra foliacea
Eucratea chelata
Scrupocellaria reptans
Bugula, sp.
Cellepora pumicosa
C. avicularis
Porella compressa
Mucronella peachii
Membranipora membranacea
M. pilosa
Alcyonidium gelatinosum
Vesicularia spinosa
Gemellaria loricata
Lichenopora hispida
Crisia eburnea
C. cornuta
Idmonea serpens
Asterias rubens

Amphiura squamata	*Antennularia ramosa*
Ophioglypha albida	*Coppinia arcta*
Tealia crassicornis	*Sertularella polyzonias*
Alcyonium digitatum	*Sertularia abietina*
Clytia johnstoni	*S. argentea*
Lafoëa dumosa	*Diphasia rosacea*
Hydrallmania falcata	*D. tamarisca*
Halecium halecinum	*Tubularia indivisa*

This was a haul—from very shallow water—which combined mere quantity of life, *i.e.*, number of *individuals*, with variety of life or *number of species and genera*. The ten species of fish were represented by 879 individuals, and we estimated that there were some hundreds of crabs and of starfishes, and some thousands of shrimps. The numbers of the Molluscs, of the hermit-crabs, of *Balanus* and of *Spirorbis* were also very large.

From these statements it is clear that whether it be a question of mere *mass* of life or of *variety* of life, haul for haul, the shallow waters can hold their own against the deep sea, and form in all probability the most prolific zone of life on this globe.

RELATIONS OF GENERA TO SPECIES.

A point which comes out in making complete lists, such as those given above, of the contents of the net on one haul is the relatively large number of genera represented by the species.* In the haul, quoted above, from the expedition of June 23rd, the 112 species were referred to 103 genera; in the haul from the Fisheries steamer on July 23rd, the 39 species obtained belong to 34 genera; on August 3rd, there were 78 species and 67 genera, and

* Dr. Murray, in the Challenger "Summary," notes this fact in the case of deep-sea hauls, but does not seem to recognise its application to shallower waters.

in the two hauls of October 27th there were 93 species in
81 genera and 111 species in 93 genera. Taking a few
instances of particular groups—on August 25th, 1894, the
15 species of Tunicata taken in one haul represented 10
genera; and Mr. Walker reports the following numbers
of species and genera in hauls of the higher Crustacea:—
March, 1893, off Rhos, shallow, 19 species in 18 genera;
May, 1893, off Rhos, two fathoms, 24 species in 21 genera;
July, 1893, off Little Orme, 5 to 10 fathoms, 31 species
in 28 genera; October, 1893, off Little Orme, 4 to 10
fathoms, 41 species in 36 genera; September, 1894, off
Little Orme, shallow, 39 species in 35 genera; and April,
1895, off Port Erin, 34 fathoms, 40 species in 35 genera.*

These figures are particularly interesting in their bearing
on the Darwinian principle that an animal's most potent
enemies are its own close allies.† Is it then the case, as
the above cited instances suggest, that the species of a
genus rarely live together; that if in a haul you get half-
a-dozen species of lamellibranchs, amphipods, or annelids
they will probably belong to as many genera, and if these
genera contain other British species these will probably
occur in some other locality, perhaps on a different bottom,
or at a greater depth? It is obviously necessary to count
the total number of genera and species of the groups in
the local fauna, as known, and compare these with the
numbers obtained in particular hauls. That has been

* These numbers refer to the Higher Crustacea only. There were many
other animals in the hauls.

† "As the species of the same genus usually have, though by no means
invariably, much similarity in habits and constitution, and always in structure, the struggle will generally be more severe between them, if they come
into competition with each other, than between the species of distinct
genera." *Darwin, The Origin of Species*, sixth edition, p. 59; see also
Wallace, Darwinism, second edition, p. 33.

done to some extent with the "Fauna" of Liverpool Bay, and the following instances may be taken as samples.

The known number of species of higher Crustacea in Vol. I. of the "Fauna" (1886) is 90, and these fall into 60 genera. But many species have been added since then, so Mr. Walker has gone over the records up to date (1895), and states that we now know in our local fauna 230 species which belong to 150 genera. This is still much the same proportion as in the former numbers, so we may take it that in our district the genera are to the species as 2 to 3, whereas in the collections quoted from Mr. Walker above the genera are to the species on the average about as 28 to 31, or nearly 7 to 8. It can also be brought out by similar series of numbers that as one extends the area investigated the number of species per genus is increased. In a single haul, in our district as we have seen, the species are to the genera about as 8 to 7, in our local fauna the proportion is about 3 species to 2 genera, while in the much wider area embraced by Sars' Amphipoda of Norway the numbers are 365 species to 157 genera or nearly 5 species to 2 genera. In other words if allied species, taking a large district, were associated together we might expect to find about twice as many species per genus in each haul as we do find.

Mr. Walker has gone carefully into this matter of the proportion of genera to species in our hauls, and in other areas, and from the figures in his notes I extract the following records, in support of the above statement:—

Rhos Bay, 13/5/93, Amphipoda, 16 sp. in 14 genera.

Little Orme, 28/7/93, Amphipoda, 24 sp. in 22 genera.

    ,,   5/10/93,   ,,   29 ,, 24 ,,

The aggregate of the species and genera of Amphipoda in the above three dredgings is 69 species in 60 genera, or an average proportion of 115 : 100. Now the total number of Amphipoda so far recorded in the L. M. B. C. District (about 5000 sq. miles) is 124 species in 78 genera, or in the proportion of 159:100; while G. O. Sars dealing with the Amphipoda of Norway—a very much more extended area—gives 365 species in 157 genera, or in the proportion of 232:100.

To sum up, the proportions of species and genera, in these Amphipoda are :—

Rhos, &c., 1 to 10 fms., 115 sp. in 100 gen., or $1\frac{1}{2}$: 1.

L.M.B.C. dist., to 70 fms., 159 sp. in 100 gen., over $1\frac{1}{2}$:1.

Norway, to 1215 fms., 232 sp. in 100 gen., or $2\frac{1}{3}$: 1.

So, it is clear that as one increases the area and depth investigated the proportion of species to genera in the fauna increases, until e.g., on the coasts of Norway it has become more than twice what it is on the north coast of Wales.

Again, the total number of recorded species of L.M.B.C. Tunicata is 46, and these are referred to 20 genera; while in the case given above (August 25th, 1894) the 12 species taken on one spot represented 10 genera, or, a little over a quarter of the species represented half the genera. These and many other series of statistics in regard to other groups which we might quote, seem to show that a disproportionately large number of genera is represented by the assemblage of species at one spot, which means that closely related species are, as a rule, not found together.

We know of some individual cases, of course, of allied species occurring together, but these do not necessarily affect the general argument. Exceptional cases may be due to some special habit which, although the species are allied forms, prevents them from being severe competitors. It is possible also that sessile animals, such as hydroids and polyzoa, may form a partial exception, and may differ from wandering forms in their method of competition. However Miss Thornely finds that in most gatherings of Polyzoa the species are less than twice the number of genera, while in our " Fauna " the average recorded number is 2·5 species in a genus. Moreover the colonies on dead shells or on stones are generally not only distinct species, but also distinct genera. As many as ten genera are sometimes represented by the Polyzoon colonies on one shell. We are accumulating further statistics on all these points.

THE SUBMARINE DEPOSITS.

In last year's report the nature of the deposits forming on the floor of the Irish Sea was discussed in a preliminary manner. During this season's work the bottom brought up on each occasion has been carefully noted and a sample kept for future study in the Jermyn Street Museum. One point which this collection of deposits from comparatively shallow shore waters seems to bring out is that the classification of submarine deposits into "terrigenous" and "pelagic," which was one of the earliest oceanographic results of the " Challenger " Expedition, and which is still adhered to in the latest " Challenger " volumes as an accepted classification, does not adequately represent or express fully the facts. Terrigenous deposits are supposed to be those formed round continents from the waste of the land, and are stated to contain on the average 68 per cent. of silica. Pelagic deposits are those formed in the open ocean from the shells and other remains of animals and plants living on the surface of the sea above, and they are almost wholly free from quartz particles.

Ordinary coast sands and gravels and muds are undoubted *terrigenous* deposits. Globigerina and Radiolarian oozes are typical *pelagic* deposits. But in our dredgings in the Irish Sea, where the deposits ought all, from their position, to be purely terrigenous, we meet with several distinct varieties of sea-bottom which are not formed mostly from the waste of the land, and do not contain anything like 68 per cent. of silica; but, on the contrary, are formed very largely of the remains of bottom haunting plants and animals, and may contain as little as 17 per cent. of silica.   Such are the nullipore bottoms, and the shell sand and shell gravel met with in some places, and the sand formed of comminuted spines and plates of echinoids which we have found off the Calf Island. These deposits are really much more nearly allied in their nature, and in respect of the kind of rock which they would probably form if consolidated,* to the calcareous oozes amongst pelagic deposits, than they are to terrigenous deposits, and yet they are formed on a continental area close to land in shallow water.   Moreover, although agreeing with the pelagic deposits in being largely organic in origin, they differ in being derived not from surface organisms, but from plants (the nullipores) and animals which lived on the bottom.   Consequently the division of deposits into "terrigenous" and "pelagic" ought to be modified or replaced by the following classification :—

1. *Terrigenous* (Murray's term, restricted)—where the deposit is formed chiefly (say, at least two-thirds, 66 %) of mineral particles derived from the waste of the land.

2. *Neritic†*—where the deposit is largely of organic origin, its calcareous matter being

* They seem closely comparable with the Coralline and Red Crag formations of Suffolk.

† Adopted from Haeckel's term for the zone of shallow water marine fauna, see "*Plankton Studien*," Jena, 1890 ; also Hickson's "Fauna of Deep Sea," 1894.

derived from the shells and other
hard parts of the animals and plants
living on the bottom.

3. *Planktonic* (Murray's pelagic)—where the greater
part of the deposit is formed of the
remains of free-swimming animals
and plants which lived in the sea
above the deposit.

The last group is Murray's "pelagic" unchanged, and
that, there can be no doubt, is a natural group of deposits;
but Murray's "terrigenous" is an unnatural or hetero-
geneous assemblage containing some deposits, such as the
gravel off Bradda Head and the sand of the Liverpool bar,
which are clearly terrigenous in their origin, along with
others such as shelly sands and nullipore deposits which
have much less to do with the waste of the land, but
are very largely organic in origin and formed by animals
and plants *in situ*. The proposal is then to recognise this
latter group of deposits by separating them from the
truly terrigenous under the name "Neritic." Probably
some of the Coral sands described by Murray and Renard
in their Challenger Report on Sea-Deposits would also
fall into this category.

Professor Johannes Walther, of Jena, who has of recent
years been working on the borderlands of geology and
bionomics, in a recent letter to me on my proposed
classification of deposits says :—"Ich meine dass der
Ausdruck *benthonisch* statt *neritisch* richtiger wäre. Denn
es kommt doch bei der Diagnose weniger daraufan,
dass die Ablagerung in der Flachsee, als dass sie durch
benthonisch Organismen (Coralline, Korallen, Echinoder-
men, Mollusken, Bryozoen, etc.) gebildet wird." With
this I can quite agree. I lay most stress on the nature
(bottom plants and animals) of the particles composing

the deposits, and I do not mind much whether they are called Neritic or Benthonic so long as the category is recognised as distinct from terrigenous.

Dr. C. Kohn has kindly analysed for me a series of fair samples of deposits from different parts of the Irish Sea, with the following results :—

	NERITIC.				TERRIGENOUS.			
	A.*	B.†	C.	D.	E.	F.	G.	H.
Silica. SiO².	16·83	46·65	54·84	23·41	84·62	83·06	77·10	78·92
Cal. Carbonate. CaCO³.	79·27	38·45	39 71	59·66	6·38	9·18	9·20	8·59
Residue (other than Silica).	3·90	14·90	6·45	19·93	9·00	13·70	13·70	12·69
	100·00	100·00	100·00	100·00	100·00	100·00	100·00	100·00

The localities and particulars are,—

A. 1 mile off Spanish Head, 16 faths., shell fragments.

B. 1 mile off Calf of Man, 20 faths., shells and spines.

C. 1 mile off Calf of Man, 18 faths., shell sand, spines.

D. 2 miles off Dalby, 15 faths., nullipores.

E. Liverpool Bar, 3 faths., sand.

F. Bahama Bank, 13 faths., muddy sand.

G. King William Bank, 5 faths., coarse sand.

H. North end of " Hole," 28 faths., mud.

It will be noticed that the four terrigenous deposits (sands and muds) all show less than 10 % of calcium carbonate; while the four neritic have all more than 38 %—well over a third—of calcium carbonate, and one (A.) has over 79 %. The silica in these neritic deposits may be less than 17 %,

* Shelly deposit. Contained 1·09 % of small stones not included in analysis.

† Contained 4·82 % of magnesium carbonate, in addition to calcium carbonate.

and does not rise in any to 55 %. In round numbers
it may be said that in these examples the silica makes up
from 20 to 50 % and the calcium carbonate from 40 to
80 %. In all the neritic deposits there are in the residue
small quantities of calcium phosphate, of iron and of
alumina.

In some of these deposits the calcareous matter is
formed almost entirely of *Lithothamnion*. Amongst the
Nullipores from our neritic deposits Professor Harvey
Gibson has identified the following species:—*Lithothamnion polymorphum*, *L. calcareum*, *L. agariciforme*, *L.
fasciculatum*, the variety *fruticulosum*, and *Lithophyllum
lenormandi*.

One of these neritic deposits (A) has its calcareous
matter formed by a large number of animals, belonging
to various groups, in addition to nullipores. One sample
(measuring 7 qts., 1½ pts., and weighing, when dry, 17 lbs.
3¼ ozs.) which I have gone over carefully for the purpose of
identifying the constituent particles contains more or less
fragmentary remains of at least the following 99 species,
all of them forms that leave calcareous remains :—

NULLIPORES :
  Lithothamnion fasciculatum
  L. calcareum
  Lithophyllum lenormandi
ECHINODERMATA :
  Echinus sphæra
  Echinocyamus pusillus
  Echinocardium cordatum
VERMES :
  Serpula, sp.
  Spirorbis, sp.
POLYZOA :†
  Cellaria fistulosa

Cellepora avicularis
C. dichotoma
C. pumicosa
Idmonea serpens
Microporella ciliata
M. malusii
M. violacea
Schizotheca fissa
*S. divisa
Mastigophora hyndmanni
*M. dutertrei
Mucronella peachii
M. variolosa

* New to district.   † Identified by Miss L. R. Thornely.

M. ventricosa
M. coccinea
Do., var. mamillata
Schizoporella auriculata
S. linearis
S. unicornis
S. simplex
*S. vulgaris
S. discoidea
*S. cristata
Membranipora catenularia
M. solidula
M. pilosa
*M. nodulosa
*M. discreta
M. aurita
M. craticula
M. flemingii
Hippothoa distans
Cribrilina annulata
C. punctata
*C. gattyæ
Crisia aculeata
Ætea recta
Amathia lendigera
Porella concinna
Do., var. belli
*P. minuta
Cribrilina radiata
Phylactella collaris
P. labrosum
Micropora coriacea
Chorizopora brongniartii
Diastopora patina

D. obelia
D. suborbicularis
Stomatopora johnstoni
*S. incurvata
Lepralia foliacea
L. pertusa
Lagenipora socialis
Smittia trispinosa
S. reticulata
Lichenopora hispida
CRUSTACEA :
Balanus balanoides
Verruca, sp.
Cancer pagurus
MOLLUSCA :
Anomia ephippium
Lima elliptica
Pecten opercularis
P. pusio
Cardium edule
Venus casina
V. ovata
Nucula nucleus
Mytilus edulis
Saxicava rugosa
Tapes, sp.
Mactra solida
Pectunculus glycimeris
Acmæa testudinalis
A. virginea
Emarginula fissura
Helcion pellucidum
Trochus magus
T. millegranus

T. cinerarius   *Buccinum undatum*
Pleurotoma, sp.   *Capulus hungaricus*
P., sp.   *Cyprœa europea*
Murex erinaceus   *Nassa incrassata*
Phasianella pullus   *Rissoa*, sp.
Natica, sp.   *R.*, sp.

From a bag of this shelly Neritic deposit (A), described above, Mr. Andrew Scott has by careful examination managed to extract the following 36 species of Copepoda, of which 4 are new records for our district and 8 others seem new to science :—*Pseudocyclops obtusatus*, B. & R.; *Ectinosoma sarsii*, Boeck; *E. melaniceps*, Boeck; *E. erythrops*, Brady; *E. gracile*, T. & A. Scott; *Tachidius brevicornis*, Müller; *Stenhelia*, n.sp.; *Stenhelia*, n.sp.; *Ameira longipes*, Brady; *A. longicaudata*, T. Scott; *A. reflexa*, T. Scott; *A. gracile*, n.sp.; *Mesochra macintoshi*, T. & A. S.; *Paramesochra dubia*, T. Scott; *Tetragoniceps consimilis*, T. Scott; *Laophonte thoracica*, Boeck; *L. curticaudata*, Boeck; *Pseudolaophonte aculeata*, n.gen. and n.sp.; *Normanella attenuata*, n.sp.; *Dactylopus stromii*, Baird; *D. tenuiremis*, B. & R.; *D. flavus*, Claus; *D. brevicornis*, Claus; *Thalestris rufocincta*, Norman; *T. peltata*, Boeck; *Harpacticus chelifer*, Müller; *Zaus spinatus*, Goodsir; *Z. goodsiri*, Brady; *Idya gracilis*, T. Scott; *Lichomolgus fucicolus*, Brady; *L. furcillatus*, Thorell; *Dermatomyzon nigripes* (B. & R.); *Ascomyzon thompsoni*, n.sp.; *Acontiophorus scutatus*, B. & R. ; and two other species which have not yet been worked out.

Mr. Thompson has also identified from a sample of the same deposit which he examined a number of the above species, and in addition the following five:—*Porcellidium*, sp., *Ameira attenuata*, *Laophonte spinosa*, *Scutellidium fasciatum*, and *Artotrogus orbicularis*, making 41 species of Copepoda in all.

These 41 species added to the 99 species from the same haul noted on p. 31 and to the following 16 species recorded from the trawl on Oct. 27th, when the haul was taken, make in all 156 species :—*Mytilus modiolus, Pecten tigrinus, Trochus zizyphinus, Fissurella græca, Eulima polita, Pagurus prideauxii, Ophiothrix fragilis, Ophicoma nigra, Adamsia palliata, Sertularia abietina, Antennularia ramosa, Hydrallmania falcata, Tubularia,* sp., *Glycera,* sp., *Amphiporus pulcher, Flustra foliacea.* It ought to be remembered, however, that a good many (by no means all) of the Mollusca and a few of the Polyzoa were dead.

———

Mr. Clement Reid, F.G.S., of the Geological Survey, has examined the samples of deposits which were sent by us to the Jermyn Street Museum, and reports as follows:—

" The series of dredgings examined since the last report is most interesting from a geological point of view. One is again struck by the common occurrence of loose angular stones at places and depths apparently well beyond the reach of any bottom drift—at least beyond the reach of currents likely to move such coarse material. This stony sea-bed is in all probability the result of submarine erosion of glacial deposits. Its occurrence renders comparison between recent marine deposits of these latitudes and Tertiary deposits a task of peculiar difficulty ; for not only is the nature of the true marine sediments masked, but the fauna also must be greatly altered. It is evident that numerous species which need a firm base on which to affix themselves will be encouraged by a stony bottom ; while in a Tertiary deposit, formed under identical conditions, except for the absence of stones, they may be entirely missing, having nothing but dead shells to which to attach themselves.

"Notwithstanding this peculiarity of most of the dredgings, a few samples may well be compared with our Older Pliocene (Coralline Crag). I would particularly draw attention to certain localities where material almost entirely of organic origin has been obtained. Of these perhaps the most interesting are some samples full of *Cellaria fistulosa* (found to the south-east of the Calf Sound, 20 fathoms). They are in many respects strikingly like certain parts of the Coralline Crag. The more ordinary type of Coralline Crag, with its extremely varied polyzoon fauna, we cannot yet match in British seas :* it was probably formed, as the mollusca indicate, in a sea several degrees warmer than ours.

"It was hoped that in the course of these dredgings some light might be thrown on the Tertiary strata under-lying the bed of the Irish Sea, for in the North Sea the dredge occasionally brings up hauls of Tertiary fossils. This expectation has not yet been realised; but possibly, by dredging in the channels where the submarine scour is greatest, such deposits may yet be reached. It is very important to obtain some knowledge of the Tertiary bed of the Irish Sea, for Irish Pleistocene deposits contain a considerable admixture of extinct forms, which may be derived from Tertiary deposits below the sea-level. The Glacial Drift of Aberdeenshire contains Pliocene Volutes and Astartes, derived from some submarine deposit off the Aberdeenshire coast. The so-called 'Middle Glacial Sands' of Norfolk are full of shells which I now believe to be derived from some older deposit, probably beneath the sea."

* See, however, the deposit described on p. 31, where nearly 60 species of Polyzoa are recorded from one haul.—W. A. H.

The important influence of the shore rocks upon the littoral fauna has not been neglected, and lists and observations are accumulating, but that subject must be left over for a fuller discussion in next year's report.

OTHER INVESTIGATIONS.

Several new lines of investigation have been started during the year, and are still in progress. One of these may be called the "larval-attachment inquiry," and consists in sinking in various parts of the bay an apparatus composed of a rope weighted at one end and buoyed at the other, and having a number of slips of glass, slate, wood, &c., attached at equal distances along its length. These ropes are hauled up and examined periodically, and may be expected, when further observations have been taken, to give information as to the times and modes of attachment of the larvæ of various species, and also as to the most suitable substances for particular kinds of larvæ to settle down upon. So far, glass seemed in the early spring (February and March) to be the favourite substance. A surprisingly large number of algæ, compared with the animals, appeared, and nearly all were on the glass slips. Later on, in the summer, Barnacles (*Balanus*) made their appearance in great numbers on the slips of wood and on the wooden buoy at the top of the apparatus, while all the upper part of the rope within a few feet of the surface became covered with algæ. A number of Ascidians (*Ascidiella virginea*) were also found, in August, to have attached themselves to the rope, and seemed to have got as far as possible in between the strands and into the coils of the knots. On the upper pieces of slate, and in one instance on a piece of glass, there were young specimens of the tubicolous Annelid, *Pomatoceros triqueter*, in no case more than $\frac{1}{2}$ to $\frac{3}{4}$ inch in length.

At the end of October another rope, which had been sunk in the bay since June with a bag of oysters, was hauled up, and the upper 4 or 5 feet, much covered with algæ, was removed for examination in the laboratory. It was found to have the following organisms adhering:—

ALGÆ (identified by Prof. Harvey Gibson):

Ceramium rubrum, C. Ag. | *P. nigrescens*, Grev.
C. strictum, Harv. | *P. elongata*, Grev.
C. deslongchampsii, Chauv. | *Dictyota dichotoma*, Lamx.
Chantransia daviesii, Thur. | *Sphacelaria cirrhosa*, C. Ag.
C. virgatula, Thur. | *Enteromorpha clathrata*, J.Ag.
Desmarestia aculeata, Lamx. | *Monostroma witrockii*, Born.
Polysiphonia urceolata, Grev. | Colonies of *Gomphonema*.

ZOOPHYTES (identified by Miss L. R. Thornely):

Obelia geniculata | *Bougainvillea muscus*
O. longissima | *Opercularella lacerata*
Cytia johnstoni |

POLYZOA (Miss Thornely):

Membranipora pilosa | *Scrupocellaria reptans*
Eucratea chelata | *Schizoporella hyalina*

TUNICATA (Miss J. H. Willmer):

Diplosoma gelatinosum | *Ascidiella virginea*

CRUSTACEA (identified by Mr. A. O. Walker):

Hippolyte varians | *Gammarus locusta*
Idotea marina | *Lilljeborgia kinahani*
Hyale nilssonii | *Amphithoe rubricata*
Apherusa bispinosa | *Podocerus falcatus*
Dexamine spinosa | *Caprella acanthifera*.

THE DRIFT BOTTLES AND SURFACE CURRENTS.

In last year's report the scheme for the distribution of drift bottles over the Irish Sea, for the purpose in helping to determine the set of the chief currents, tidal or otherwise, which might influence the movements of fish food and fish embryos, was fully explained. Since then the work has been going on actively, and now at the end of about twelve months one thousand drift bottles in all have been set free. Many have been let out at intervals of ten minutes, or quarter of an hour, or twenty minutes (corresponding to distances of from 3 to 6 miles apart) from the Isle of Man boats when crossing between Liverpool and Douglas—a very convenient line of 75 miles across the middle of the widest part of our area, traversing the "head of the tide" or meeting place of the tidal currents entering by St. George's Channel and the North Channel. Others have been let off from Mr. Alfred Holt's steamers, in going round from Liverpool to Holyhead and in coming down from Greenock. Mr. Dawson on the Fisheries steamer "John Fell" has distributed a number along the coast in various parts of the district, and the Fisheries bailiffs have let off some dozens from their small boats. Other series have been set free at stated intervals during the rise and fall of the tide from the Morecambe Bay Light Vessel in the northern part of our area, north of the "head of the tide;" and, through the kindness of Lieutenant M. Sweny, R.N., a similar periodic distribution has taken place from the Liverpool North-West Light Vessel, to the south of the "head of the tide." Others, finally, have been despatched by Mr. R. L. Ascroft, by Mr. Andrew Scott and by various members of the Committee in other parts of the area from small boats and on our dredging expeditions, in some cases between the Isle of Man and Ireland. Altogether we have pretty well covered this northern area of the Irish Sea in our distribution of floating bottles.

The first bottles, and the printed paper they contained, were described last year. We afterwards adopted a rather larger size of bottle, 8·5 cm. in length ; and, after various postal difficulties and experiments, we hit upon a convenient size and thickness of private post card, which, ready stamped and addressed, and marked with a distinguishing letter or number, is rolled up in its bottle and has printed on its back the following :—

For scientific enquiry into the currents of the Sea.

Whoever finds this is earnestly requested to write distinctly the DATE and LOCALITY, with full particulars, in the space below, and to put the card in the nearest post office.

[No. here]

LOCALITY, where found................................

....

....

DATE, when found.......................................

Name and address of sender.............................

....

[No. here]

The number is marked with blue and black pencils in duplicate on opposite corners of the card, in case of one edge of the card getting worn by moisture ; and the card is so rolled in the bottle that one of these numbers can be read through the glass, in order that a record may be kept of when and where each bottle is set free. Mr. Andrew Scott, Fisheries Assistant at University College, has collated these records with the particulars of finding of those bottles which have been recovered, and I am indebted to him for the details upon which the following statement of results is based.

Altogether out of the 1000 bottles distributed, over 340 or 34 per cent., more than one in three, have been subsequently picked up on the shore, and the paper or post card has been duly filled up and returned to me. I beg to thank the various finders of the bottles for their kindness in filling and posting the cards. They come from various parts of the coast of the Irish Sea—Scotland, England, Wales, Isle of Man, and Ireland. Some of the bottles have gone quite a short distance, having evidently been taken straight ashore by the rising tide; while others have been blown ashore by the wind, *e.g.*, two (post cards 211 and 214) let off near New Brighton stage on 9th October, 1895, the tide ebbing and the wind N.N.W., were found next day near the Red Noses, 1 mile to the west. Others have been carried an unexpected length, *e.g.*, one (No. 35), set free near the Crosby Light Vessel, off Liverpool, at 12.30 p.m., on October 1st, was picked up at Saltcoats, in Ayrshire, on November 7th, having travelled a distance of at least 180 miles* in thirty-seven days; another (H. 20) was set free near the Skerries, Anglesey, on October 6th, and was picked up one mile north of Ardrossan, on November 7th, having travelled 150 miles in thirty-one days; and bottle No. 1, set free at the Liverpool Bar on September 30th, was picked up at Shiskin, Arran, about 165 miles off, on November 12th. On the other hand, a bottle (J. F. 34) set free on November 7th, in the Ribble Estuary, was picked up on November 12th at St. Anne's, having only gone 4 miles.

It may be doubted whether our numbers are sufficiently large to enable us to draw very definite conclusions. It is only by the evidence of large numbers that the vitiating effect of exceptional circumstances, such as an unusual gale, can be eliminated. Prevailing winds, on the other hand, such as would usually affect the drift of surface organisms, are amongst the normally acting causes which we are trying to ascertain. Mr. W. E. Plummer of the Bidston Observatory has kindly given us access to his

* More probably, very much further, as during that time it would certainly be carried backwards and forwards by the tide.

records of weather for the last twelve months, and we have noted opposite the bottles, from whose travels we are drawing any conclusions, an approximate estimate of the wind influences during the period when the bottle may have been at sea. There have been a few rather extraordinary journeys, e.g., one let off in the middle of Port Erin Bay on April 23rd was found at Fleetwood on July 6th ; another let off at Bradda Head on June 3rd was found on Pilling Sands (near Fleetwood) on July 24th.

It is important to notice that the bottles may support one another's evidence, those set free about the same spot often being found in the same locality, e.g., out of a batch of 6 set free off New Brighton, on Oct. 9th, 1895, 5 have come back and all were found at about the same place.

Dr. Fulton, who is conducting a similar inquiry by means of drift bottles, in the North Sea, for the Scottish Fishery Board, writes to me that he is now having large numbers of his bottles returned to him from the Continent, chiefly Schleswig and Jutland. And he draws the conclusion, "There is no doubt that the current goes across, down as far as Norfolk—none of the bottles have been found south of Lincoln and none in Holland—and this will explain the presence of banks and shallows in the south and east and the immense nurseries of immature fish there." No detailed account of these experiments on the Scottish coast has yet appeared, and it will be interesting to compare our results with them at some future period.*

What is already well-known† in regard to the tidal

---

* Since the above was in type an account has been published.

† All the accounts I have had access to seem based upon Admiral Beechey's observations published in the Philosophical Trans. for 1848 and 1851. Admiral Wharton, F.R.S., the present Hydrographer to the Navy, has kindly informed me that Admiral Beechey took his observations by the direct method of anchoring his ship in various places and then observing the direction and force of the tide.

streams or currents in the Irish Sea is that for nearly six hours after low-water at, say, Liverpool, two tidal streams pour into the Irish Sea, the one from the north of Ireland, through the North Channel, and the other from the southward, through St. George's Channel. Parts of the two streams meet and neutralise each other to the west of the Isle of Man, causing the large elliptical area, about 20 miles in diameter and reaching from off Port Erin to Carlingford, where no tidal streams exist, the level of the water merely rising and falling with the tide. The remaining portions of the two tidal streams pass to the east of the Isle of Man and eventually meet along a line extending from Maughold Head into Morecambe Bay. This line is the "head of the tide." During the ebb the above currents are practically reversed, but in running out the southern current is found to bear more over towards the Irish coast.

There is some reason to believe that, as a result of the general drift of the surface waters of the Atlantic and the shape and direction of the openings to the Irish Sea, more water passes out by the North Channel than enters that way, and more water enters by the South (St. George's) Channel than passes back, and that consequently there is, irrespective of the tides, a slow current passing from south to north through our district. The fact that so many of our drift bottles have crossed the "head of the tide" from S. to N., and that of those which have gone out of our district nearly all have gone north to the Clyde Sea-area supports this view, which I learn from Admiral Wharton is *a priori* probable, and which is believed in by some nautical men in the district from their experience of the drift of wreckage.

It may be objected to our observations by means of drift bottles that they are largely influenced by the wind and waves, and are not carried entirely by tidal streams. Well, that is an advantage rather than any objection to the method. For our object is to determine not the tidal currents alone but the resulting effect upon small surface organisms such as floating fish eggs, embryos and fish food, of *all* the factors which can influence their movements, including prevalent winds. The only factors which can vitiate our conclusions are unusual gales or any other quite exceptional occurrences, and the only way to eliminate such influences is (1) to allow for them so far as they are known from the weather reports, and (2) to employ a large number of drift bottles and continue the observations over a considerable time. We have carefully considered the bearing of the weather records, and we think that the large number of bottles we have made use of, during the year, ought to enable us to come to some definite results. Our conclusions so far (November) then are :—

(1) A large number (over 34 %) have been stranded and found and returned, (2) only a small proportion (13 %) have been carried out of our part of the Irish Sea, (3) nearly 12 % have crossed the " head of the tide " showing the influence of wind in carrying floating bodies over from one tidal system into another, (4) most of the bottles set free to the west of the Isle of Man have been carried across to Ireland, only a small number (3·8 %) of them have got round to the eastern side of the Island and been carried ashore on the English coasts, (5) the majority of the bottles set free off Dalby have gone to the Co. Down coast, (6) a considerable number of bottles have been set free over the deep water to the east of the Isle of Man, where our more valuable flat fish spawn, and of those that have been returned the majority had been carried to the Lancashire, Cheshire and Cumberland coasts. So we may reasonably conclude that the embryos

of fish spawning off Dalby would tend to be carried across
to the Irish Coast, while those of fish spawning in the
deep water on this eastern side of the Isle of Man would
go to supply the nurseries in the shallow Lancashire and
Cheshire bays, and very few would be carried altogether
out of the district.

OTHER FAUNISTIC WORK.

Mr. Arnold Watson writes as follows in regard to the
Annelids at which he is working :—" The most interesting
item is probably the capture by Mr. R. L. Ascroft, in
May last year, and subsequently at intervals, of larval
Pectinaria in the waters near Blackpool. The specimens
were taken with a tow-net attached to the beam of a
trawl, and show that in the first instance the animal
secretes a minute free tube of organic matter of somewhat
cellular appearance. This tube is about $\frac{1}{25}$ of an inch
long, $\frac{1}{100}$ of an inch in its widest diameter, tapering to
about $\frac{1}{300}$ at the narrow end. To the wider end of this
membranous tube the worm attaches very minute grains
of sand, course after course, forming a sand tube about
$\frac{1}{50}$ of an inch in external diameter. The length of the
larval worm from tip of tail to the outer margin of the
minute headbristles, or combs, is about $\frac{1}{25}$ of an inch.
Last spring Mr. Ascroft was good enough to send me some
living specimens of these larvæ which, for a few days,
survived their journey, and were very active. At this
stage of the animal's existence a pair of eye-spots are
visible. He also sent me in March last, for identification,
a specimen of *Autolytus alexandri* (with its egg sac)
taken by surface tow-netting in the daytime off the
Bahama Light Ship, near Ramsey, Isle of Man. Hornell
recorded in 1892 a MALE specimen of this worm taken by
tow-netting off Puffin Island, which was the first recorded
from British waters."

Mr. Watson has completed his work on the tube-building habits of *Panthalis oerstedi*, referred to in last report, and his paper on the subject, with two plates, has been published in Vol. IV. of the "Fauna." One of the specimens of *Panthalis* from Port Erin lived in Mr. Watson's Aquarium at Sheffield from September 30th, 1894, to October 8th, 1895, when as it seemed ailing he killed it with corrosive.

Dr. J. D. F. Gilchrist who has paid several short visits to the Biological Station, and worked there for some time at Easter, has sent in the following report upon his work to the Director :—" During my stay at the Marine Station at Port Erin, I was chiefly concerned with the Mollusca, but found the frequent shore collecting and dredging excursions very profitable for general work. *Aplysia* was found in abundance by dredging and I took this opportunity of trying various methods of killing the animal in an expanded condition. After trying several, I found the following the only method which could be depended on with certainty. A few drops of a 5 % solution of cocain were mixed with the water in which the Aplysias were. After a time they expanded fully. They were then left in the solution (12 hours or more) till no contraction took place when removed and put into weak alcohol. If contraction took place they would be put back into the cocain solution when they again expanded. This was repeated till no contraction took place, when they could, after a time, be put into stronger alcohol. Other methods though simpler, and not so tedious, were less dependable and at best gave a somewhat abnormally inflated appearance.

"At Prof. Herdman's suggestion a solution of formol was tried as a preserving fluid for *Aplysia* and *Pleurobranchus*. In both cases a considerable amount of colouring matter was dissolved out of the integument and stained the surrounding fluid of a reddish colour.

"A series of experiments on the method of feeding in Lamellibranchs was begun, to show in what manner the gills exercised the function of collecting food material and the labial palps of discriminative selection of food particles. I hope to be able to give the results of these experiments after further observations.

"I also procured at the Station specimens of Opisthobranchs which will form material for future work."

The Rev. T. S. Lea (who has kindly presented a large ordinance map, 6 inches to the mile, of the S.W. of the Isle of Man, to the Biological Station) has continued this summer his series of observations upon the zones of algæ on the shore, and has taken a number of photographs of species *in situ* on the rocks and in pools.

The work done by Mr. Browne and by Mr. Beaumont at the Station is sufficiently dealt with in their reports, upon the Medusæ and the Nemertea respectively, which have now been published in Vol. IV. of the "Fauna." Mr. Walker and Mr. Thompson have discussed the results of their work on Crustacea some pages back (p. 10); and the investigations of Prof. Boyce and Prof. Herdman on oysters under various normal and abnormal conditions is work of a very special nature—partly bacteriological and partly experimental—which is still in progress, and will be reported upon in full to the British Association and to the Lancashire Sea-Fisheries Committee.

In a brief report from the Curator giving an outline of the work of the summer the following observations which seem worthy of permanent record occur:—"The dredgings in May produced amongst other nudibranchs *Cuthona arenicola*, new to the fauna round the Isle of Man; and two Polychætes, which though common here have not, as yet, been recorded in the L. M. B. C. lists, were found on the shore—these were *Amphitrite johnstoni* and *Arenicola ecaudata*. The latter species seems to take the place of *A. marina* amongst stones and muddy shingle where it is invariably found, while *A. marina* is confined to the sand...The dredgings in June brought to light several interesting animals, some of the more important finds being :—*Cratena olivacea*, new to the Isle of Man, and which has proved to be not uncommon in the upper Coralline zone off Port Erin ; *Embletonia pulchra*, new to the district, which during June and July appeared in almost every dredging that was taken ; *Coryphella landsburgi*, new to the Isle of Man, taken several times ; *Oscanius membranaceus*, dredged in 15 fathoms off Port Erin ; and *Eolis concinna*. At Whitsuntide, *Polygordius* was dredged from a gravel bottom off Bradda Head, and it has since turned up in two different localities, from similar ground... About this time of the year (July) and for the rest of the summer the bay was full of young fish known to the fishermen as "Gilpins." These are chiefly young cod and pollach—mostly the latter. Some were caught and put into one the tanks, several are still alive and have grown considerably since their capture...Towards the end of the month the annual cleaning of the buoy took place. This year there were several tubes and worms of what is probably *Sabella penicillus*...There were, as before, great quantities of Caprellids, *Ascidiella virginea* and *Ciona intestinalis* were present, along with the nudibranchs *Facclina drummondi*, *Cuthona aurantiaca*, and *Dendronotus arborescens*...During the spring months, until towards the middle of June, *Aplysia punctata* was one of the commonest animals in the bay. It came up in quantities in the dredge, it was to be found commonly on the shore amongst rocks, and after a westerly wind the shore was covered with masses of its spawn. After June *Aplysia* almost entirely disappeared; and was not found again until the end of September...The following common animals could usually be supplied alive to laboratories and museums at any time without much delay :—*Actinia mesembryanthemum, Tealia crassicornis, Bunodes gemmaceus, Actinoloba dianthus, Alcyonium digitatum, Echinus esculentus, E. miliaris, Ophiothrix fragilis, Ophiocoma nigra, Ophiura ciliaris, Arenicola marina* and *A. ecaudata, Nereis pelagica, Pecten maximus* and *P. opercularis, Doris tuberculata, Aplysia punctata* (spring), *Cancer pagurus, Carcinus moenas, Nephrops norvegicus, Galathea squamifera.*"

There are of course many other animals, both common species and rarer forms, which could be obtained alive by giving a little notice, or preserved in spirit, on applying to the Hon. Director, at University College, Liverpool.

The little reference library at the Station is growing gradually, but is still badly in want of many common books and pamphlets. Any works on Marine Zoology, on British Animals, or on the structure and development of marine invertebrates will be thankfully received. The Committee are much indebted to Prof. G. B. Howes who has kindly presented to the Station a series of 7 volumes of collected fishery papers—the result of the Fisheries Exhibition of 1883. Other books and pamphlets have been received from members of the Committee, and the following books have been purchased during the year :— Baird's British Entomostraca, Johnston's British Museum Catalogue of Worms, and Jeffrey's British Conchology (5 vols.).

PUBLICATIONS.

Since the last Annual Report, we have issued Vol. IV. of the "Fauna of Liverpool Bay." It is the largest volume of the series and contains about 475 pages and 53 plates, several of which are coloured. In addition to the reports and papers which had been already announced as forthcoming in this volume, Vol. IV. contains a note upon the yellow variety of *Sarcodictyon* by Prof. Herdman, a paper on the structure of the cerata of *Dendronotus* by Mr. J. A. Clubb, a revision of the Amphipoda of the L. M. B. C. District by Mr. Walker, and a supplementary report on the Port Erin Nemertines by Mr. Beaumont.

It ought to be noticed that although the primary objects of the Committee were originally faunistic and speciographic, yet observations on habits and life-histories, and bionomics in general, have not been neglected ; and now some of our papers in this Vol. IV., such as Mr. Chadwick's on the Vascular Systems of the Starfishes, and Mr. Clubb's on the Cerata of Nudibranchs, are coming to deal with purely structural and morphological questions.

The other Reports in this volume deal, some of them— such as Mr. Gamble's on Turbellaria, Mr. Beaumont's on Nemertea, and Mr. Browne's on Medusæ—with fresh groups of animals which had not been adequately discussed in the previous volumes ; while others, such as Mr. Thompson's and Mr. Walker's reports, are welcome revisions of these authors' own previous work on the Crustacea. Dr. Hanitsch has furnished us with a paper on the Classification and Nomenclature of British Sponges, which it may be said does not come strictly within the scope of the L.M.B.C. Reports. Still the subject matter is of such importance to anyone working systematically at our sponge fauna, and the treatment seems so well adapted to render the lists an indispensable working addition to Bowerbank's Monograph, that I had no hesitation in asking Dr. Hanitsch to allow the paper to be included in our series of reports upon the Fauna of Liverpool Bay.

As to the future, there are a number of reports upon groups, and other pieces of work, in progress. The " List of Fishes " is still in hand. Mr. Andrew Scott has undertaken to collect and report upon the Ostracoda, Dr. Hurst has still charge of our Pycnogonida; while Prof. Boyce and Prof. Herdman are engaged on an extensive investigation on Oysters in healthy and in diseased conditions which has been partly laid before the British Association, but ought to be published in full next year after some further series of observations and experiments have been made.

The Infusoria of all kinds, some of the parasitic groups of Crustacea, the marine Rotifera, and some of the lower worms are still not allotted to workers; while there is plenty for many hands to do in working out the detailed distribution of genera and species, and in tabulating and discussing the results of dredging in various depths and localities.

There is no need to dwell upon the large number of species now recorded, and the additions that have been made by our explorations both to the British fauna and to science; such results, though very necessary, are no longer the sole, perhaps not even the chief objects which the Committee have in view. I think all who are engaged in this L. M. B. C. work feel that it is growing steadily under their hands in every direction. Not only are there many animals and whole groups of animals in our sea awaiting examination and record, but there are many points of view, the speciographic, distributional, anatomical, physiological, embryological, bionomical and others, from which even the best known forms would well repay further and more detailed investigation; and wider problems such as the association of animals together on particular sea-bottoms and at particular depths, and other questions of bionomics and of oceanography—some of them having important bearings upon Geology and upon Fishery questions—are now opening up before us and pressing for solution.

We are a small body, the Naturalists of Liverpool, our laboratory at Port Erin is a modest establishment with but scanty equipment, we have no State, County or Municipal subsidies, and our available funds (private subscriptions) are barely sufficient for the necessary expenses of steamer and apparatus in our explorations, and for the publication of our results ; but fortunately there is no lack of work for us to do, work which is interesting in the doing, and work which, if we seek it earnestly and do it honestly, we cannot but believe will be of value to science, and may, through its connection with the fishing industries, be of direct benefit to mankind.

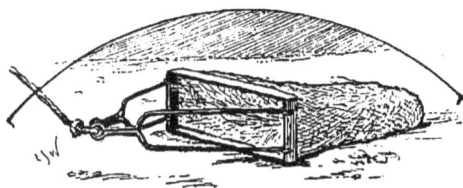

APPENDIX A.

LIVERPOOL MARINE BIOLOGICAL STATION
at PORT ERIN.

REGULATIONS.

I.—This Biological Station is under the control of the Liverpool Marine Biology Committee, the executive of which consists of the Hon. Director (Prof. Herdman, F.R.S.) and the Hon. Treasurer (Mr. I. C. Thompson, F.L.S.).

II.—In the absence of the Director, and of all other members of the Committee, the Station is under the temporary control of the Resident Curator or Laboratory Assistant, who will keep the keys, and will decide, in the event of any difficulty, which places are to be occupied by workers, and how the tanks, collecting apparatus, &c., are to be employed.

III.—The Resident Assistant will be ready at all reasonable hours and within reasonable limits to give assistance to workers at the Station, and to do his best to supply them with material for their investigations.

IV.—Visitors will be admitted, on payment of a small specified charge, to see the Aquarium and the Station, so long as it is found not to interfere with the scientific work.

V.—Those who are entitled to work in the Station, when there is room, and after formal application to the Director, are:—(1) Annual subscribers of one guinea or upwards to the funds (each guinea subscribed entitling to the use of a work place for four weeks), and (2) others who are not annual subscribers, but who pay the Treasurer

10s. per week for the accommodation and privileges. Institutions, such as Colleges and Museums, may become subscribers in order that a work place may be at the disposal of their staff for a certain period annually : a subscription of two guineas will secure a work place for six weeks in the year, a subscription of five guineas for four months, and a subscription of £10 for the whole year.

VI.—Workers at the Station can always find comfortable and convenient quarters at the closely adjacent Bellevue Hotel ; but lodgings can readily be had by those who prefer them.

VII.—Each worker is entitled to a work place opposite a window in the Laboratory, and may make use of the microscopes, reagents, and other apparatus, and of the boats, dredges, tow-nets, &c., so far as is compatible with the claims of other workers and with the routine work of the Station.

VIII.—Each worker will be allowed to use one pint of methylated spirit per week, free. Any further amount required must be paid for. All dishes, jars, bottles, tubes, and other glass may be used freely, but must not be taken away from the laboratory. If any workers desire to make, preserve, and take away collections of marine animals and plants, they must make special arrangements with the Director or Treasurer in regard to bottles and preservatives. Although workers in the Station are free to make their own collections at Port Erin, it must be clearly understood that (as in other Biological Stations) no specimens must be taken for such purposes from the laboratory stock, nor from the Aquarium tanks, nor from the steam-boat dredging expeditions, as these specimens are the property of the Committee. The specimens in the Laboratory stock are preserved for sale, the animals in the tanks are for the instruction of visitors to the

Aquarium, and as all the expenses of steam-boat dredging expeditions are defrayed by the Committee the specimens obtained on these occasions must be retained by the Committee (*a*) for the use of the specialists working at the Fauna of Liverpool Bay, (*b*) to replenish the tanks, and (*c*) to add to the stock of duplicate animals for sale from the Laboratory.

IX.—Each worker at the Station is expected to lay a paper on some of his results—or at least a short report upon his work—before the Biological Society of Liverpool during the current or the following session.

X.—All subscriptions, payments, and other communications relating to finance, should be sent to the Hon. Treasurer, Mr. I. C. Thompson, F.L.S., 19, Waverley Road, Liverpool. Applications for permission to work at the Station, or for specimens, or any communications in regard to the scientific work should be made to Professor Herdman, F.R.S., University College, Liverpool.

APPENDIX B.

SUBSCRIPTIONS and DONATIONS.

	Subscriptions.			Donations.		
	£	s.	d.	£	s.	d.
Ayre, John W., Ripponden, Halifax ...	1	1	0	—		
Banks, Prof. W. Mitchell, 28, Rodney-st.	2	2	0	—		
Beaumont, W. I., Cambridge	2	2	0	—		
Bickersteth, Dr., 2, Rodney-st.	2	2	0	—		
Boulnois, H. P., 7, Devonshire-rd. ...	1	1	0	—		
Brown, Prof. J. Campbell, Univ. Coll....	1	1	0	—		
Browne, Edward T., B.A., 141, Uxbridge-road, Shepherd's Bush, London ...	1	1	0	—		
Boyce, Prof., University College ...	1	1	0	—		
Caton, Dr., 31, Rodney-street ...	—			1	1	0
Clague, Dr., Castletown, Isle of Man ...	1	1	0	—		
Clague,Thomas, Bellevue Hotel, Port Erin	1	1	0	—		
Comber,Thomas,J.P.,Leighton,Parkgate	1	1	0	—		
Crellin, John C., J.P., Ballachurry, Andreas, Isle of Man	0	10	6	—		
Darbishire,R.D.,Victoria-pk.,Manchester	1	1	0	4	4	0
Dawkins, Professor W. Boyd, Owens College, Manchester...	1	1	0	—		
Derby, Earl of, Knowsley ...	5	0	0	—		
Delius, W. Meyer, Hamburg	2	2	0	—		
Dumergue, A. F., 7 Montpellier-terrace	0	10	6	—		
Gair, H. W., Smithdown-rd., Wavertree	2	2	0	—		
Gamble,Col.C.B.,Windlehurst,St.Helens	2	0	0	—		
Gamble,F.W.,OwensCollege,Manchester	1	1	0	—		
Gaskell, Frank, Woolton Wood... ...	1	1	0	—		
Gaskell, Holbrook, J.P., Woolton Wood	1	1	0	—		
Gell, James S., High Bailiff of Castletown	1	1	0	—		
Gibson, Prof. R. J. H., 41, Sydenham-av.	1	1	0	—		

Gifford, J., Whitehouse-terrace, Edin. ...	1	0	0	—		
Gilchrist, Dr. J. D. F., Edinburgh Univ.	1	1	0	—		
Glynn, Dr., 62, Rodney-street	2	2	0	—		
Greening, Linnæus, 5, Wilson Patten-st., Warrington	1	1	0	—		
Gotch, Prof., Museum, Oxford ...	1	1	0	—		
Halls, W. J., 35, Lord-street	1	1	0	—		
Henderson, W. G., Liverpool Union Bank	1	1	0	—		
Herdman, Prof., University College ...	2	2	0	—		
Holder, Thos., 1, Clarendon-buildings, Tithebarn-street	1	1	0	—		
Holland, Walter, Mossley Hill-road ...	2	2	0	—		
Holt, Alfred, Crofton, Aigburth ... '...	2	2	0	—		
Holt, George, J.P., Sudley, Mossley Hill	1	0	0	—		
Howes, Prof. G. B., Royal College of Science, South Kensington, London	1	1	0	—		
Hoyle, W. E., Museum, Owens College, Manchester	1	1	0	—		
Isle of Man Natural History and Antiquarian Society	1	1	0	—		
Jones, C.W.,J.P., Field House,Wavertree	1	0	0	—		
Kermode, P. M. C., Hill-side, Ramsey...	1	1	0	—		
Lea, Rev. T. Simcox, 3, Wellington-fields	1	1	0	—		
Leicester, Alfred, Harlow, Essex	1	1	0	—		
Liverpool Museum Committee ...	2	2	0	—		
Macfie, Robert, Airds	1	0	0	—		
Meyer, Dr. Kuno, University College ...	0	5	0	—		
Meade-King, H. W., J.P., Sandfield Park	1	1	0	—		
Meade-King, R. R., 4, Oldhall-street '...	0	10	0	—		
Melly, W. R., 90, Chatham-street ...	1	1	0	—		
Miall, Prof., Yorkshire College, Leeds ...	1	1	0	—		
Michael, Albert D., Cadogan Mansions, Sloane Square, London, S.W. ...				1	1	0
Monks, F. W., Brooklands, Warrington	1	1	0	—		
Muspratt, E. K., Seaforth Hall	5	0	0	—		
Newton, John, M.R.C.S., 44, Rodney-st.	0	10	6	—		

Poole, Sir James, Tower Buildings ...	2	2	0	—
Rathbone, S.G.,Croxteth-drive, Sefton-pk.	2	2	0	—
Rathbone, Mrs. Theo., Backwood, Neston	1	1	0	—
Rathbone, Miss May, Backwood, Neston	1	1	0	—
Rathbone, W., Greenbank, Allerton	2	2	0	—
Roberts, Isaac, F.R.S., Crowborough ...	1	1	0	—
Shaw, Prof. H. S. Hele, Ullet-road ...	1	1	0	—
Shepheard, T., Kingsley Lodge, Chester	1	1	0	—
Simpson,J.Hope,Annandale,Aigburth-dr	2	2	0	—
Smith, A. T., junr., 24, King-street ...	1	1	0	—
Talbot, Rev. T. U., 4, Osborne-terrace,				
Douglas, Isle of Man	1	1	0	—
Thompson, Isaac C., 19, Waverley-road	2	2	0	—
Thornely, James, Baycliff, Woolton ...	1	1	0	—
Thornely, The Misses, Baycliff, Woolton	1	1	0	—
Toll, J. M., 340, Walton Breck-road ...	1	1	0	—
Turnbull, Thos. S., 18, Spring-gardens,				
Manchester	1	1	0	—
Walker, A. O., Nant-y-glyn, Colwyn Bay	3	3	0	—
Walker, Horace, South Lodge, Princes-pk.	1	1	0	—
Walters, Rev. Frank, B.A., King William				
College, Isle of Man...	1	1	0	—
Watson, A. T., Tapton-cresent, Sheffield	1	1	0	—
Weiss, Prof. F. E.,Owen's College, Man'tr.	1	1	0	—
Westminster, Duke of, Eaton Hall ...	5	0	0	—
White, Prof., University College, Bangor	2	0	0	—
Wiglesworth, Dr., Rainhill	1	1	0	—
	109	6	6	6 6 0

THE LIVERPOOL MARINE BIOLOGY COMMITTEE.

In Account with ISAAC C. THOMPSON, Hon. Treasurer.

Dr.

1895.	£	s.	d.
To Printing Reports, Plates, &c.,	20	17	9
„ Printing and Stationery,	1	12	8
„ Expenses of Dredging Expeditions,	28	15	7
„ Boat Hire,	0	7	6
„ Books and Apparatus at Port Erin Biological Station	10	13	3
„ Postage, Carriage of Specimens, &c.,	3	16	11
„ Salaries, Curator and Laboratory Boy	50	3	4
„ Rent of Port Erin Biological Station,	15	0	0
„ Repairs, „ „ „	3	17	2
„ Sundries	0	10	5
	£135	14	7

Cr.

1895.	£	s.	d.
By Balance in hand, Dec. 31st, 1894,	10	16	2
„ Subscriptions and Donations actually received	105	8	6
„ Dividend, British Workman's Public House Co., Ltd., Shares	5	18	9
„ Sale of Reports	4	13	6
„ Bank Interest	0	19	6
„ Admissions to Aquarium	2	4	6
„ Balance due Treasurer	5	13	5
	£135	14	7

Endowment Investment Fund :—
British Workmans' Public House Co's. Shares ...£173 1 0

ISAAC C. THOMPSON,
Hon. Treasurer.

Audited and found correct,

A. T. SMITH, Junr.

LIVERPOOL, *December 31st*, 1895.

[WORK FROM THE PORT ERIN BIOLOGICAL STATION.]

DESCRIPTION and NOTES of some NEW and RARE COPEPODA from LIVERPOOL BAY.

By Mr. ANDREW SCOTT.

FISHERIES ASSISTANT, UNIVERSITY COLLEGE.

With Plates I. to V.

[Read March 13th, 1896.]

THE copepoda which form the subject of the following notes were obtained by washing dredged material, trawl refuse and mud collected in various parts of the L.M.B.C. area by Professor Herdman, Dr. Hanitsch and myself during the past year and latter part of the previous year (1894) while on expeditions to the dredging grounds, trawling stations and mussel beds. They are all additions to the fauna of Liverpool Bay and may form a useful appendix to the valuable papers published by Mr. I. C. Thompson, F.L.S.

Family HARPACTICIDÆ.

Sunaristes paguri, Hesse.

This rather peculiar and interesting species was obtained by washing the shells of *Buccinum* inhabited by the hermit crabs *Pagurus bernhardus*, collected in the trawl-net of the steamer while working at the mouth of the Mersey estuary on the 23rd of July, 1895. It seems to be a comparatively rare species and so far as is known this is only the third time it has been found in British waters. From our present knowledge of its distribution it appears to be confined to areas having large volumes of brackish water passing over the bottom, and has not been found in pure sea-water.

Sunaristes paguri is not unlike *Canuella perplexa* in general appearance but is readily distinguished from that species by the structure of the various appendages, especially the antennules and second pair of swimming feet of the male.

Stenhelia herdmani, n. sp. Pl. I., figs. 1—11.

Description of the species.—Female. Length 1·43 millim. ($\frac{1}{17}$th of an inch. Body moderately stout; rostrum prominent and curved. Antennules long and slender, eight-jointed; the first, second, fourth and eighth joints longer than the others, the fifth joint being the smallest of the series; the second, third and fourth joints have each a tuft of setæ on their upper distal margins. The proportional lengths of the various joints are as follows:—

$$\frac{14 \;.\; 14 \;.\; 8 \;.\; 10 \;.\; 5 \;.\; 6 \;.\; 7 \;.\; 11}{1 \quad 2 \quad 3 \quad 4 \quad 5 \quad 6 \quad 7 \quad 8}$$

Antennæ moderately stout, secondary branch small and slender, two jointed; basal joint elongate narrow with one seta on its upper distal end, second joint short, about one third of the length of the first and furnished with two terminal setæ. Mandibles large and well developed, the broad biting part armed with a few large teeth and a number of smaller ones; mandible palp comparatively large, consisting of a one-jointed basal part which carries at its lower extremity two branches, one large and one small, the smaller of the two being two-jointed, whilst the larger one is composed of a single joint. Masticatory portion of the maxillæ furnished with a number of strong teeth, palp two branched, the outer one bearing three setiferous lobes. Anterior foot-jaws furnished with one large terminal claw and three digitiform setose tubercles. Posterior foot-jaws stout, of moderate length and furnished with a strong, slightly curved terminal claw at the base of which are two setæ; the basal joint of the foot-jaw has four small ciliated tubercles on its lower side, while the second joint has a row of fine cilia on its upper margin and a row of stronger cilia on its lateral surface a little way down from the upper margin, there are also two plumose setæ on the upper margin of the joint. First pair of swimming feet somewhat similar to those of *Stenhelia ima*, Brady; basal joints of the inner branches nearly as long as the entire outer branches, second joint about half the length of the third which is less than one third the length of the long basal joint. Outer branches of the second, third and fourth pairs elongate, inner branches much shorter, those of the fourth pair only reaching to the end of the second joint of the outer branches. Fifth pair of feet large and well developed, inner branches considerably larger than the outer ones, with a subtriangular apex bearing five plumose setæ, two on the outer angle close together and three arranged at regular intervals along the inner margins; outer branches subovate, bearing six setæ on the external distal margins, the second seta from the inside is considerably longer than any of the others. Caudal stylets about as long as broad and about half the length of the last abdominal segment.

Habitat, 1 mile off Spanish Head, Isle of Man, in neritic material dredged from a depth of 16 fathoms, October 27th, 1895.

Remarks.—This large and well marked species though somewhat like *Stenhelia ima* in general appearance is readily distinguished from it and the other known members of this genus, by the form and armature of the fifth pair of feet, and by the structure and proportional lengths of the antennules.

Stenhelia similis, n. sp. Pl. I., figs. 12—25.

Description of the species.—Female. Length 1 millim. ($\frac{1}{25}$ of an inch). Body elongate, moderately robust; rostrum prominent and curved with a bifid apex. Antennules long and slender, sparingly setiferous, the second joint longer than any of the others and slightly contracted near the middle, but expanding again towards the distal end,

third, fifth, sixth and seventh joints small, the others of
moderate length as shown by the formula :—

$$\frac{13 \ . \ 24 \ . \ 7 \ . \ 12 \ . \ 5 \ . \ 6 \ . \ 6 \ . \ 10}{1 \quad 2 \quad 3 \quad 4 \quad 5 \quad 6 \quad 7 \quad 8}$$

Antennæ well developed, secondary branch three-jointed,
second joint very small, terminal joint fully half the
length of the basal one and furnished with two setæ on
its apex, one large and spiniform and one very small;
one seta springs from near the middle of the upper
margin of the terminal joint, the basal joint bears one
seta on its upper distal angle.   Mandibles furnished with
several strong and serrated teeth on the biting parts,
mandible palp consisting of a basal part carrying two
branches, the inner branch which is smaller than the
outer is two-jointed, both branches are furnished with a
number of setæ on the apex and upper margins, the basal
part has three terminal plumose setæ, and a curved row
of short spines on its lateral surface.   Maxillæ somewhat
similar to those of *Stenhelia herdmani*.   Posterior foot-
jaws slender and furnished with a short curved claw,
basal joints short and furnished with three small plumose
setæ on the upper distal margin, second joint fully three
times longer than broad and bearing a few cilia and one
seta on its upper margin, there are also a few spines on
its lateral surface.   The first four pairs of swimming feet
are nearly as in *Stenhelia ima*, the joints of the outer
branches of the first pair are subequal, basal joint of the
inner branches nearly as long as the entire outer branch,
second joint small and about half the length of the third
which is about half the length of the basal joint, the apex
of the third joint is furnished with one short stout spine
and two plumose setæ, one long and one short.   Fifth
pair of feet large, inner branches broad and triangular,
bearing five short plumose setæ from the middle of the

inner margin to the apex ; outer branches elongate ovate, about two-thirds the length of the inner, proximal half of the outer margin ciliated, inner margin slightly ciliated towards the distal end, apex and distal half of the outer margin furnished with six setæ, the second from the inner part of the apex considerably longer than the others. Caudal stylets rather shorter than broad and about one-third the length of the last abdominal segment.

Male. Antennules ten-jointed, fourth and sixth joints very small. Swimming feet, with the exception of the second pair, similar to those of the female. Inner branches of the second pair two-jointed, second joint bearing at the apex two strong and slightly curved spines, the inner spine which is slightly longer than the outer one, becomes distinctly bifid at the middle. The form of the fifth pair of feet is somewhat similar to those of the female, but smaller and furnished with fewer setæ, the inner branches have only two setæ which are placed on the apex, the outer branches have two setæ on the outer distal margin, the lower one being stout and spiniform, two setæ on the middle of the inner margin and one seta on the apex.

Habitat, 1 mile off Spanish Head, Isle of Man, in neritic material dredged from a depth of 16 fathoms. A considerable number of specimens were obtained.

Remarks.—This species comes near *Dactylopus tenui-remis,* but can easily be distinguished from it by the structure and proportional lengths of the antennules, the length and armature of the inner branches of the first feet, and also by the structure of the fifth feet.

Stenhelia reflexa, T. Scott.

[T. Scott, Thirteenth An. Rep. Fish. Board for Scot., pt. III., p. 166, 1895.]

A few specimens of this *Stenhelia* were obtained from dredged material collected off Port Erin in June, 1895.

Ameira gracile, n. sp.   Pl. II., figs. 1.—11.

Description of the species.—Female.   Length · 5 millim.
($\frac{1}{50}$th of an inch).   Body elongate and slender, rostrum
small and inconspicuous.   Antennules long and very
slender, seven-jointed; second and fifth joints longer than
any of the others, fourth joint very short, the second,
third and fourth joints have each a tuft of long setæ on
the upper distal margins, the following formula shows
the proportional lengths of the joints :—

$$\frac{9 \cdot 18 \cdot 10 \cdot 4 \cdot 13 \cdot 8 \cdot 9}{1 \quad 2 \quad 3 \quad 4 \quad 5 \quad 6 \quad 7}$$

Antennæ slender, three-jointed, secondary branch small,
two-jointed, the second joint very small.   Mandibles
elongate narrow, apex obliquely truncate and armed with
a number of teeth, mandible palp with a distinct basal
part, narrow at the base but somewhat dilated towards
the apex to which is attached a one-jointed elongate
narrow branch.   Posterior foot-jaws moderately robust
and armed with a strong terminal claw, lower margin of the
second joint furnished with a row of fine cilia.   First pair
of swimming feet elongate and slender, basal joint of the
inner branches nearly as long as the entire outer branch,
second joint about one-fourth the length of the basal
joint and fully half the length of the third joint.   Outer
branches of the second, third and fourth pairs elongate
three-jointed, inner branches also three-jointed but shorter
than the outer branches.   Fifth pair of feet foliaceous,
the inner branch produced into a subtriangular lobe
which reaches to about the middle of the outer branch and
furnished at the apex with a stout setiform spine and a
small seta, outer branch oblong ovate in shape, the
greatest breadth being very nearly half the length,
furnished with three setæ on the outer margin, one on
the apex and one on the inner distal margin, both the

inner and outer margins are clothed with fine cilia. Caudal stylets long and narrow, being about five times longer than broad and nearly twice the length of the last abdominal segment.

Male. Antennules ten-jointed, fifth and sixth joints very small, hinged between the third and fourth joints and also between the seventh and eighth joints. The form of the fifth pair of feet is somewhat similar to those of the female, but the inner branch is much smaller.

Habitat, 1 mile off Spanish Head, Isle of Man, in neritic material dredged from a depth of 16 fathoms, a number specimens were obtained.

Remarks.—This species in general appearance is not unlike *Ameira longicaudata* but is readily distinguished from it by the shape of the cephalothoracic segment and on dissection by the characters described above. Nearly all the specimens obtained had the last three joints of the antennules broken off.

Ameira reflexa, T. Scott.

[T. Scott, Twelfth An. Rep. Fish. Board for Scot., pt. III., p. 240, 1894.]

One or two specimens of this *Ameira* were obtained from the shelly deposit dredged 1 mile off Spanish Head, Isle of Man, depth 16 fathoms. The species is easily distinguished from the other members of this genus by the structure of the inner branches of the first pair of swimming feet and also by the fifth pair of feet.

Canthocamptus palustris, Brady. Pl. II., figs. 12—23.

[Brady, Monograph Brit. Copep., Vol. II., p. 53, 1880.]

A considerable number of specimens of a copepod apparently belonging to this species were washed from mud adhering to samples of Mussels (*Mytilus edulis*) sent from the St. Annes Mussel beds near Lytham, one of the samples was from that part of the bed which never becomes dry at low-water, and was obtained by means of a "mussel rake," it was from this sample that the first specimens were obtained, other samples sent later on in the year also contained numbers of specimens.

The specimens differ a little from the figures given by Dr. Brady in his "monograph," especially in the length of the basal joint of the first pair of swimming feet and also in the shape of the fifth pair of feet of the female.

Mesochra macintoshi, T. and A. Scott.

[T. & A. Scott, An. & Mag. Nat. Hist., Ser. 6, Vol. XV., p. 53, 1895.]

A number of specimens of this species were obtained from the shelly material dredged 1 mile off Spanish Head, Isle of Man, from a depth of 16 fathoms. The slender appearance of the species along with the structure and armature of its various appendages, enable it to be readily distinguished from the other members of the genus.

Tetragoniceps trispinosus, n. sp. Pl. II., figs. 24 and 25; III., figs. 1—6.

Description of the Species.—Female. Length ·5 millim. ($\frac{1}{50}$th of an inch). Body elongate cylindrical, tapering gently towards the posterior end, rostrum small and triangular in shape. Antennules long and slender, six-jointed and sparingly setiferous, the basal joint is considerably longer than any of the others, fifth joint very small, about half the length of the fourth; the proportional lengths of the joints are as shown by the following formula:—

$$\frac{28 \cdot 13 \cdot 14 \cdot 8 \cdot 4 \cdot 16}{1 \quad 2 \quad 3 \quad 4 \quad 5 \quad 6}$$

Antennæ of moderate length and three-jointed, secondary branch small and rudimentary, consisting of a single seta attached to the lower margin of the second joint of the primary branch at a distance of about one-third from the base. Posterior foot-jaw small, with a strong curved claw as long as the joint to which it is attached. Both branches of the first pair of swimming feet two-jointed, outer branches small, the joints subequal and reaching to about the middle of the basal joint of the inner branch ; inner branches long and slender, basal joint nearly twice the length of the entire outer branch and fully seven times longer than broad, a moderately long seta springs from near the base of the inner margin. Second joint short and narrow, fully one-fourth the length of the basal joint, furnished at its apex with a short curved seta, a seta of considerable length springs from near the middle of the inner margin. Outer branches of the second, third and fourth pairs of feet elongate, three-jointed, inner branches short and narrow, one-jointed, in the fourth pair the inner branches are only about one-third the length of the basal joint of the outer branches and furnished at the apex with three short setæ. Fifth pair of feet small, one branched and divided into two distinct portions, an inner which is produced into an elongate curved spiniform apex devoid of setæ and an outer tubercle-like process which arises from near the base of the elongate portion furnished with two short stout setæ and one long slender hair. Caudal stylets elongate narrow, slightly divergent, tapering to an acute apex and about twice the length of the last abdominal segment ; on the inner margin of each stylet at a distance of about one-third from the apex there arises a single seta which is fully two-thirds the length of the animal and having a slighly thickened base. Anal operculum semi-circular in shape and produced into three spines, a median and two lateral.

Habitat, 1 mile off Spanish Head, Isle of Man, in neritic material, dredged from a depth of 16 fathoms. Only two specimens were observed.

Remarks.—This species though placed in the genus

Tetragoniceps differs somewhat from the generic description given in the Monograph of the British Copepoda, especially in the number of joints in the outer branches of the first pair of feet and in the inner branches of the second, third and fourth feet, but as the mouth organs have not been satisfactorily worked out, it is perhaps better meanwhile to place it under the genus *Tetragoniceps* its nearest ally rather than institute a new genus for its reception.

Tetragoniceps consimilis, T. Scott.

[T. Scott, Twelfth An. Rep. Fish. Board for Scot., pt. III., p. 244, 1894.]

A few specimens of this species were obtained from the material dredged 1 mile off Spanish Head, Isle of Man, from a depth of 16 fathoms, it closely resembles *Tetragoniceps bradyi* in general appearance as well as in a few structural details, but differs from it in the absence of the strong hook on the second joint of the antennules, in the inner branches of the first pair of feet being three-jointed and in the fifth pair being composed of two distinct branches.

Laophonte propinqua, T. and A. Scott.

T. & A. Scott, An. & Mag. Nat. Hist., Ser. 6, Vol. XV., p. 460, 1895.

A few specimens of this species were obtained from material washed from sponges collected by Dr. Hanitsch at Port Erin, Isle of Man, in August, 1894; it is not unlike *Laophonte denticornis* at first sight but on closer examination is found to differ very markedly, not only from that species, but from any of the other known members of the genus.

Laophonte intermedia, T. Scott.

T. Scott, Thirteenth An. Rep. Fish. Board for Scot., pt. III., p. 168, 1895.

This species was obtained from the same material as the last, and also from the mussel beds at Duddon and Morecambe, it appears to be intermediate between *Laophonte lamellata* and *Laophonte hispida* but is quite distinct from either of them, the sub-conical form of the stylets alone enable it to be easily recognised when mixed up in a collection of Copepoda along with *L. lamellata* and *L. hispida*.

Pseudolaophonte, n. gen.

Description of the genus.—*Pseudolaophonte* resembles *Laophonte*, Philippi, in the structure of the antennules and antennæ; the mandibles, maxillæ and foot-jaws, and the first pair of swimming feet, but differs from that genus in the structure of the second and third pairs; the second pair of swimming feet consist of a single one-jointed branch, and the outer and inner branches of the third pair are each composed of two joints. The fourth and fifth pairs of feet are somewhat similar to those of *Laophonte*.

Pseudolaophonte aculeata, n. sp. Pl. III., figs. 7—23.

Description of the species.—*Female.* Length 1 millim. ($\frac{1}{25}$th of an inch). Body seen from above elongate narrow, of nearly equal breadth throughout, all the segments are more or less angular in shape and furnished with a row of short teeth on their posterior margins; surface of all the segments clothed with minute cilia; rostrum small and inconspicuous, with a small hair on each side of the base. Antennules moderately stout, four-jointed, first and third joints longer than the other two, the fourth joint being the smallest, the basal joint has a row of blunt pointed teeth on its upper margin and three rows on its lateral aspect, the middle row being the longest; a stout tubercle with a quadri-dentate apex arises from near the middle of the lower margin; second joint furnished on its lower margin with a strong slightly curved tooth which reaches to near the middle of the basal joint, and forms with the dentate tubercle of that joint, a powerful grasping apparatus; the third joint is covered with minute spines for about three-fourths of its length, the remaining fourth being covered with fine cilia, the fourth joint is also covered with cilia and has the lower distal part produced into a strong spine, the following formula shows the proportional lengths of the joints :—

$$\frac{17 \cdot 11 \cdot 16 \cdot 6}{1 \quad 2 \quad 3 \quad 4}$$

Antennæ two-jointed and of moderate size, with a small one-jointed secondary branch arising from near the middle of the lower margin of the basal joint and furnished with four setæ. Mandibles small, with a few serrated teeth on the truncate apex, mandible palp very small, with ciliated margins and bearing three setæ on the apex. Maxillæ and foot-jaws somewhat similar to those of a typical *Laophonte*, the second joint of the posterior foot-jaw long and slender, being about four times longer than broad, the terminal claw is also long and slender and is considerably longer than the second-joint. First pair of swimming feet similar to those of a typical *Laophonte*, outer branch composed of two joints. Second pair of swimming feet rudimentary, consisting of a single one-jointed branch, bearing three setæ at the apex, the innermost being longer than the other two. Both branches of the third pair of feet two-jointed, the inner branch being slightly shorter than the outer. The fourth pair of feet has the outer branch three-jointed and the inner two, the basal joint of the outer branch is nearly as long as broad and is equal to the combined lengths of the second and third joints, the first and second joints have each one stout ciliated spine on the outer distal angle, the second joint

which is very narrow, is produced on the inner margin
into a hook-like process furnished with a short seta, the
third joint has three strong spines on the outer margin
and apex, inner branches short, reaching to about the
middle of the outer branch, the second joint is furnished
with three short setæ on its apex. Fifth pair of feet large
and foliaceous, inner branch triangular in shape, ciliated
on the inner margin and covered with a number of more
or less curved rows of cilia, the branch is also furnished
with five moderately stout plumose setæ on its inner
margin and apex; outer branch broadly ovate, and fully
half the size of the inner branch, it is also covered with
rows of cilia and bears five short stout plumose setæ on
its apex. Caudal stylets elongate narrow, of moderate
length, about three times longer than broad and slightly
longer than the last abdominal segment; bearing on the
inner angles of the apex, a short stout curved spine and
near the middle the dorsal surface, a slightly shorter spine
and a seta, the outer margins are furnished with two
short setæ, the apex also bears two setæ, one of which is
very long. Anal operculum produced into a short stout
spine.

Male. Antennules six-jointed, first and second joints
like those of the female, third and sixth joints very small,
fourth joint considerably dilated. Mouth organs similar
to those of the female. The first and second feet are also
similar to to those of the female. The basal joint of the
outer branches of the third pair of feet has a strong
curved spine on its outer distal angle which is nearly
twice the length of the joint itself and extends considerably
beyond the end of the second joint, second joint of the
inner branches produced into a curved spine which reaches
to beyond the end of the outer branch, both branches of
the third pair two-jointed. The fourth pair of feet has

the outer branch three-jointed and the inner two; the basal joint of the outer branches is longer than the combined lengths of the second and third joints and bears a strong spine on its outer distal angle, second and third joints of the outer branch of about equal length; inner branches very short reaching to about the middle of the basal joint of the outer branch, basal joint of the inner branch very small and only about one-fourth the length of the second joint. Fifth pair small, inner branch not produced, furnished with two plumose setæ on its apex, the inner one being three times longer than the outer; outer branch elongate narrow, bearing at its apex three stout setæ.

Habitat, 1 mile off Spanish Head, Isle of Man, in neritic material dredged from a depth of 16 fathoms; a number of specimens were obtained.

Remarks.—This species comes very near *Laophonte spinosa*, I. C. Thompson, especially in the structure of the antennules and mouth organs, but differs considerably in the structure of the second, third and fourth pairs of swimming feet; the outer branches of the second, third and fourth feet in *Laophonte spinosa* are two jointed and the inner three, whilst in *Pseudolaophonte aculeata* the second pair of feet consists of a single one-jointed branch, in the third pair each branch is composed of two joints and in the fourth pair the outer branch consists of three joints and the inner of two, the fifth feet also differ somewhat. The appendages of the male differ also from those of the male *Laophonte spinosa*.

Laophontodes bicornis, n. sp.   Pl. III., figs. 24—25; IV., figs. 1—7.

Description of the species.—*Female.*   Length ·5 millim. (¹⁄₅₀th of an inch).   Body seen from above elongate narrow, the breadth gradually decreasing towards the posterior end; all the segments are more or less angular in shape and with the exception of the cephalic segment, bear each a row of short teeth on the distal margin. Cephalo-thoracic segment broadly triangular in outline, the frontal portion being produced into a small rostrum, and the lateral margins near the distal end into strong curved spines directed backwards and extending slightly beyond the middle of the second segment. Antennules short, five-jointed, all the joints are of moderate length except the fourth which is very short; the proportional lengths of the joints are as shown in the following formula:—

$$\frac{13}{1} \cdot \frac{17}{2} \cdot \frac{22}{3} \cdot \frac{3}{4} \cdot \frac{13}{5}$$

Antennæ small, two-jointed without any secondary appendage. Mandibles and other mouth organs nearly as in *Laophonte*. The first pair of swimming feet are similar to those of *Laophontodes typicus*, and the second, third and fourth pairs are also similar to the corresponding feet of that species. The fifth pair are large and prominent and project outwards from the sides of the fifth segment; each foot consists of a single narrow elongate branch, composed of two-joints, furnished with one seta on the inner distal angle of the first joint and two on the outer angle, the second joint has two setæ on the inner margin, two on the apex and one on the outer margin, the basal joint has also a row of cilia on its inner margin. Caudal stylets long and narrow, about equal to the combined lengths of the last two abdominal segments.

Habitat, Off Port Erin, from dredged material collected June, 1895; only one specimen has been observed.

Remarks.—This species is easily distinguished from *Laophontodes typicus* the only other member of the genus, by the lateral projections of the cephalothoracic segment, the proportional lengths of the joints of the antennules and the length of the caudal stylets; the fifth feet also differ, in this species they are two-jointed whilst in *Laophontodes typicus* they are composed of a single joint only.

Normanella attenuata, n. sp. Pl. IV., figs. 8—20.

Description of the species.—Female. Length 1 millim. ($\frac{1}{25}$th of an inch). Body elongate cyclindrical, slender. Antennules nine-jointed; the second much longer than the others, seventh and eighth joints very small, the others are of moderate length as shown by the formula :—

$$\frac{9 \cdot 15 \cdot 10 \cdot 7 \cdot 4 \cdot 5 \cdot 1 \cdot 1 \cdot 5}{1 \quad 2 \quad 3 \quad 4 \quad 5 \quad 6 \quad 7 \quad 8 \quad 9}$$

Antennæ three-jointed, stout and of moderate length, a small one-jointed secondary branch arises from the lower distal end of the basal joint of the primary branch and is furnished with two setæ; the lower one of which appears to be articulated to the apex of the joint. Mandibles slender with a serrated apex, basal portion of the mandible palp considerably dilated and bearing two one-jointed branches, the outer branch being much longer than the inner. Maxillæ and foot-jaws nearly as in *Normanella dubia*. Inner branches of the first pair of swimming feet long and slender, two-jointed, basal joint longer than the entire outer branch, second joint about one-third the length of the basal joint, bearing one curved spine and two setæ on the apex, outer branches three-jointed, shorter than the basal joint of inner branches. In the second and third pairs of feet, the inner branches are short, and two-jointed; the outer branches are considerably longer than the inner and three-jointed. Inner branches of the fourth pair of feet three-jointed and very short, only reaching to about the middle of the second joint of the outer branches. Fifth pair of feet foliaceous, two branched, inner branch large and subtriangular bearing two setæ on the inner distal margin and two on the apex, outer branch pyriform, arising from the middle of the outer margin and extending considerably beyond the apex of the inner branch, bearing four setæ on its outer distal margin and two on the apex. Caudal stylets of moderate length, about twice as long as broad and fully half the length of the last abdominal segment.

Male. Antennules nine-jointed, sixth joint very short, the others of moderate length, hinged between the fourth and fifth joints and also between the seventh and eighth, all the other appendages with the exception of the fifth pair of feet are similar to the corresponding appendages of the female. The inner branch of the fifth pair sub-triangular in form bearing one stout plumose spine and two plumose setæ on its apex, the outer branch pyriform, bearing three setæ on its outer distal margin, and two on the apex, with a strong plumose spine between the two apical setæ.

Habitat, 1 mile off Spanish Head, Isle of Man, in neritic material dredged from a depth of 16 fathoms; very few specimens were obtained.

Remarks.—This species differs considerably in shape from *Normanella dubia* but the structural details are almost similar to those on which the genus was founded, the only differences being that the antennules have nine joints instead of seven, and the inner branches of the fourth pair of feet have three joints instead of two. These differences are not considered to be of sufficient importance to warrant the establishment of a new genus for its reception.

Cletodes similis, T. Scott.

[T. Scott, Thirteenth An. Rep. Fish. Board for Scot. Pt. III., p. 168, 1895.]

A few specimens were obtained from material washed

from sponges collected by Dr. Hanitsch at Port Erin,
Isle of Man, in August, 1894. This species is very like
Cletodes lata in general appearance but is easily distinguished from it on dissection by the structure of the
antennules, the proportional lengths and armature of
the outer and inner branches of the first pair of swimming
feet, and also by the form of the fifth pair of feet.

Nannopus palustris, Brady.

Several specimens of this species were obtained in the
mud collected from the Mussel beds near Duddon and
from mud sent to the laboratory from the Fleetwood
Oyster beds. It seems to be a brackish water species and
in general appearance is very like *Platychelipus littoralis*
another brackish water copepod, it can be distinguished
from that species however, even without dissecting, by
making an examination of the fifth pair of feet and also
of the inner branches of the third and fourth pairs of feet.
Nannopus palustris has two ovisacs and *Platychelipus
littoralis* one only.

Idya elongata, n. sp. Pl. IV., figs. 21—24; Pl. V.,
figs. 1—5.

Description of the species.—Female. Length ·74 millim.
($\frac{1}{35}$th of an inch). Body seen from above elongate narrow,
tapering rapidly towards the posterior end, the length
being nearly equal to four times the greatest breadth;
rostrum prominent with a bluntly rounded apex. Antennules short and comparatively stout; shorter than the
cephalothoracic segment, eight-jointed; second and third
joints longer than any of the others, as shown in the
following formula:—

$$\frac{11 \cdot 16 \cdot 17 \cdot 13 \cdot 6 \cdot 8 \cdot 5 \cdot 12}{1 \quad 2 \quad 3 \quad 4 \quad 5 \quad 6 \quad 7 \quad 8}$$

Antennæ, mandibles and maxillæ nearly as in *Idya
gracilis*, T. Scott. Foot-jaws also similar to those of that

species but shorter and stouter. Inner branches of the
first pair of swimming feet slender and of moderate
length, basal joint nearly as long as the entire outer
branch, and furnished with a plumose seta arising from
the lower half of the inner margin and extending to
slightly beyond the end of the branch, second joint fully
two-thirds the length of the basal joint also furnished
with a plumose seta arising from near the middle of its
inner margin, third joint very small, bearing on its apex
two stout spines and one short plumose seta; outer
margins and proximal halves of the inner margins of the
first and second joints fringed with short hairs, the joints
of the outer branches are short and broad, the second
joint is slightly shorter than the first and the third joint a
little shorter than the second, the armature of the joints
is somewhat similar to that of the first pair in *Idya
furcata*; the spines are furnished with a row of moderately
long cilia on the upper margins. Second, third and fourth
pairs of swimming feet similar to those of *Idya furcata*.
Fifth pair of feet very short being little more than half
the length of the joint to which they are attached and
extending only a little way beyond the base of the first
segment of the abdomen, the length of each foot is about
equal to twice the breadth, the secondary joint is furnished
with three setæ on the apex, the innermost one being
longer than either of the other two, outer very short; a
short seta is attached to the outer margin a little way
from the apex. Caudal stylets narrow and slightly
divergent, length equal to about twice the breadth and
nearly as long as the last segment of the abdomen.

Male. Antennules nine-jointed, hinged between the
third and fourth joints and also between the seventh and
eighth joints, fourth joint very small; the other appendages
are similar to those of the female, fifth feet also similar to
the fifth feet of the female but smaller.

Habitat, obtained from the mud collected on the Mussel beds between Morecambe and Heysham; only a few specimens were obtained.

Remarks.—This species is very distinct from *Idya furcata* and also from two other species recently described—*Idya longicornis,* T. and A. Scott, and *Idya gracilis,* T. Scott—and can easily be recognised from either of them by the elongate form of the animal, the short antennules and the small fifth feet.

Idya gracilis, T. Scott.

[T. Scott, Thirteenth An. Rep. Fish. Board for Scot., pt. III., p. 171, 1895.]

A number of specimens of this species were obtained from the shelly material dredged 1 mile off Spanish Head, Isle of Man, from a depth of 16 fathoms; it is easily recognised by the long and slender inner branches of the first pair of swimming feet and also by the shape and arrangement of the setæ on the fifth pair of feet.

Family SAPPHIRINIDÆ, Thorell.

Modiolicola insignis, Aurivillius.

Living as a messmate within the mantle of the "horse mussel," *Mytilus modiolus.* A number of specimens were found in the examples of this Mollusc which were brought up in the trawl-net of the steamer, while working in the vicinity of the north end of "the Hole" on March 23rd, 1895. This appears to be a widely distributed species of Copepod, its range being probably co-extensive with that of the Mollusc. It has been recorded from the Firth of Forth, the Moray Firth, and from the vicinity of Mull. It has also been obtained in specimens of the same species of Mollusc dredged by Dr. Norman in 1893, off Trondhjem in Norway.

Family Ascomyzontidæ, Thorell (1859).

Dermatomyzon gibberum, T. and A. Scott.

[T. & A. Scott, An. & Mag. Nat. Hist., Ser. 6, Vol. XIII., p. 144, 1894.]

A considerable number of specimens of this species were obtained by washing the common starfish (*Asterias rubens*) in weak methylated spirit and afterwards examining the sediment. It was taken from starfish collected at Hilbre Island and afterwards from the same species of starfish taken in other parts of the district; both males and females were found, many of the latter with ovisacs attached.

Collocheres elegans, n. sp.   Pl. V., figs. 6—15.

Description of the species.—Female. Length 1 millim. ($\frac{1}{25}$th of an inch). Body elongate, subpyriform, anterior segment large and somewhat triangular in outline and equal to twice the combined lengths of the second, third and fourth segments, rostrum small and inconspicuous. Antennules moderately long, slender and sparingly setiferous, twenty-jointed; the first, eighteenth and twentieth joints of about equal length and longer than any of the others, the second and tenth joints slightly smaller than the others; a sensory filament springs from the end of the third last joint. The following formula shows the proportional lengths of the joints :—

$$\frac{9\,.\,2\,.\,3\,.\,3\,.\,3\,.\,3\,.\,3\,.\,4\,.\,4\,.\,2\,.\,3\,.\,6\,.\,5\,.\,6\,.\,6\,.\,6\,.\,6\,.\,9\,.\,3\,.\,8}{1\;\;2\;\;3\;\;4\;\;5\;\;6\;\;7\;\;8\;\;9\;\;10\;11\;12\;13\;14\;15\;16\;17\;18\;19\;20}$$

Antennæ three-jointed, basal joint long and narrow, bearing near the middle of the lower margin a small secondary branch, which consists of a single joint, nearly oval in outline and furnished with three small setæ on the apex and one near the middle of the upper margin, second joint of the antennæ about half the length of the first, third joint about two-thirds the length of the second and bearing at the apex a long slender spine having a slightly thickened base, and a small hair; a short seta also springs from near the base of the upper margin. Mandibles elongate narrow, denticulated on the oblique apex, palp rudimentary and consisting of a single moderately long hair. Maxillæ two-lobed, both lobes of about equal length, but one is slightly narrower than the other and is furnished with one seta at the apex, the broad lobe has four setæ on its apex. Foot-jaws somewhat similar to those of *Collocheres gracilicauda* (Brady). First four pairs of swimming feet also similar to those of that species; the outer branches of all the four pairs are armed with short dagger shaped spines and the terminal joint of the inner branch of the fourth pair is furnished with one stout dagger shaped spine on the apex and a smaller one near the middle of the outer margin. Fifth pair of feet somewhat rudimentary, two-jointed, basal joint broadly triangular in shape, the second joint which is attached to near the middle of the outer margin of the basal joint is elongate, curved, and bluntly serrated at its apex, the length being about equal to three and one-half times the breadth; it is furnished with three setæ, one on the apex and two a little lower down on the outer margin and slightly separated from each other. Abdomen slender, four-jointed, genital segment elongate narrow, length nearly equal to twice the breadth, and longer than the combined lengths of the next three segments, second joint about one-third the length of the first, third joint slightly smaller than the second, fourth joint smaller than the third. Caudal stylets about four times as long as broad and nearly equal to the length of the last two segments of the abdomen.

Habitat, off Port Erin, from dredged material collected June, 1895, only one specimen has been observed.

Remarks.—This species is not unlike *Collocheres gracili-*

cauda and may perhaps have been passed over for that species, but it can be readily distinguished from it by the much shorter caudal stylets and also by the shape of the fifth pair of feet.

Ascomyzon thompsoni, n. sp. Pl. V., figs. 16—26.

Description of the species.—Female. Length 1 millim. ($\frac{1}{25}$th of an inch). Body broad, suborbicular in shape, cephalothorax broadly ovate, last segment of thorax and abdomen much narrower, rostrum not prominent. Antennules slender, twenty-one-jointed, the first being the largest and ciliated on its upper margin ; second to eighth joints small and of about equal length, ninth joint smaller than any of the others, eighteenth joint furnished with a short sensory filament. The proportional lengths of the joints are shown in the following formula :—

$$\frac{48\,.\,7\,.\,7\,.\,5\,.\,6\,.\,6\,.\,6\,.\,7\,.\,8\,.\,4\,.\,7\,.\,8\,.\,11\,.\,8\,.\,12.11.14.15\,.\,7\,.\,8\,.\,7}{1\quad 2\quad 3\quad 4\quad 5\quad 6\quad 7\quad 8\quad 9\quad 10\,11\quad 12\,13\,14\,15\,16\,17\,18\,19\,20\,21}$$

Antennæ four-jointed, first joint long and bearing near the distal end of the lower margin, a small one-jointed secondary branch, which bears at the apex a moderately long seta, a small hair also springs from near the middle of the upper margin; second joint of the antennæ shorter and narrower than the first and having its lower margin ciliated, third joint very small, fourth joint about as long as broad and bearing at its apex one strong curved spine and two setæ. Mandibles slender, and stylet shaped ; palp elongate narrow, two-jointed, second joint about one-third the length of the first and bearing at its apex, one long and one short plumose seta. The maxillæ consist of a short basal joint bearing two lobes of about equal length, but one is considerably narrower than the other, each lobe is furnished with four plumose setæ ; one of the setæ on the broad lobe is much stouter and longer than the others, two of the other setæ on the same lobe are also comparatively stout but are only about half the length of the long seta. Anterior foot-jaws simple, bearing a strong curved apical claw. Posterior foot-jaws elongate slender, four-jointed, resembling those of *Dermatomyzon nigripes* (B. and R.). Both branches of the first four pairs of swimming feet short and stout, three-jointed and nearly equal in length. Fifth pair of feet rudimentary, two-jointed, inner joint short and broad, furnished with one plumose seta on its upper distal angle, outer joint elongate, length about equal to twice the breadth and bearing at its apex two moderately long plumose setæ and one small spine, both margins of the joint ciliated. Abdomen three-jointed, genital segment about as long as broad and nearly equal to the combined lengths of the next two segments and caudal stylets, second joint about half the length of the first, third joint about two-thirds the length of the second. Caudal stylets slightly longer than the last abdominal segment, length about equal to twice the breadth.

Habitat, 1 mile off Spanish Head, Isle of Man, in neritic material dredged from a depth of 16 fathoms; a few specimens only were obtained. A number of specimens have since been found in material washed from Ophiuroids (*Ophioglypha* and *Ophiothrix*) taken in the trawl-net off Blackpool, and sent to us by Mr. Ascroft.

Remarks.—This species is readily distinguished from the other members of the *Ascomyzontidæ* by the almost oval outline of the cephalothorax and on dissection by the structure of the mandible palp and maxillæ, the stout setæ on the larger lobe of the maxillæ appears to be a well marked character. Dr. W. Giesbrecht of the Zoological Station Naples, is preparing a monograph on this interesting family and an abstract which appeared in the Ann. and Mag. of Natural History for August, 1895, shows a number of changes in the nomenclature and classification of the genera and species.

_____

EXPLANATION OF THE PLATES.

___

PLATE I.

Stenhelia herdmani, n. sp. (A. Scott).

Fig. 1. Female seen from the side, × 27. 2. Antennule, × 63. 3. Antenna, × 85. 4. Mandible, × 85. 5. Maxilla, × 85. 6. Anterior foot-jaw, × 127. 7. Posterior foot-jaw, × 90. 8. Foot of first pair of swimming feet, × 85. 9. Foot of fourth pair, × 85. 10. Foot of fifth pair, × 127. 11. Abdomen and caudal stylets, × 170 .

Stenhelia similis, n. sp.

Fig. 12. Female seen from the side, × 40. 13. Antennule, × 127. 14. Antennule of male, × 127. 15. Antenna, × 125. 16. Mandible, × 253. 17. Maxilla, × 253. 18. Posterior foot-jaw, × 253. 19. Rostrum, × 253. 20. Foot of first pair of swimming feet, × 127. 21. Foot of fourth pair, × 127. 22. Foot of second pair, male, × 127. 23. Foot of fifth pair, × 125. 24. Foot of fifth pair, male, × 125. 25. Abdomen and caudal stylets, × 53

PLATE II.

Ameira gracile, n. sp.

Fig. 1. Female seen from the side, × 64. 2. Antennule, × 152. 3. Antennule, male, × 152. 4. Antenna, × 253. 5. Mandible, × 380. 6. Pos-

terior foot-jaw, × 380. 7. Foot of first pair
of swimming feet, × 170. 8. Foot of fourth
pair, × 170. 9. Foot of fifth pair, × 380. 10.
Foot of fifth pair, male, × 380. 11. Abdomen
and caudal stylets, × 80.

Canthocamptus palustris, Brady.

Fig. 12. Female seen from the side, × 50. 13. Antennule, × 200. 14. Antennule, male, × 152.
15. Antenna, × 253. 16. Mandible, × 300.
17. Posterior foot-jaw, × 380. 18. Foot of
first pair of swimming feet, × 170. 19. Foot of
fourth pair, × 170. 20. Foot of fifth pair,
× 253. 21. Foot of fifth pair, male, × 253.
22. Appendage to the first abdominal segment,
male, × 253. 23. Abdomen and caudal stylets,
× 80.

Tetragoniceps trispinosus, n. sp.

Fig. 24. Posterior foot-jaw, × 300. 25. Foot of fifth
pair, × 380.

PLATE III.

Tetragoniceps trispinosus, n. sp.

Fig. 1. Female seen from above, × 80. 2. Antennule,
× 170. 3. Antenna, × 253. 4. Foot of first
pair of swimming feet, × 253. 5. Foot of
fourth pair, × 253. 6. Abdomen and caudal
stylets, × 253.

Pseudolaophonte aculeata, n. gen. and n. sp.

Fig. 7. Female seen from above, × 106. 8. Antennule,
× 170. 9. Antennule, male, × 170. 10. Antenna, × 125. 11. Mandible, × 253. 12.
Maxilla, × 253. 13. Anterior foot-jaw, × 253.
14. Posterior foot-jaw, × 253. 15. Foot of
first pair of swimming feet, × 253. 16. Foot

of second pair, × 253. 17. Foot of third pair,
× 253. 18. Foot of fourth pair, × 253. 19.
Foot of fifth pair, × 170. 20. Foot of third
pair, male, × 253. 21. Foot of fourth pair,
male, × 253. 22. Foot of fifth pair, male, ×
380. 23. Abdomen and caudal stylets, × 80.

Laophontodes bicornis, n. sp.

Fig. 24. Mandible, × 500. 25. Anterior foot-jaw, × 500.

PLATE IV.

Laophontodes bicornis, n. sp.

Fig. 1. Female seen from above, × 120. 2. Antennule,
× 253. 3. Antenna, × 253. 4. Posterior foot-
jaw, × 380. 5. Foot of first pair of swimming
feet, × 253. 6. Foot of fourth pair, × 253.
7. Foot of fifth pair, × 253.

Normanella attenuata, n. sp.

Fig. 8. Female seen from the side, × 50; 9. Antennule,
× 127. 10. Antennule, Male, × 127. 11.
Antenna, × 150. 12. Mandible, × 253. 13.
Maxilla, × 253. 14. Posterior foot-jaw, × 253.
15. Foot of first pair of swimming feet, × 150.
16. Foot of second pair, × 150. 17. Foot of
fourth pair, × 150. 18. Foot of fifth pair, ×
300. 19. Foot of fifth pair, male, × 300. 20.
Abdomen and caudal stylets, × 90.

Idya elongata, n. sp.

Fig. 21. Posterior foot-jaw, × 380. 22. Foot of fifth
pair, female, × 115. 23. Foot of fifth pair,
male, × 115. 24. Appendage to the first abdominal segment, × 115.

PLATE V.

Idya elongata, n. sp.

Fig. 1. Female seen from above, × 64. 2. Antennule,

× 253.   3. Antennule, male, × 253.   4. Foot
first pair of swimming feet, × 190.   5. Foot
fourth pair, × 190.

Collocheres elegans, n. sp.

Fig. 6. Female seen from above, × 52.   7. Antennule,
× 133.   8. Antenna, × 170.   9. Mandible, ×
253.   10. Maxilla, × 170.   11. Anterior foot-
jaw, × 170.   12. Posterior foot-jaw, × 170.
13. Foot of first pair of swimming feet, × 125.
14. Foot of fourth pair, × 125.   15. Foot of
fifth pair, × 190.

Ascomyzon thompsoni, n. sp.

Fig. 16. Female seen from above, × 50.   17. Antennule,
× 133.   18. Antenna, × 190.   19. Mandible,
× 190.   20. Maxilla, × 125.   21. Anterior
foot-jaw, × 127.   22. Posterior foot-jaw, × 127.
23. Foot of first pair of swimming feet, × 127.
24. Foot of fourth pair, × 127.   25. Foot of
fifth pair, × 200.   26. Abdomen and caudal
stylets, × 85.

Andrew Scott del.

COPEPODA.

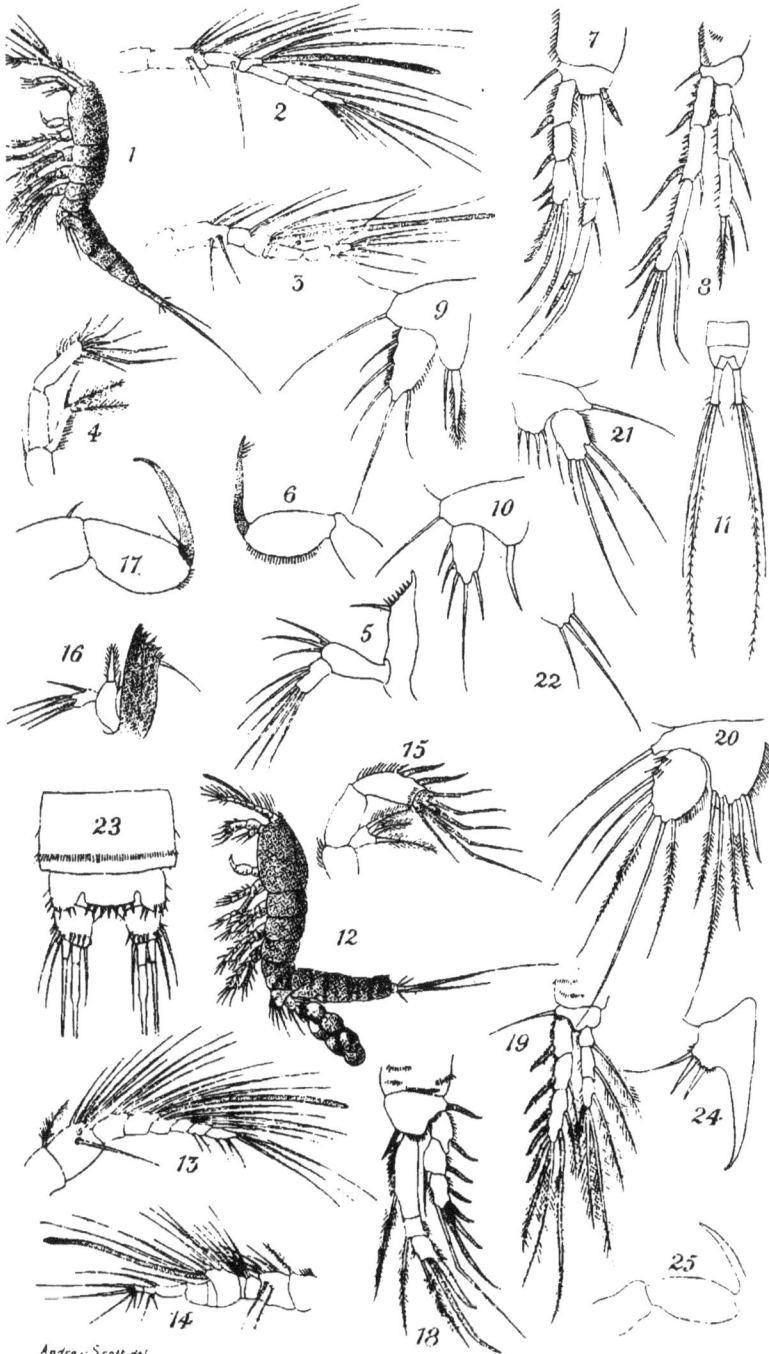

Andrew Scott. del

COPEPODA.

Vol. X., Pl. III.

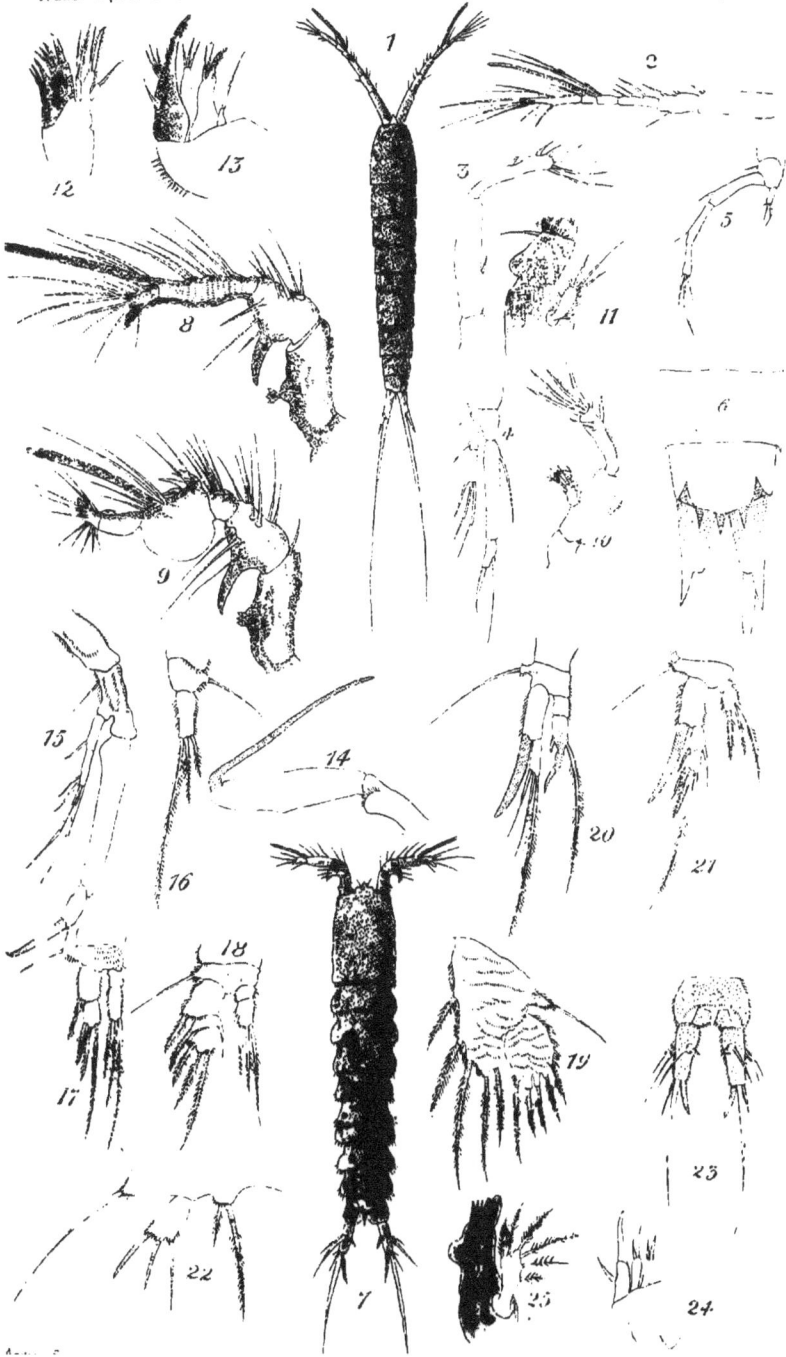

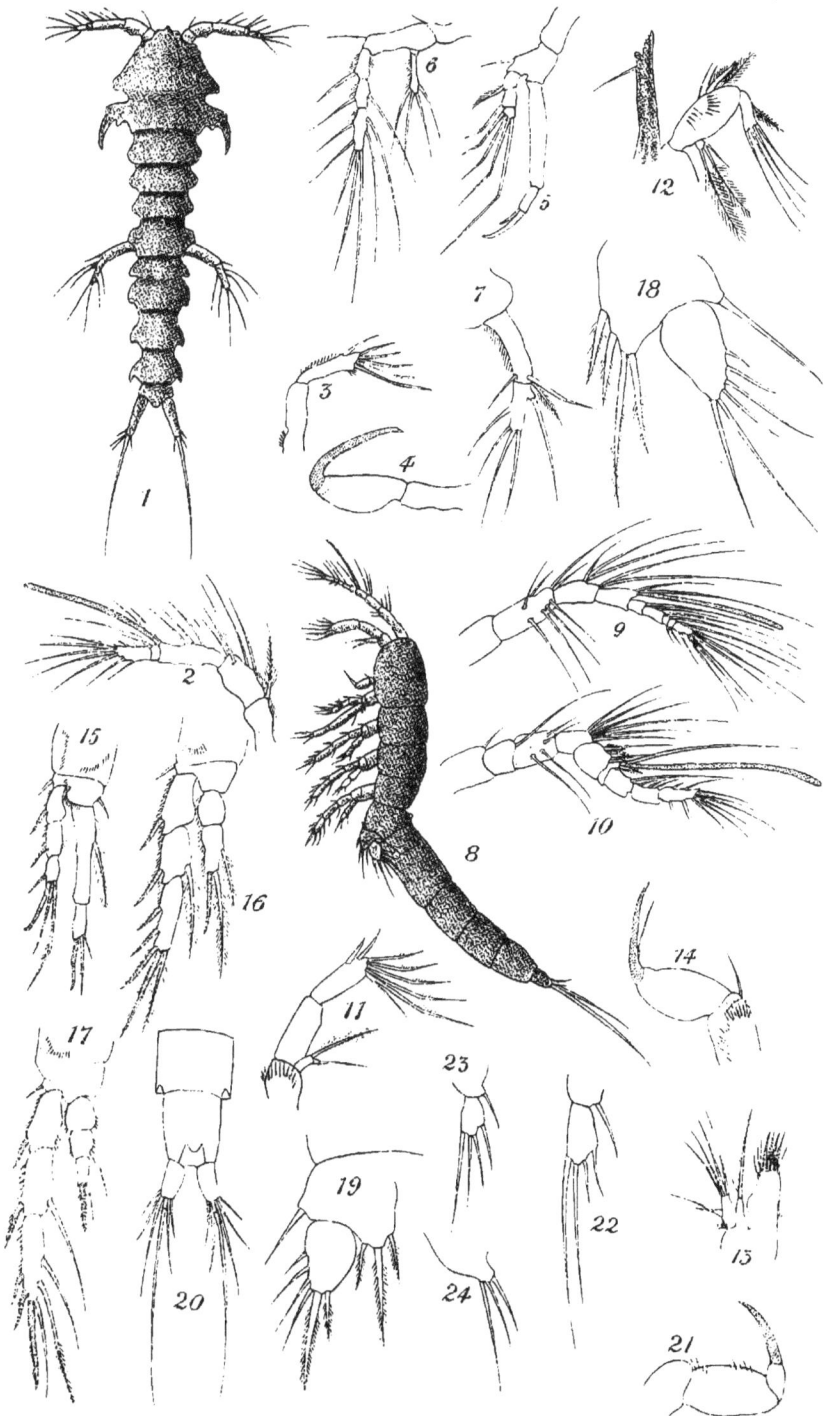

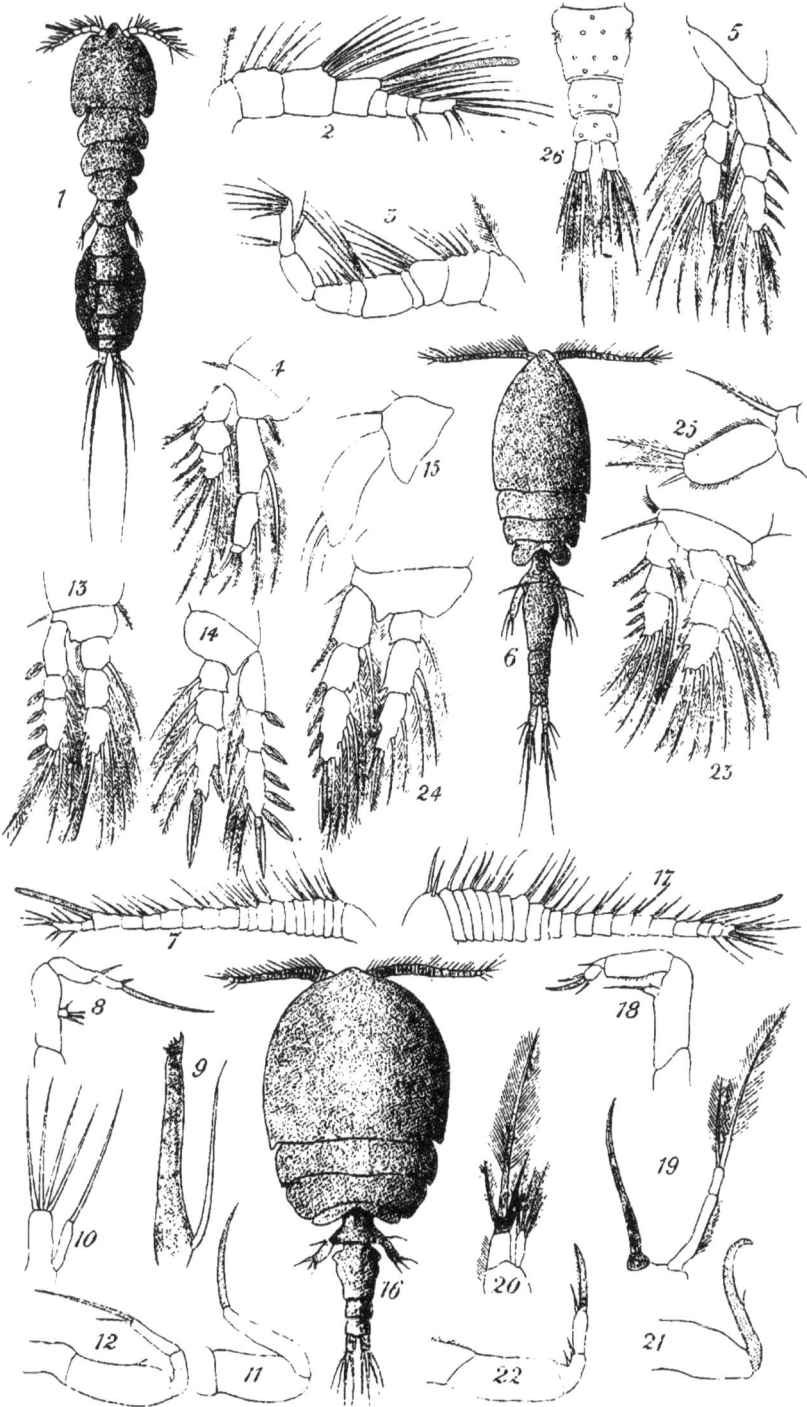

[From Trans. Biol. Soc., L'pool. Vol. XI.]

TENTH ANNUAL REPORT of the LIVERPOOL MARINE BIOLOGY COMMITTEE and their BIOLOGICAL STATION at PORT ERIN.

By W. A. HERDMAN, D.Sc., F.R.S.,

DERBY PROFESSOR OF NATURAL HISTORY IN UNIVERSITY COLLEGE, LIVERPOOL ;
CHAIRMAN OF THE LIVERPOOL MARINE BIOLOGY COMMITTEE,
AND DIRECTOR OF THE PORT ERIN STATION.

[Read December 11th, 1896.]

AMONG the events of the past year which should be specially mentioned are 1°, the visit of the British Association, which involved a dredging expedition in Liverpool Bay, and an excursion of several days duration to Port Erin at the conclusion of the Liverpool meeting ; and 2°, the compilation of an index list of all the species of marine animals and plants recorded by the L.M.B.C. during the first ten years of their work.

The Station at Port Erin has been made use of by various members of the Committee and other naturalists for varying periods during the year ; the tanks in the aquarium have also enabled some experiments in sea-fish hatching to be carried out for the Lancashire Sea-Fisheries Committee ; and several students from both Owens College, Manchester, and University College, Liverpool, have occupied work-tables in the laboratory during the Easter, Whitsuntide, and Summer vacations.

STATION RECORD.

The following naturalists have worked at the Port Erin Laboratory during the past year :—

DATE.	NAME.	WORK.
February.	Mr. I. C. Thompson, Liverpool ...	Copepoda.
—	Prof. W. A. Herdman, Liverpool	Collecting.
March.	Mr. R. A. Dawson, Preston ⎱ ...	Sea-Fisheries.
—	Mr. Andrew Scott, Liverpool ⎰	

March.	Prof. W. A. Herdman, Liverpool	Tunicata.
—	Mr. I. C. Thompson, Liverpool ...	Copepoda.
—	Mr. Charles E. Jones, Liverpool	Algæ.
—	Mr. A. H. Burtt, Manchester	Algæ.
April.	Prof. F. E. Weiss, Manchester ...	Diatoms.
—	Mr. Charles E. Jones, Liverpool	Algæ.
—	Mr. A. H. Burtt, Manchester ...	Algæ.
—	Prof. W. A. Herdman, Liverpool	Tunicata.
—	Mr E. T. Browne, London ...	Medusæ.
—	Mr. Arnold T. Watson, Sheffield	Annelids.
—	Mr. F. W. Gamble, Manchester...	Turbellaria.
—	Mr. J. A. Clubb, Liverpool ...	Collecting.
—	Prof. Gustave Gilson, Louvain	Annelids.
—	Mr. P. M. C. Kermode, Ramsey ...	General.
—	Mr. Harold Murray, Manchester	Collecting.
—	Mr. Terry, Manchester	Geology.
—	Mr. E. J. W. Harvey, Liverpool	Studying Fauna.
—	Mr. A. O. Walker, Colwyn Bay...	Amphipoda.
—	Mr. R. A. Dawson, Preston ⎫	
—	Mr. R. L. Ascroft, Lytham ⎬	Sea-Fish Hatching.
—	Mr. Andrew Scott, Liverpool ⎭	
—	Miss L. R. Thornely, Liverpool ...	Polyzoa.
—	Mr. I. C. Thompson, Liverpool...	Copepoda.
May.	Mr. E. T. Browne, London	Medusæ.
—	Mr. Andrew Scott, Liverpool	Fish.
—	Mr. W. Narramore, Liverpool ...	General.
June.	Rev. T. S. Lea, Liverpool ...	Algæ, &c.
July.	Mr. A. Dwerryhouse, Liverpool...	General.
—	Rev. T. S. Lea, Liverpool ...	Algæ, &c.
—	Prof. W. A. Herdman, Liverpool	General.
—	Mr. A. Chopin, Manchester ...	Actiniaria.
August.	Mr. J. H. Ashworth, Manchester	Annelids.
—	Prof. W. A. Herdman, Liverpool	Tunicata.
—	Mr. W. Wadsworth, Manchester	General.
—	Prof. R. Boyce, Liverpool ...	General.
—	Miss L. R. Thornely, Liverpool...	Polyzoa.
—	Mr. J. W. Woodall, Scarborough	General.
September.	Mr. H. C. Chadwick, Bootle ...	Echinoderms.
—	Prof. W. A. Herdman, Liverpool	Tunicata.
—	Mr. I. C. Thompson, Liverpool ...	Copepoda.
—	Mr. Alfred Leicester, Kent	Mollusca.
—	Prof. E. B. Poulton, Oxford	General.

September.	Dr. Johan Hjort, Christiania	...	Ascidians.
—	Dr. J. G. de Man, Iersekc	...	General.
—	Dr. J. F. Gilchrist, Cape Town	...	General.
—	Mr. Conrad Cooke, London	...	General.
—	Mr. Bedford, Cambridge	...	General.
—	Prof. Magnus, Berlin		
—	Prof. Pfitzer, Heidelberg		
—	Prof. Chodat, Geneva	...	Algæ.
—	Prof. Weiss, Manchester		
—	Prof. Zacharias, Hamburg		
October.	Mr. Herbertson, Edinburgh	...	General.
November.	Mr. I. C. Thompson, Liverpool	...	Copepoda.
—	Professor W. A. Herdman, Liverpool	..	Tunicata.

In addition to these, a number of other members of the British Association party visited the Station between September 24th and 28th, including the Lieutenant Governor (Lord Henniker), Sir James Gell, Dr. Munro, Dr. Montelius, Professor Haddon, and others.

This list shows an increase in the number of workers. It is also satisfactory in containing the names of several senior students preparing for science degrees in Victoria University. The Councils of both the Owens College, Manchester, and University College, Liverpool, have arranged with the L.M.B.C. to engage by the year a "table" or work-place each in the Laboratory, to be constantly at the disposal of any members of the Biological departments of the Colleges—either staff or senior students—who may be nominated by the Professors. The "tables" in question have been labelled with the names of their respective Colleges, and the Committee sincerely hope that a succession of students will occupy these places and take advantage of the opportunity thus afforded them of becoming acquainted with marine animals and plants in their native haunts. The very perfection of modern Laboratory methods in our Colleges renders it the more necessary that all students of natural

science should, during some part of their course, study a
living Fauna and Flora so as to realise the natural
appearance, mode of occurrence and environment of the
forms which they know otherwise only from books,
lectures, and preserved specimens.

The Committee of the Liverpool Free Public Museum
has also become a subscriber to the Station in order that
one or other of the officials of the museum may have the
use of a work place in the Laboratory during the period
of weeks covered by the annual subscription. Other public
bodies, or local scientific societies are invited to become
subscribers in a similar manner (see Regulations, p. 49).

THE AQUARIUM.

Over 300 visitors paid 3d. for admission to the aquarium
during the summer when it was on exhibition, and many
other visitors, including members of the Isle of Man
Natural History and Antiquarian Society, and of the
British Association party who visited the island in
September were taken round the establishment. The
usual animals, including most of the common invertebrates of the shore and shallow water, and a few fishes,
have been on exhibition. Some of the commonest forms,
such as the sea-anemones, the hermit crabs, shrimps and
prawns, and the flat-fish, are those that are most admired
and excite most interest. People, as a rule, are pleased
to see something they can recognise, and like to be told
or shown something new about an old friend.

Our visitors to the Aquarium have ranged from the
Governor (Lord Henniker) and his party to the fisher lads
of the neighbourhood and the "trippers" from the inland
towns of Lancashire and Yorkshire. We hear many
curious remarks, and extraordinary opinions, in regard to
the lower living things in the sea, when we take visitors

round the Aquarium. It is remarkable how prevalent the idea still is, even amongst people with some education, that sea-anemones, medusæ, and sponges are not living things, but are merely " growths " as they express it; and I have more than once, when talking about some well known fish which was swimming before us in the tank, been interrupted with the incredulous—almost indignant —exclamation, " But a fish is not an animal—is it ? " Some are incredulous in regard to the real life-histories and habits we tell them of, or show them ; others are extraordinarily credulous as to impossible stories they " have heard " of our doings on dredging expeditions. The boatmen of the neighbourhood apparently regale the summer visitors with sensational accounts of the wonders we find in the deep sea. A party of ladies and children came to the Aquarium on one occasion, and after looking anxiously round the tanks said, " But where are the pigmy elephants ? We were told that you had dredged up some sea-elephants three inches long with tusks and trunk complete." Whether this was a highly-coloured version of our having found some large living specimens of *Dentalium*, the so-called " tusk shell " or " elephants' tooth shell," or was wholly imaginary, we failed to discover.

Children are usually much interested in our tanks, and with a little encouragement and help become keen collectors and quick observers. We have had many specimens of anemones, crabs, worms, and small fish brought to us from the rock-pools by boys and girls, who then take an additional interest in the tank or dish they have helped to stock, and bring their companions to see " the ones I caught." If we are able by some little experiment, or some observations, to demonstrate to them from their own specimens a fact in Natural History or an example of some principle of Biology, not only does it add to their gratification but I am sure has considerable educational influence by opening their minds to new realms of thought, new methods of questioning nature.

For example, if a boy brings us a light-coloured shanny caught in a shallow exposed pool, we can place the little fish in a deep vessel in semi-darkness under a table, or cover it with some brown sea-weed, the result being that when the boy comes next day to look for his specimen, he has been known to exclaim, "Hullo! where is my shanny? There is only a black one here." It is then easy, by putting the fish into a shallow white dish in the bright sunlight, in a short time to turn the black shanny into what he recognises as the light-coloured one he caught. You can then tell him of the beautiful pigment cells of the skin and show them to him under the microscope in a small living fish, in a watch-glass full of sea-water. You can show him a speckled shrimp hiding in sand and a mottled shrimp in gravel, and the little prawn *Virbius* which may be almost any colour according as you change its surroundings from green to red or to dark brown sea-weeds. You explain the difference in pigmentation on the upper and lower sides of a flat fish, you remind him of the Chamæleon, tell of Sir Joseph Lister's observations on the change of colour in the skin of the Frog, and—most beautiful experiment of all—show him the "blushing" of the newly-born cuttle-fish. From this there opens up a wide range of physiology, of the influence of light and the controlling action of nerves, not to mention natural selection and evolution in general.

This is only one of many examples that might be taken. Almost any of the common marine animals, if carefully watched as to structure and habits, show us interesting cases of adaptation to their surroundings and mode of life.

We are often asked about the sea-anemones of Port Erin. It may be useful to give here a list of the different kinds we have found round the southern end of the Isle of Man, between Fleshwick Bay and Port St. Mary. One of the best places to collect anemones is the Calf Sound, especially on the group of rocks called "The Cletts," at low water. Mr. A. Chopin, of Manchester, has kindly given me some additions to the list. I give first the correct modern* name of each species, then after each is placed in brackets the name used by Gosse in the "Actinologia Britannica," the book generally consulted, or any other name that has been much employed, and finally is given in each case the English name used by Gosse and other writers.

SEA-ANEMONES (ACTINIARIA) OF PORT ERIN.

Family I. PROTANTHIDÆ.

Corynactis viridis, Allm. (do. Gosse)—The Globehorn.
Capnea sanguinea, Forb. (do.)—The Crock.

Family II. HEXACTINIDÆ.

Halcampa chrysanthellum, Peach (do.)—the Sand Pintlet.
Anemonia sulcata, Penn. (*Anthea cereus*)—the Opelet.
Actinia equina, Linn. (*A. mesembryanthemum*)—the Beadlet.
Bunodes verrucosa, Penn. (*B. gemmacea*)—the Gemlet.
Urticina crassicornis, Müll. (*Tealia crassicornis*)—The Dahlia Wartlet, or Crass.
Sagartia miniata, Gosse (*Heliactis miniata*)—the Scarlet-fringed Anemone.
Sagartia venusta, Gosse (*Heliactis venusta*)—the Orange-disked Anemone.

* Professor Haddon has kindly revised the nomenclature and classification.

Sagartia nivia, Gosse (*Heliactis nivea*)—the Snowy Anemone.

Sagartia rosea, Gosse (*Heliactis rosea*)—the Rosy Anemone.

Sagartia sphyrodeta, Gosse (*Sagartia sphyrodeta*)—the Sandalled Anemone.

Sagartia lacerata, Dall. (? not distinguished by Gosse).

Cylista viduata, Müll. (*Sagartia viduata*)—the Snake-locked Anemone.

Cylista undata, Müll. (*Sagartia troglodytes*)—the Cave-dwelling Anemone.

Cereus pedunculatus, Penn. (*Sagartia bellis*)—the Daisy Anemone.

Adamsia palliata, Boh. (*Adamsia palliata*)—the Cloak Anemone.

Metridium dianthus, Ellis (*Actinoloba dianthus*)—the Plumose Anemone.

Paraphellia expansa, Hadd. (not known to Gosse).

Family III.   ZOANTHIDÆ.

Epizoanthus arenaceus, D. Ch. (*Zoanthus couchii*)—the Sandy Creeplet.

Family IV.   CERIANTHIDÆ.

Cerianthus lloydii, Gosse (do.)—the Vestlet.

Of many of these species several marked varieties occur.

SEA-FISH HATCHING.

In several previous reports we have discussed the suitability of the Port Erin Biological Station as a hatching establishment for sea-fish. No public body seems willing to move in the matter to the extent of erecting the necessary building and plant; but, with the help of a small grant to fit up temporary wooden tanks, we under-took, during the last hatching season (Easter 1896), a series of experiments for the Lancashire Sea-Fisheries Com-mittee. As the result of these experiments we success-fully fertilised the eggs (obtained from the parent fish caught with the trawl) of the grey Gurnard (*Trigla gurnardus*), the lemon Sole (*Pleuronectes microcephalus*), and the Witch (*Pleuronectes cynoglossus*), and kept them in the tanks until they hatched out as young larvæ. We were not prepared in this first season to proceed with the rearing; but we propose, with additional tanks and an improved circulation of water, to carry the experiments a stage further next year.

Last spring we fitted up, in the lower floor of the Aquarium house, three wooden hatching tanks, each 5 ft. by 3 ft. by 1 ft., and so arranged like steps that water could flow by bamboo spouts covered with a fine silk net through the series of tanks. From the lowest wooden tank the water fell into the concrete floor tank, into which dipped an endless chain formed of an india-rubber belt bearing numerous little buckets. This chain of buckets revolved on a drum, octagonal in section, which was kept in motion by india-rubber belting passing from its axle to a pulley on a large water wheel actuated by the fresh water tap.*

Consequently, by turning the tap the whole apparatus was set in motion, and the sea-water from the concrete floor tank was raised by the little buckets and emptied into a sloping wooden trough which guided it to the upper hatching tank. Thus the same water was used over again,

---

* The tanks and the water motor apparatus were made most carefully and ingeniously, from my plans, by Mr. R. Garner, superintendent of the wood-working department at University College, Liverpool.

a couple of gallons of fresh sea-water being added to the system every day.

During the period when the apparatus was working the temperature and the specific gravity of the water in the tanks kept fairly constant, the extremes of the range being:—

Temperature from 50° to 53° F., and

Specific gravity from 0·0265 to 0·0270.

Each of the three tanks had a partition 1 foot from its outflow end which stopped 2 inches from the top, and a second partition 6 inches nearer the end, which reached the top but stopped short 2 inches from the bottom of the tank. In the two compartments imperfectly separated by this last partition, clean washed sand was placed so as to reach to about 4 inches from the bottom. Consequently all water escaping from the tank had to flow over the first partition and *under* the second, filtering through the bed of sand as it went. The object of this was to form a sand trap which would let the water pass through, but keep back the suspended fish eggs and embryos. By this method the same water can be used to circulate through several tanks containing different kinds of embryos.

A more detailed account of these fish-hatching experiments at Port Erin will be given in the Annual Report of the Sea-Fisheries Laboratory to the Lancashire Committee.

DREDGING EXPEDITIONS.

Since the last report, the Committee have organised eight dredging expeditions, nearly all in steamers, as follows:—

I. November 24, 1895.—Small boats. Localities dredged:—Port Erin Bay, in depths up to 7 fathoms.

II. February 2, 1896.—Hired steamer 'Rose Ann.'

Localities dredged and trawled:—Through the Calf Sound, and off its eastern and western ends, at depths of 16 to 20 fathoms.

III. March 14, 1896.—Sea-Fisheries steamer 'John Fell.' Off Port Erin.

IV. April 5, 1896.—Hired steamer 'Rose Ann.' Localities trawled :—Out in the deep channel, 12 miles S.W. of Calf; bottom reamy mud, with many spawning fish ; depths 40 to 50 fathoms.

V. April 21-24, 1896.—Sea-Fisheries steamer 'John Fell.' Localities trawled:—Deep channel, 12 miles S.W. of Calf, and further north to opposite Port Erin ; also west of Dalby, 8 miles off; reamy bottom ; depths 20 to 40 fathoms.

VI. May 29 and 30, 1896.—Sea-Fisheries steamer 'John Fell.' Localities:—Estuary of the Wyre and around Piel Island, in Barrow Channel ; shallow water.

VII. August 31, 1896.—Mr. Woodall's S.Y. 'Vallota.' Localities dredged and trawled :—Between Port Erin and Calf Island ; depth 17 to 22 fathoms.

VIII. September 19, 1896. — Sea-Fisheries steamer 'John Fell.' Localities :—Liverpool Bay, Hilbre Swash and the Rock Channel ; 4 to 10 fathoms.

Two of these expeditions—those at Easter in the 'Rose Ann,' and at the end of April in the 'John Fell'—were particularly successful, and resulted in the capture of a number of new and interesting species. Amongst these is a large green Gephyrean worm, which is either *Thalassema gigas*, M. Müller, or a new species of *Thalassema* with a remarkable pigment (see p. 108) ; and a Cumacean, for which a new genus is necessary.

Additions have been made during the year to most of the groups of invertebrate animals, and these will be found noted in the lists below; in addition Mr. A. O. Walker has prepared the following special account of the higher Crustacea obtained on these expeditions:—

CRUSTACEA.

The following species of MALACOSTRACA have been added to the fauna since the last report. Nearly all were dredged off the S. end of the Isle of Man in the 'John Fell' expedition, from April 22 to 24, 1896.

PODOPHTHALMA:—*Portunus corrugatus* (Pennant).— S.E. of Calf Sound, 26 fathoms.

Nika edulis, Risso.—Co. Down Coast (Ascroft); and from stomach of whiting, 12 m. S.W. of Chicken Rock, 33 fathoms.

SCHIZOPODA:—*Erythrops serrata*, G. O. Sars; 12 m. S.W. of Chicken Rock, 33 fathoms.

Siriella armata (M. Edw.).   Port Erin harbour, April.

CUMACEA:—Family Leuconidæ—*Leuconopsis*, n. gen.

Female with a distinct two-jointed appendage to the fourth pair of feet, not furnished with natatory setæ. Lower antennæ short, with the third joint conical, with a minute one-jointed rudimentary flagellum. Rami of uropoda subequal.

Male with the third pair of feet each provided on the second joint with a pair of curved blade-like processes.

Remaining characters as in *Leucon*.

Leuconopsis ensifer, n. sp.

Female:—Carapace about as long as the free thoracic segments, dorsal crest of fourteen teeth beginning about the middle of the upper margin, and curving down to the base of the rostrum; a small tooth on the upper and near the posterior margin; lower margin with the anterior half coarsely toothed, and forming with the anterior margin an acute angle the upper portion of which is finely toothed. Rostrum about quarter the length of the carapace, obliquely truncate; almost horizontal lower margin with two or three teeth near the extremity and two or three near the base.

Fourth pair of legs with an exopodite or imperfect natatory appendage, two-jointed, reaching nearly to the end of the first joint, which is almost as long as the remaining four.

Telson triangular, as in *Leucon*.

Uropoda with peduncle and both rami subequal in length; peduncle almost spineless, inner ramus with six unequal spines on the inner and two on the outer side of the first joint; second joint with two very short and slender spines on the inside; outer ramus obliquely truncate, with five plumose setæ on the inner side and four at the end. Length 5½ mm.

Male :—Upper margin of carapace as long as the free segments; lower margin with five or six teeth on the anterior half increasing in size anteriorly, forming a right angle with the anterior margin which has five teeth just below the rostrum, the second from the rostrum being the largest; rostrum horizontal, blunt, about one-sixth the length of the carapace, with five small teeth on the lower margin.

First pair of legs with seven teeth on the lower margin of the first joint. Second pair with a large spine at the distal end of the second, and two unequally long spines at the end of the third joint. Third pair with an append- age on the second joint, consisting of two parallel curved blades, twice as long as the succeeding three joints. Length 8½ mm.

The above interesting species has a general resemblance to *Leucon*, from which genus, however, it may be at once distinguished by the appendages on the fourth pair of legs in the female and the third pair in the male. It was taken in the tow net attached to the back of the trawl net on April 22, 12 m. S.W. of Chicken Rock, 33 fathoms.

Eudorella emarginata (Kröyer).—One female. Same locality as last.

Campylaspis glabra, G. O. Sars.—Three specimens, from same locality as last. A Mediterranean species, not previously recorded from British Seas. I have specimens taken by Mr. Ascroft off the Ile d'Yeu.

AMPHIPODA:—*Normanion quadrimanus* (Bate and Westwood).—One small specimen; length 2 mm., 6 miles W.S.W. of Calf, 23 fathoms.

Stenothoë crassicornis, n. sp.

Three males. Same locality as last.

Mandibles without a palp.

Maxillipedes with the basal lobe very small, divided to its base.

Antennæ stout, the flagellum of the lower but little longer than the last joint of the peduncle; its first joint almost as long as the remaining four together.

First gnathopods as in *S. marina*.

Second gnathopods with the palm of the propodos defined near the base by a triangular tooth, the distal extremity expanded and cut into four blunt lobes, of which the proximal is much the largest; dactylus with a prominence on the inner margin, coinciding with the palmar lobus.

Peræopods short and strong, the third (meros) joint in the last three pairs much produced backwards, as in *Probolium calcaratum*, G. O. Sars.

Third uropods with four spines on the upper surface of the peduncle, which is twice as long as the first joint of the ramus.

Telson with three pairs of dorsal spines on its proximal half, the first pair the smallest. Length 2 mm.

In the form of the hand of the second gnathopods this species approaches *S. tenella*, G. O. S., and *S. dollfusi*, Chevreux; but both these (perhaps identical) species are remarkable for the length and slenderness of their antennæ and peræopods.

Halimedon parvimanus, Sp. Bate.—Five or six specimens, 12 m. S.W. of Chicken Rock, 33 fathoms.

Argissa hamatipes (Norman) = *Syrrhoë hamatipes*, Norman, 'Brit. Ass. Rep.,' 1868 (1869). p. 279. Same locality as last.--Two females, one with ova, 2 mm. long.

FIG. 1.—Plan of the L.M.B.C. District.

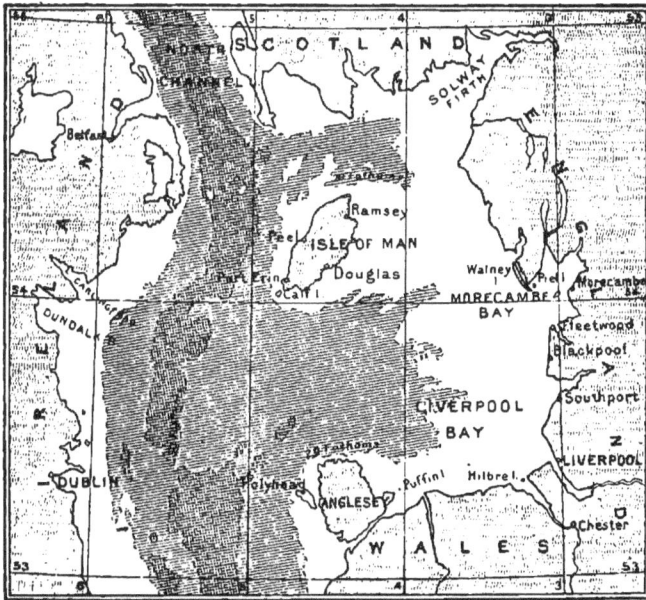

Prof. G. O. Sars, with some hesitation, follows Boeck in placing *Argissa* among the Pontoporeiidæ, but there can be little doubt that Canon A. M. Norman was right in classing it with the Syrrhoidæ.

Gammarus campylops, Leach.—Brackish pond near Colwyn Bay ; also Port Erin harbour.

Fig. 2.—Section across the Irish Sea through Douglas.

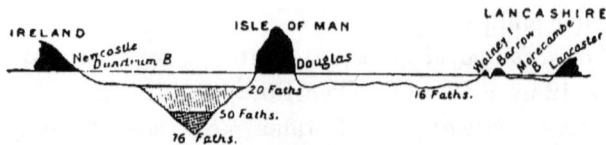

OTHER ADDITIONS TO THE FAUNA.

In making a thorough revisal of our lists of the Local Fauna and Flora for the purpose of presenting a complete report to the British Association at the recent Liverpool meeting, it was found that some species we had discovered in the past had escaped record, a few others had been recorded under names that have now been superseded, while a considerable number of previously unknown forms have turned up as the result of the year's work. It was stated in the British Association Report that all these species would be duly recorded, with their localities, in this the Tenth Annual Report. They are as follows :—*

FORAMINIFERA :—*Hyperammina arborescens*, Norm., off Peel, Isle of Man.

Gypsina vesicularis, P. & J., off the Isle of Man.

COELENTERATA :—*Coryne vaginata*, Hincks, I. of Man S.

Corymorpha nutans, Sars, off Port Erin.

Hybocodon prolifer, Agassiz, off Port Erin.

Podocoryne carnea, Sars, off Port Erin.

Phialidium temporarium, Browne, off Port Erin.

---

* For these additional records we are indebted to a number of the Naturalists who are now working at the Fauna of the Irish Sea—especially to Dr. G. W. Chaster, Mr. I. C. Thompson, Mr. A. Leicester, Mr. E. T. Browne, Mr. A. O. Walker Miss L. R. Thornely, and Mr. Andrew Scott.

Tiaropsis multicirrata (Sars), off Port Erin.

Agalmopsis elegans, Sars, off Port Erin.

Virgularia mirabilis, Lamk., off the west of the Isle of Man, deep water.

VERMES :—*Planaria littoralis*, Van Ben., Port Erin.

Thalassema, sp. (? n. sp.), deep water off Port Erin.

Polynoe reticulata, Clap., this is recorded in Vol. III. of " Fauna," p. 136, as *P. extenuata*, Grube.

P. semisculpta, John. Recorded in Vol. III, p. 137, as *P. propinqua*, Mgrn.

Chætopterus variopedatus, Ren. Recorded as *C. insignis*, Baird, in Vol. III., p. 158.

Arenicola ecaudata, John., Port Erin, &c.

Autolytus longisetosus (Orsted), Port Erin.

Flabelligera affinis, Mgrn., recorded in Vol. III., p. 159, as *Siphonostoma diplochaitos*, Otto.

POLYZOA :—*Membranipora spinifera*, Johnst., off Garwick Head, and off South Coast, Isle of Man.

Schizoporella alderi, Busk, off the West Coast of the Isle of Man.

Smittia cheilostoma, Manz., on shells, dredged off the Calf of Man.

Cylindræcium giganteum, Busk, Puffin Island.

Loxosoma phascolosomatum, Vogt, on *Phascolosoma vulgare*, from near Puffin Island.

CRUSTACEA :—*Portunus corrugatus* (Penn.), S.E. Calf Sound, 26 faths.

Nika edulis, Risso, Co. Down Coast, 12 m. S.S.W. of Chicken Rock, in whiting's stomach.

Erythrops serrata, G. O. Sars, 12 m. S.W. of Chicken Rock, 33 faths.

Siriella armata (M. Edw.), Port Erin Harbour, April.

Nebalia bipes, (M. Edw.), 12 m. S.W. of Chicken Rock, 33 faths.

Leuconopsis ensifer, A. O. Walker, 12 m. S.W. of
Chicken Rock, 33 faths. (new genus—see p. 98).

Eudorella emarginata, (Kröyer), 12 m. S.W. of
Chicken Rock, 33 faths.

Campylaspis glabra, G. O. Sars, 12 m. S.W. of
Chicken Rock, 33 faths.

Normanion quadrimanus (Bate & Westwood), 6 m.
W.S.W. of Calf, 23 fathoms.

Stenothöe crassicornis, A. O. Walker, same locality.

Halimedon parvimanus, Sp. Bate, 12 m. S.W. of
Chicken Rock.

Argissa hamatipes (Norman), same locality.

Gammarus campylops, Leach, brackish Pond, near
Colwyn Bay; and Port Erin Harbour.

OSTRACODA :—(?) *Argillœcia cylindrica*, Sars, off Peel,
Isle of Man, 50 faths.

Bairdia acanthigera, Brady, off Southport, 25 faths.

Cythere albomaculata, Baird, Southport, and Port St.
Mary.

C. globulifera, Brady, near Nelson buoy, off Ribble,
14 faths.

C. lutea, O. F. M., in *Laminaria*, Port St. Mary.

C. robertsoni, Brady, Southport shore.

Eucythere declivis, Norman, off Peel, 50 faths.; and
Nelson buoy, 14 faths.

Cytherura nigrescens, Baird, Roosebeck Mussel Bed.

C. acuticostata, Sars, Southport, and Nelson buoy, 14
faths.

Cytheropteron punctatum, Brady, off Peel, 50 faths.

Paradoxostoma flexuosum, Brady, Roosebeck Mussel
Bed.

Cytheridea elongata, Brady, Morecambe Mussel Beds.

C. torosa, Jones, Nelson buoy, 14 faths.

Cytherideis subulata, Brady, Roosebeck Mussel Bed.

Bythocythere simplex, Norman, off Peel, 50 faths.

COPEPODA :—*Scolecithrix hibernica*, A. Scott, between Isle of Man and Ireland, off County Down Coast.

Centropages typicus, Kröyer, off Port Erin.

Pseudocyclops crassiremis, Brady, in Neritic material, off Spanish Head.

P. obtusatus, Brady and Rob., in Neritic material, off Spanish Head.

Lamippi proteus, Clap., from *Alcyonium digitatum*, off Peel, Isle of Man.

L. forbesi, T. Scott, from *Alcyonium digitatum*, trawled off Peel, Isle of Man.

Stenhelia herdmani, A. Scott, in Neritic material, off Spanish Head.

Stenhelia similis, A. Scott, in Neritic material, off Spanish Head.

Canthocamptus palustris, Brady, Ainsdale Mussel Bed.

Tetragoniceps trispinosus, A. Scott, in Neritic material, off Spanish Head.

Laophontodes bicornis, A. Scott, in Neritic material, off Spanish Head.

Thalestris forficuloides, T. and A. Scott, off Port Erin.

Pseudothalestris major, T. and A. Scott, off Port Erin.

Idya elongata, A. Scott, Morecambe Mussel Beds.

Parartotrogus richardi, T. and A. Scott, off Peel, amongst refuse.

Collocheres elegans, A. Scott, off Port Erin.

Chondracanthus merlucci, Holt, from Hake, caught off Calf of Man.

Lerneonema sprattæ, Sow., from Sprats caught in the Mersey.

Lerneopoda galei, Kr., from dogfish caught in Menai Straits.

PYCNOGONIDA :—*Nymphon rubrum*, Hodge, Turbot Hole,
Puffin Island.

MOLLUSCA :—*Mytilus phaseolinus*, Phil., Isle of Man, S.
Leda minuta, Müll., var. *brevirostris*, Jeff., Port Erin.
Mactra solida, var. *truncata*, Mont., Puffin Island.
Mactra subtruncata, var. *striata*, Brown, Southport.
,, var. *inæqualis*, Jeff., Southport.
· *Saxicava rugosa*, var. *arctica*, L., Isle of Man South.
Panopea plicata, Mont., Southport.
Teredo megotara, Han., and var. *mionota*, Southport.
T. norvegica, Speng., var. *divaricata*, Desh., South-
port.
Siphonodentalium lofotense, Sars, off Peel, I. of Man.
Trochus zizyphinus, var. *humilior*, Jeff., Port Erin.
Emarginula fissura, var. *elata*, Jeff., central area.
Rissoa striatula, Mont., Waterloo.
R. zetlandica, Mont., Isle of Man, South.
R. striata, var. *arctica.*, Lov., Puffin Island, Port
Erin, and Port St. Mary.
,, var. *distorta*, Mar., Puffin Island.
Hydrobia ventrosa, Mont., Colwyn Bay and South-
port.
Odostomia minima, Jeff., Isle of Man South; and
Puffin Island.
O. clavula, Lov., Southport.
O. rissoides, var. *glabrata*, off Port Erin.
O. albella, Lov., Port St. Mary.
,, var. *subcylindrica*, Marsh., Port Erin.
O. insculpta, Mont., Puffin Island, Southport and
Isle of Man South.
O. turrita, var. *nana*, Jeff., Port Erin.
O. interstincta, var. *suturalis*, Phil., Southport, Puffin
Island, and Isle of Man.
Cerithium perversum, L., Isle of Man South.

Buccinum undatum, L., var. *littoralis*, King, Southport.

,, var. *jordoni*, Chaster (MS.), Southport.

,, monstr. *acuminatum*, Brod., Southport.

Trophon truncatus, Str., var. *alba*, Jeff., Port Erin.

Fusus propinquus, Ald., var. *jeffresiana*, Fisch. (see Fauna, vol. I., p. 244).

. *F. antiquus*, L., var. *alba*, Jeff., Isle of Man South.

F. gracilis, Da C., var. *convoluta*, Jeff., Port Erin.

Defrancia linearis, Mont., and var. *æqualis*, Jeff., Isle of Man.

Pleurotoma attenuata, Mont., Isle of Man South.

Utriculus truncatulus, var. *pellucida*, Bro., Puffin I.

U. mammillatus, Phil., Isle of Man South.

Philine catena, Mont., Isle of Man South.

Melampus bidentatus, var. *alba*, Turt., Isle of Man S.

TUNICATA :—*Fritillaria*, sp., Port Erin.

CEPHALOCHORDA :—*Branchiostoma lanceolatum*, Pall.

PISCES:—*Zeugopterus norvegicus*, Gunth., S.W. of Chicken Rock.

MAMMALIA :—*Phoca vitulina*, Linn., Mersey.

Balænoptera musculus, Linn., North Coast of Wales.

OTHER INVESTIGATIONS.

The Rev. T. S. Lea has been continuing his observations on the distribution of the species of sea-weeds and
associated animals on the shore at Port Erin. He has
produced a very beautiful series of photographs which
were exhibited in the Loan Museum at the British
Association Meeting, and were also shown as lantern
slides to section D. at one of the forenoon sittings.
Besides photographs of Algæ, natural size, *in situ*, and as

microscopic objects, Mr. Lea has a number of views into rock pools, taken with his vertical camera, showing anemones fully expanded and fish lying on variously coloured floors. Some of the foreign Biologists were much pleased with these photographs, and Mr. Lea has supplied Prof. Chodat with a number of reproductions of the lantern slides for use in lectures at the University of Geneva. Mr. Lea has kindly presented the complete series of his photographs as lantern slides to the New Museum of Zoology at University College, where they will be permanently on exhibition, classified and labelled, so as to illustrate the littoral fauna and flora at the south end of the Isle of Man.

The large green *Thalassema* of which several specimens, all more or less mutilated, were trawled from the deep water to the S.W. of Port Erin at Easter, seems to be an undescribed form. It must be, when perfect, about 20 cm. in length over all, and 10 or 12 mm. in average thickness. The extended proboscis measures about 10 cm. in length, and 15 to 20 mm. in breadth. In appearance it most nearly resembles *T. gigas*, M. Müller, but differs from that species in the relative proportions of body and proboscis, in the greater breadth of the proboscis, and in the shape of its extremity. The colour is a rich green. Prof. Lankester, who has seen one of the specimens, calls it a " beautiful chrome green," and says " it is exactly the colour of my specimen of *Hamingia*."

Our species differs from *Hamingia* (as defined by Lankester) in having strong setæ present at the genital pores in the female, and from *Bonellia* (another allied, green form) in the shape of the proboscis and other particulars. It is, in its anatomical characters, a member of the genus *Thalassema*, but differs in some points from all the known species. The ciliated funnels of the cloacal nephridia are borne on tangled twigs of a snow white colour given off by the dark brown central tube of the organ. There is only a single pair of anterior nephridia. These contain ova, but no rudimentary males were found. All our specimens are females. A full description, with figures, of this new species will be published shortly.

Prof. Sherrington and Dr. Noël Paton have independently investigated the green pigment spectroscopically. They report that it is a very remarkable and apparently unknown pigment which is not allied to hæmoglobin or chlorophyll. It is not a respiratory pigment and is apparently nearer to " bonellein," described by Dr. Sorby from the Gephyrean *Bonellia viridis*, than to any other known pigment, but differs markedly in some respects and cannot be identical with it. Prof. Sherrington gives the spectral characters as follows :—" The solution of the pigment in formol (5% solution) exhibited considerable absorption of the violet end of the spectrum (nearly as far as solar line F, to λ 468), less of the red end (to solar a, λ 716), and a single broad band of absorption in the red between C and D with its centre at λ 617 and extending from λ 602 to λ 630*. No other absorption band existed. Hæmoglobin in formol solution exhibits the spectrum of *reduced* hæmoglobin. There is no similarity between the spectrum of the pigment here examined and that of hæmoglobin. On the other hand the position of the band recalls that of the strong band given by bonellein λ 643 to λ 617 (Sorby). But bonellein was not examined in formol solution. No other definite absorption band was given by the Thalassema pigment in formol."

Mr. J. H. Ashworth, Demonstrator of Zoology at the Owens College, spent some weeks at Port Erin in August, and besides collecting, preserving, and examining various

---

* Dr. Noël Paton gives the centre of the band at λ 640.

marine forms devoted himself specially to the investigation of the Lug-worms, *Arenicola*. The following extract is from a letter written by Mr. Ashworth to the Hon. Director on leaving the Laboratory :—" During the last weeks I was engaged nearly all the time upon *Arenicola*. I took your advice and went over to Bay-ny-Carrickey last Monday, and in about one-and-a-half-hour's searching obtained five specimens of *Arenicola ecaudata*, and on Wednesday I went to the same place again and obtained about a dozen more. I have got the ova and sperms from them, the latter almost ripe, and I have made several dissections and find many points of difference between this species and *Arenicola piscatorum*. I intend to follow this work up after my return to College in October. I have enjoyed my visit to Port Erin very much, and have found the laboratory very convenient for work. I am greatly obliged to you for the many valuable suggestions and help you have given me while I have been there, and I thank you most sincerely for them."

Mr. Ashworth is preparing a paper on *Arenicola piscatorum* and *A. ecaudata*, which will be laid before the Liverpool Biological Society during the present session.

Professor Weiss reports that he investigated the Diatoms of the plankton during the month of April, with special reference to the variation in the preponderance of the various forms at different times. In this connection he observed the breaking-up of the protoplasm of *Chætoceros* and *Coscinodiscus* into eight or sixteen nucleated masses within the parent frustule, as recently described by Mr. George Murray before the Linnean Society of London (June 18th). He also collected a large quantity of the Coralline Algæ, both shore and deep water forms, an account of which he is preparing for the Liverpool Biological Society.

Mr. Hiern, with the assistance of several other Botanists present at the Biological Station at the end of September, compiled a list of Manx plants which will appear shortly in the " Journal of Botany."

Dr. C. H. Hurst reports as follows in regard to some specimens of *Nymphon* dredged from the " Turbot hole," near Puffin Island :—

" There were 14 specimens :—

1 was a typical *N. gracile*.

3 were spiny *N. gracile*.

2 „ *N. gracile*, but with *flat* ocular tubercle.

2 „ *N. gracile*, but with the tarsus of *N. brevitarse*.

3 „ young, and doubtfully *N. gracile*.

3 „ recorded as *N. rubrum*—but *none were red*.

" The average species-monger would make *six* species of those 14 specimens out of one " hole." In spite· of the decision of Sars that *N. gracile* and N. "*rubrum*" (which is *not* red) are distinct, I do not believe they are. Typical specimens of both were found as well as some specimens possessing some characters of *N. brevitarse* : but there were also found, *in the same hole*, other forms which bridge over the gap between the supposed species. The differences between the most extreme forms were less than those between individuals of *Bombus terrestris* (workers) found in a single nest and far less than the differences between *Araschnia* (*Vanessa*) *prorsa* and *A. levana*, which are now known to be a single species. The species (*N. gracile-rubrum-brevitarse*) may be a polymorphic species, but I believe it is *one* species and not three."

Mr. James Hornell has supplied the following notes as being supplementary to his Report on the Polychætous Annelids of the L.M.B.C. District published in vol. iii. of the " Fauna " (and Trans. L'pool Biol. Soc., vol. v., p. 223).

" The following remarks are intended to bring the above Report up to the present date, so far as the writer (who has been absent from the district for several years) is able.

p. 233. *Polynoë extenuata*, here mentioned, should be referred to *P. reticulata*, Claparède.

p. 234. *Polynoë propinqua*, Mgrn. is the *Lepidonotus semisculptus* of Johnston's Brit. Museum Catalogue, and hence should appear now as *Polynoë semisculpta*.

p. 235. Undoubtedly the *P. floccosa* of Prof. M'Intosh's list is also a synonym for the last named species, *P. semisculpta*.

p. 255. Joyeux Lafuie (Archiv. d. Zool. Exp. (Ser. 2), viii., p. 244, 1890) shows that only one European species of *Chætopterus* exists, viz., *C. variopedatus*, Ren., hence this name replaces that of *C. insignis*.

p. 256. The form here entered as *Siphonostoma diplochaïtos* has been shown by further investigation to be the *Flabelligera affinis* of Malmgren.

p. 248. Here is given a note on the embryology of *Arenicola* and of *Scoloplos*, and while the remarks relating to the latter have been found to require no correction, my experience on the Jersey Coast has brought up facts which show that an error was made as to the parent species of those larvæ described as belonging to *Arenicola*; hence this note must be corrected by the substitution of the name *Phyllodoce maculata* for that of *Arenicola*. With this alteration of name the description holds good. In the description of pl. xiv., figs. 12 to 21 refer therefore to the embryology of *Phyllodoce maculata*, and not to *Arenicola*."

Since Mr. Hornell left Liverpool for Jersey three additional species of Polychæta have been added, viz., *Arenicola ecaudata*, Johnston, *Magelona papillicornis*, Müll., and *Autolytus longisetosus*, Örst.

Professor Boyce has continued his important investigations into the bacteriology of the oyster and its possible connection with disease in man. He has drawn up a report upon the subject which was read before the Liverpool Meeting of the British Association. As an account of the present state of the question, and a summary of Prof. Boyce's bacteriological work will be given, in a few weeks, in the Annual Report of the Lancashire Sea-Fisheries Laboratory, it is unnecessary to do more here than to state that the fresh experiments on inoculating Oysters with the typhoid Bacillus and keeping them under observation, both in stagnant and in running sea-water, show (1) that the typhoid organism does not multiply in the stomach or tissues of the Oyster, (2) that Oysters fresh from the sea contain fewer bacteria (chiefly the common colon bacillus) than those that have been stored or kept in shops, and (3) the power of the Oyster to get rid of bacterial infection when placed in a stream of running water, there being a great diminution or total disappearance of the *Bacillus typhosus* in from one to seven days.

Mr. Edward T. Browne has sent me the following notes on the species of Medusæ, and other constituents of the pelagic fauna, taken by him at Port Erin during his work there in April 1896 :—

" This visit to Port Erin in April, 1896, was specially made to obtain more specimens of the interesting medusa *Hybocodon prolifer*, for the completion of my work on the development of its ova. This medusa was fairly plentiful in 1893, and very abundant in April 1894, but was unfortunately absent in 1896.

" The pelagic fauna throughout the whole of April 1896. was conspicuously poor in medusæ and other pelagic animals usually found in the spring of the year. This scarceness is difficult to account for, as the previous winter had been mild and the spring favourable for an early fauna.

"The temperature of the sea was 48°F. at the beginning of April, two degrees higher than in 1894, when the temperature did not reach 48° until 26th of April. The fauna, nevertheless, was more like that usually recorded for February than for April. Diatoms throughout the month were exceedingly abundant and aided by the gelatinous algæ quickly clogged the meshes of the tow-net. Often when the can at the end of the net was emptied into a glass bottle, the contents had the appearance of thick pea-soup, so great was the abundance of diatoms.

"The medusæ showed a decrease in the number of species compared with 1894, and a great decrease in quantity, especially in the case of *Margellium octopunctatum*, which swarmed in the Bay in 1894, but of which only four specimens were taken in 1896. Another noticeable feature was that all the medusæ, except *Obelia*, were young forms and usually belonged to the earliest free-swimming stage.

"The ctenophores usually plentiful in Port Erin Bay in the spring were entirely absent. A species of *Fritillaria* made its first appearance on 21st of April, and a single specimen of the larval *Magelona* on 29th of April.

"*Agalmopsis elegans*, Sars (recorded as *Halistemma*, sp.? Fauna, iv. p. 279), first taken in April 1894, did not make its appearance in 1896."

Mr. Browne will communicate this session to the Biological Society a revised list of the L.M.B.C. Medusæ.

The Committee have lately purchased from Mr. M. Treleaven Reade, the inventor, one of his folding "Shell-bend" boats for the use of the Biological Station.

Although workers at Port Erin will no doubt in the future, as in the past, make considerable use of the ordinary pleasure boats of the bay, still it frequently happens that one, two, or three desire to go out tow-netting, or collecting round the rocks or the break-water at low tide on occasions when it is inconvenient or impossible to hire a boat. Under these circumstances the " Shell-bend " will be most handy. It is a flat-bottomed dinghy 10 feet in length, with plenty of room for three men to work tow-nets and other collecting implements. When hauled ashore, the sides fold down on the bottom, and then one or two men can easily carry the boat for a considerable distance.

THE VISIT OF THE BRITISH ASSOCIATION.

At the conclusion of the Meeting in Liverpool last September, about 100 members of the British Association crossed to the Isle of Man for the purpose of spending five days in exploring the Natural History and Antiquities of the Island. The party broke up into four sections, of which two (the Archæologists and the Geologists) made their headquarters at Douglas, while the other two (Zoologists and Botanists) went on to Port Erin and lived at the Bellevue Hotel. The weather throughout the visit was very unsuitable for Biological work. The steam-trawler " Rose Ann " was in attendance, but it was impossible to go to sea in her, although attempts were made both from Port Erin and Port St. Mary. The time was spent in shore-collecting at various parts of the coast, and in searching for the rarer Algæ and encrusting animals amongst the banks of *Laminaria* and other coarser seaweeds cast ashore by the storm.

The foreign Botanists were much pleased with the marine flora, and several of the Zoologists were especially interested in the abundant supply of Compound Ascidians, belonging to the genera *Botryllus*, *Botrylloides*, *Lepto-clinum*, *Amaroucium* and *Diplosoma*, which were found attached to the stems and roots of the *Laminaria*. Dr. Johan Hjort was anxious to see the buds in the colonies of as many species as possible. He was also desirous of examining the stolons of *Clavelina*, of which specimens can usually be procured on the side walls of a deep shore pool near Spaldrick; and we were able to show him the hibernating condition of the buds in the stolons which Professor Giard has lately discussed,* and which have been known for some years at Port Erin. Several of the Zoologists and Botanists preserved and carried off collections, and Prof. Chodat since his return to Geneva has given two public lectures before his University on what he saw of the Marine Fauna and Flora of the Isle of Man during our Expedition.

THE BRITISH ASSOCIATION FUND.

The surplus of the Local Fund, collected for the purpose of meeting the expenses of the visit of the British Association to Liverpool, was, by a resolution of the Local Committee at their final meeting on November 30th, 1896, placed in the hands of trustees to be invested for the purpose of promoting the work of the Liverpool Marine Biology Committee. The view of the British Association Executive Committee in recommending this allocation of the fund, was that the money had been subscribed *locally* for the purposes of a meeting *for the Advancement of Science*, and that consequently the surplus should be devoted to some investigation which would result in the advancement of local science. In recommending the L.M.B.C. as a suitable body to receive the fund and carry on the researches, the Executive Com-

* Comptes-rendus, Aug. 3, 1896.

mittee drew up and circulated amongst the subscribers and the members of the Local Committee the following statement :—

"Memorandum on Behalf of the Executive Committee."

"In considering the best allocation for the surplus which the Treasurers are happily able to report, the Committee have sought to select the object which would most commend itself to Subscribers as *local*, as *representative*, and as *permanently conducive* to the great aim of the British Association, the Advancement of Science. After considering various suggestions, the Committee are unanimous in recommending that the fund should be entrusted to the Liverpool Marine Biology Committee for administration under trust.

"The group of Sciences promoted by the Liverpool Marine Biology Committee includes those which most admit, and indeed require, local investigation.

"Its composition is widely representative, combining members from Liverpool with representatives of North Lancashire, Manchester, North Wales, and the Isle of Man. It is intimately associated with the work of the Lancashire Sea-Fisheries Committee, so important to the industrial prosperity and development of Liverpool and the neighbourhood. It has, moreover, been remarkably successful in eliciting the enthusiasm and support of non-professional as well as professional workers in Science.

"Its scope is comprehensive, and papers directly emanating from its action have been read before several different Sections of the British Association. Alike in quality and quantity, the work of the Committee has done honour to Liverpool. The Annual Reports and the four published volumes of ' Fauna and Flora of Liverpool Bay' are notable contributions to Marine Biology; and altogether it has helped to inspire more than a hundred papers contributed to Scientific Journals. At its Laboratory, first at Puffin Island, now at Port Erin, laborious researches have been carried out, such as in other countries rest on State support, but in England are left to the public spirit and enterprise of individuals or communities.

" A small endowment of the kind contemplated will be invaluable for securing fruit and permanence to the activities of the Liverpool Marine Biology Committee. The publication of Proceedings, and the conduct of scientific investigations, make a continuous and heavy drain upon the resources of a voluntary society. A grant which has been allotted for the last four years to the Liverpool Marine Biology Committee from the funds of the British Association, expired with the present year, and the annual income from this endowment will at a fortunate moment make good the loss.

" The close association of the Liverpool Marine Biology Committee with two of the British Association Honorary Local Secretaries, whose exertions contributed so much to the success of the recent Meeting, is no mere coincidence. In connection with the Liverpool Marine Biology Committee, Prof. Herdman and Mr. Isaac Thompson established not only their enthusiasm for Science, but also their capacity and resource in organisation. By happy fortune the British Association Meeting, which owed so much to their energy, can make an apt return by permanently forwarding that local development of Science which they have most at heart."

The recommendation of the Executive Committee met with universal approval; and only 2 subscribers out of a total of 300 took advantage of the opportunity which was given to them of withdrawing their share of the surplus. The formal resolutions which were unanimously passed by the Local Committee on November 30th were:—

1°. "That the balance remaining after the settlement of all expenses connected with the Meeting of the Association be handed to Trustees, the income of the fund to be applied in or towards the publication of Scientific Proceedings and the prosecution of Scientific Research; the Trustees to pay such income to the Treasurer of the Liverpool Marine Biology Committee for the above purposes, and the receipt of such Treasurer shall be a sufficient discharge to the Trustees for such payment."

2°. "That the Vice-Presidents, the Treasurers and the Secretaries be empowered to select Trustees, and to define and settle the exact terms of the Trust."

At the subsequent meeting of these Local Officers, the Trust deed, drawn up by Mr. J. W. Alsop, was submitted and Mr. W. E. Willink, J.P., Mr. Charles Booth, Jun., and Professor Herdman were appointed Trustees.

The fund which amounts, after the payment of all expenses, to about £950 will be invested, and the annual proceeds will be available as a small fixed income for the advancement of our work.

The L.M.B.C. while gratefully accepting this welcome addition to their means, and while they appreciate highly this mark of confidence in their work, cannot but feel that they are accepting also increased responsibility. They have constantly in the past kept before them the view that in undertaking to report upon the Marine Biology and Geology of the district, and in asking for subscriptions to defray the necessary expenses of the work, they were incurring responsibilities both to the scientific world and to the public of Liverpool. They have been responsible to the latter for the wise administration of such funds as are entrusted to them, and to the former both for the energetic and careful prosecution of the scientific work and also for the due recognition and encouragement of all those workers, amateur as well as professional, whose contributions to knowledge come within the scope of their investigation. This sense of responsibility is only increased and justified by this trust which has been created for the benefit of the Marine Biological work; and the British Association Local Committee and the Subscribers to the fund may rest assured that the L.M.B.C. Officials will use their utmost endeavour to so direct the investigations that they may be a credit to Liverpool and " permanently conducive to the great aim of the British Association, the Advancement of Science."

It is the view of the Trustees, and also the desire of the L.M.B.C., that the annual interest should not be merely added to the income of the Committee, but should so far as possible be expended either upon the publication of results or upon some definite line of investigation, such as the hire of steamers for dredging explorations, or series of experiments in the Biological Station, so that the name of the fund may from time to time be publicly associated with some tangible result in such a way as to keep alive in Liverpool the memory of the British Association meeting of 1896.

It ought to be borne in mind by our own subscribers, by those of the Liverpool public who have in the past so generously helped the L.M.B.C. work, that if this British

Association Fund is to be a real advantage and bear fruit, it is absolutely necessary that our ordinary income derived from subscriptions should be in no way diminished. We confidently appeal to all those interested in any way in the Natural History of our neighbourhood to co-operate with us. Those who work with the microscope, who are collectors, who have any aptitude for practical work, will be gladly welcomed at the Biological Station or on the expeditions ; while those who feel that they can only appreciate the work of others, but are interested in the extension of our knowledge of nature, can most effectively help and encourage us by adding to the slender annual income of the Committee, which is barely sufficient to meet the necessary expenses of the work at Port Erin and in Liverpool Bay.

LIST OF L.M.B.C. WORKERS.

We think it useful to give here a list of those Naturalists who are definitely working at special groups of organisms in the L.M.B.C. district, and who, as a rule, undertake the identification of the animals reported upon, and contribute information about their groups to the L.M.B.C. Reports.

BACTERIA.—Professor R. Boyce.

DIATOMACEÆ.—Dr. H. Stolterfoth.

ALGÆ.—Professor Harvey Gibson and Professor Weiss.

FORAMINIFERA.—Dr. G. W. Chaster.

DINOFLAGELLATA.—Mr. R. L. Ascroft.

INFUSORIA, &c.—Vacant.

PORIFERA.—Dr. R. Hanitsch.

HYDROID ZOOPHYTES.—Miss L. R. Thornely.

MEDUSÆ.—Mr. E. T. Browne, B.A., F.Z.S.

ACTINIARIA.—Mr. J. A. Clubb, B.Sc.

ECHINODERMATA.—Mr. H. C. Chadwick.

TURBELLARIA.—Mr. F. W. Gamble, M.Sc.

TREMATODA.—Vacant.

NEMERTIDA.—Mr. W. I. Beaumont, B.A.

ROTIFERA.—Vacant.

NEMATODA.—Vacant.

GEPHYREA, HIRUDINEA, AND OLIGOCHÆTA.—Vacant.

POLYCHÆTA.—Mr. J. Hornell and Mr. Arnold T. Watson.

POLYZOA.—Miss L. R. Thornely.

CIRRIPEDIA.—Vacant.

COPEPODA.—Mr. Isaac C. Thompson, F.L.S.

OSTRACODA.—Mr. Andrew Scott.

HIGHER CRUSTACEA.—Mr. A. O. Walker, F.L.S.

PYCNOGONIDA.—Dr. C. H. Hurst.

MOLLUSCA (TESTACEOUS).—Mr. Alfred Leicester.

NUDIBRANCHIATA.—Professor Herdman, F.R.S., and Mr. J. A. Clubb, B.Sc.

CEPHALOPODA.—Mr. W. E. Hoyle, M.A.

TUNICATA.—Prof. Herdman, F.R.S., and Miss J. H. Willmer.

FISHES.—Prof. Herdman, Mr. R. A. Dawson, and Mr. A. Scott.

SEA-BIRDS.—Dr. H. O. Forbes.

SUBMARINE GEOLOGY.—Mr. J. Lomas, Mr. G. W. Lamplugh, and Mr. Clement Reid.

PHYSICS AND CHEMISTRY OF THE SEA.—Vacant.

One of our greatest needs is a young Chemist or Physicist who would join our expeditions with the object of reporting upon the condition of the sea water at the various localities, depths and seasons.

THE LIBRARY.

The Committee consider it advisable to publish here a list of the books forming their nucleus of a working library at Port Erin, first, for the purpose of letting workers and students know what books they will find in the laboratory, and secondly in the hope that the short list will suggest to members of the Committee, other naturalists, sub- scribers, and friends, some deficiencies in our library which might be made good by contributions from their own shelves. It may be convenient to state that what the Committee aim at is merely a small working library of Marine Biology, and that the most important books for their purposes—after a few standard text books and works of reference—are monographs or important papers on British Marine animals and plants.* In addition to the books in the following list, there is also in the book-case a considerable number of pamphlets kindly sent by authors, and dealing mostly with the Marine Biology of the neighbourhood. We are always glad to have such author's reprints.

ALDER and HANCOCK.—Monograph of the British Nudibranchiate Mollusca.—in seven parts. Ray Society, 1845-55.

BAIRD.—British Entomostraca. Ray Society, 1850.

BALFOUR.—Comparative Embryology. 2 vols., 1880.

BELL.—British Stalk-Eyed Crustacea, 1853.

BRADY and NORMAN.—Monograph of British Ostracoda Part I. Trans. R. Dublin Soc., 1889.

BRADY.—Monograph of the Free and semiparasitic Copepoda of the British Islands. 3 vols., Ray Soc., 1878.

* A few books and papers which have from time to time been kindly sent to the L.M.B.C., but which have no particular bearing upon British Marine Biology, have been deposited temporarily in the library of the Biological Society in Liverpool, where they will be more used and more appreciated than at Port Erin.

CAMBRIDGE NATURAL HISTORY, vol. II., 1896.

CARUS.—Prodromus Faunæ Mediterraneæ. 4 vols., 1885-93.

CUNNINGHAM.—Marketable Marine Fishes, 1896.

DAY.—Fishes of Great Britain and Ireland. 2 vols., 1880.

FRAIPONT.—Recherches sur les Acinétiens, &c.

FORBES.—Monograph of the British Naked-Eyed Medusæ. Ray Soc., 1848.

GOSSE.—Manual of Marine Zoology for the British Isles. 2 vols., 1855.

GOSSE.—Handbook to Marine Aquarium, 1856.

GOSSE.—British Sea-Anemones and Corals, 1860.

HAUCK.—Meeres-Algen.

HARVEY.—Manual of British Marine Algæ.

HELLER.—The "Novara" Crustacea, 1865.

HOLMES and BATTERS.—Revised List Brit. Mar. Algæ.

HERDMAN.—Phylogenetic Classification of Animals, 1885.

HERDMAN and LESLIE.—Marine Invertebrate Fauna of the Firth of Forth, 1881.

HERDMAN.—Annual Reports upon the Biological Station at Puffin Island, 1888-92. (see also Liverpool Marine Biology Committee).

HINCKS.—British Hydroid Zoophytes. 2 vols., 1868.

HINCKS.—British Marine Polyzoa. 2 vols., 1880.

HUGHES.—Principles and Management of the Marine Aquarium, 1875.

INTERNATIONAL FISHERIES EXHIBITION in London, 1883, by Huxley, Hubrecht, Holdsworth, Walpole, &c. Addresses and papers read at the Conferences, &c. 9 vols., 1883.

JEFFREYS.—British Conchology. 5 vols., 1862-69.

JOHNSTON.—Brit. Museum Catalogue of Worms, 1865.

KENT.—Manual of the Infusoria. 3 vols., 1880-82.

KORSCHELT and HEIDER.—Text Book of Embryology of Invertebrates. 2 vols., 1895.

LANG.—Text-book of Comparative Anatomy. 2 vols., 1891-96.

LEE.—Microtomist's Vade-Mecum, 1893.

LIVERPOOL MARINE BIOLOGY COMMITTEE.—Reports upon the Fauna of Liverpool Bay, &c., vols. I., II., III., IV., 1886-95.

MALMGREN.—Annulata Polychæta, 1867.

MARSHALL and HURST.—Practical Zoology, 1895.

M'INTOSH.—Monograph of the British Annelids, Part I., Nemerteans, 2 vols., 1873.

MURRAY.—Introduction to study of seaweeds.

NICHOLSON.—Text-book of Zoology. 7th Edn., 1887.

PENNINGTON.—British Zoophytes, 1885.

PIZON.—Blastogénèse des Botryllides, 1892.

POUCHET.—Changements de Coloration, &c., 1876.

ROLLESTON.—Forms of Animal Life. 2nd Edition, by W. Hatchett Jackson, 1888.

SARS.—Crustacea of Norway, 2 vols.

THOMPSON.—Revised Report on the Copepoda.

VOGT and YUNG.—Traité d'Anat. Comp. Prat. 2 vols.

PUBLICATIONS.

The fifth volume of the " Fauna and Flora " will not be ready for a couple of years ; but a revision of all the groups already reported upon has been carried out during the summer and as the result a complete list, brought up to date, was laid before Section D of the British Association and is published in the report of that meeting. Copies of this list have been reprinted, and will be issued in our next volume. As references to the literature are given after

the name of each species, the list forms a useful index to
all the L.M.B.C. publications up to date; and in addition
contains a record of all the FISHES we have obtained in
the district—a group not yet reported on. The list in
question appeared before the British Association as the
fourth and final report of that Committee of the Association which has for some years been co-operating in the
L.M.B.C. work. That report contains the following
" Concluding Remarks :—

"Although this is put forward as a final report of the
present Committee, they do not desire thereby to indicate
that the work of exploring the zoology, botany, and geology
of the Irish Sea is finished. Probably such an investigation can never be finished; but the Committee feel that
the occasion of the British Association meeting in Liverpool is one that they ought to take advantage of to present
a report which is final, in the sense that it completes the
present series of reports, and brings together and sums
up the results of all previous marine biological work in
the district.

" For the future, they feel that the work will be carried
on actively by the Liverpool Marine Biology Committee,
the body of investigators by whom most of the work has
been done in the past. The Port Erin Biological Station
is equipped for such work, and the British Association
can best render effective help by supporting the general
investigations carried on at that station, or by giving
grants for special researches."

APPENDIX A.

THE LIVERPOOL MARINE BIOLOGY COMMITTEE (1896).

R. D. DARBISHIRE, Esq., B.A., F.G.S., Manchester.

PROF. R. J. HARVEY GIBSON, M.A., F.L.S., Liverpool.

PROF. W. A. HERDMAN, D.Sc., F.R.S., F.L.S., Liverpool, Chairman of the L.M.B.C., and Hon. Director of the Biological Station.

ALFRED LEICESTER, Esq., formerly of Southport and Liverpool.

SIR JAMES POOLE, J.P., Liverpool.

DR. ISAAC ROBERTS, F.R.S., formerly of Liverpool.

I. C. THOMPSON, Esq., F.L.S., Liverpool, Hon. Treasurer.

JOHN VICARS, Esq., formerly of Bootle.

A. O. WALKER, Esq., F.L.S., J.P., Colwyn Bay.

DR. SPENCER WALPOLE, formerly Governor of the Isle of Man.

CONSTITUTION OF THE L.M.B.C.

(Established March 1885.)

I.—The OBJECT of the L.M.B.C. is to investigate the Marine Fauna and Flora (and any related subjects such as submarine geology and the physical condition of the water) of Liverpool Bay and the neighbouring parts of the Irish Sea; and if practicable to establish and maintain a Biological Station on some convenient part of the coast.

II.—The COMMITTEE shall consist of not more than 12 and not less than 10 members, of whom 3 shall form a quorum ; and a meeting shall be called at least once a year for the purpose of arranging the Annual Report, passing the Treasurer's accounts, and transacting any other necessary business.

III.—During the year the AFFAIRS of the Committee shall be conducted by an HON. DIRECTOR, who shall be Chairman of the Committee, and an HON. TREASURER, both of whom shall be appointed at the Annual Meeting and shall be eligible for re-election.

IV.—Any VACANCIES on the Committee, caused by death or resignation, shall be filled up by the election, at the Annual Meeting, of those who, by their work on the Marine Biology of the district, or by their sympathy with science, seem best fitted to help in advancing the work of the Committee.

V.—The EXPENSES of the investigations, of the publication of results, and of the maintenance of the Biological Station shall be defrayed by the Committee, who for this purpose shall ask for subscriptions or donations from the public, and for grants from scientific funds.

VI.—The BIOLOGICAL STATION shall be used primarily for the Exploring work of the Committee, and the SPECIMENS collected shall, so far as is necessary, be placed in the first instance at the disposal of the members of the Committee and other specialists who are reporting upon groups of organisms ; but, in order to add to the funds, some of the workplaces in the Biological Station may be rented by the week or year to students and others, and duplicate specimens which, in the opinion of the Committee, can be spared may be sold to museums and laboratories.

LIVERPOOL MARINE BIOLOGICAL STATION
at PORT ERIN.

REGULATIONS.

I.—This Biological Station is under the control of the Liverpool Marine Biology Committee, the executive of which consists of the Hon. Director (Prof. Herdman,F.R.S.) and the Hon. Treasurer (Mr. I. C. Thompson, F.L.S.).

II.—In the absence of the Director, and of all other members of the Committee, the Station is under the temporary control of the Resident Curator or Laboratory Assistant, who will keep the keys, and will decide, in the event of any difficulty, which places are to be occupied by workers, and how the tanks, collecting apparatus, &c., are to be employed.

III.—The Resident Assistant will be ready at all reasonable hours and within reasonable limits to give assistance to workers at the Station, and to do his best to supply them with material for their investigations.

IV.—Visitors will be admitted, on payment of a small specified charge, to see the Aquarium and the Station, so long as it is found not to interfere with the scientific work.

V.—Those who are entitled to work in the Station, when there is room, and after formal application to the Director, are :—(1) Annual subscribers of one guinea or upwards to the funds (each guinea subscribed entitling to the use of a work place for four weeks), and (2) others who are not annual subscribers, but who pay the Treasurer 10s. per week for the accommodation and privileges. Institutions, such as Colleges and Museums, may become subscribers in order that a work place may be at the disposal of their staff for a certain period annually ; a subscription of two guineas will secure a work place for six weeks in the year, a subscription of five guineas for four months, and a subscription of £10 for the whole year.

VI.—Each worker* is entitled to a work place opposite a window in the Laboratory, and may make use of the microscopes, reagents, and other apparatus, and of the boats, dredges, tow-nets, &c., so far as is compatible with the claims of other workers and with the routine work of the Station.

VII.—Each worker will be allowed to use one pint of methylated spirit per week, free. Any further amount required must be paid for. All dishes, jars, bottles, tubes, and other glass may be used freely, but must not be taken away from the laboratory. If any workers desire to make, preserve, and take away collections of marine animals and plants, they must make special arrangements with the Director or Treasurer in regard to bottles and preservatives. Although workers in the Station are free to make their own collections at Port Erin, it must be clearly understood that (as in other Biological Stations) no specimens must be taken for such purposes from the laboratory stock, nor from the Aquarium tanks, nor from the steam-boat dredging expeditions, as these specimens are the property of the Committee. The specimens in the Laboratory stock are preserved for sale, the animals in the tanks are or the instruction of visitors to the Aquarium, and as all the expenses of steam-boat dredging expeditions are defrayed by the Committee the specimens obtained on these occasions must be retained by the Committee (*a*) for the use of the specialists working at the Fauna of Liverpool Bay, (*b*) to replenish the tanks,

* Workers at the Station can always find comfortable and convenient quarters at the closely adjacent Bellevue Hotel; but lodgings can readily be had by those who prefer them.

and (c) to add to the stock of duplicate animals for sale from the Laboratory.

VIII.—Each worker at the Station is expected to lay a paper on some of his results—or at least a short report upon his work—before the Biological Society of Liverpool during the current or the following session.

IX.—All subscriptions, payments, and other communications relating to finance, should be sent to the Hon. Treasurer, Mr. I. C. Thompson, F.L.S., 53, Croxteth Road, Liverpool. Applications for permission to work at the Station, or for specimens, or any communications in regard to the scientific work should be made to Professor Herdman, F.R.S., University College, Liverpool.

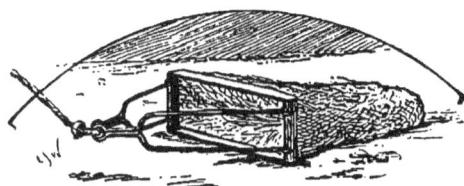

APPENDIX B.

SUBSCRIPTIONS and DONATIONS.

	Subscriptions.			Donations.		
	£	s.	d.	£	s.	d.
Ayre, John W., Ripponden, Halifax ...	1	1	0	—		
Banks, Prof. W. Mitchell, 28, Rodney-st.	2	2	0	—		
Bateson, Alfred, Harrop-road Bowdon ...	1	1	0	—		
Beaumont, W. I., Cambridge	1	1	0	—		
Bickersteth, Dr., 2, Rodney-st.	2	2	0	—		
Brown, Prof. J. Campbell, Univ. Coll....	1	1	0	—		
Browne, Edward T., B.A., 141, Uxbridge-						
road, Shepherd's Bush, London ...	1	1	0	—		
Brunner, Sir J. T., Bart., Druids Cross...	5	0	0	—		
Boyce, Prof., University College ...	1	1	0	—		
Caton, Dr., 86, Rodney-street	—			1	1	0
Clague, Dr., Castletown, Isle of Man ...	1	1	0	—		
Clague, Thomas, Bellevue Hotel, Port Erin	1	1	0	—		
Comber, Thomas, J.P., Leighton, Parkgate	1	1	0	—		
Crellin, John C., J.P., Ballachurry, An-						
dreas, Isle of Man	0	10	6	—		
Darbishire, R.D., Victoria-pk., Manchester	1	1	0			
Dawkins, Professor W. Boyd, Owens						
College, Manchester...	1	1	0	—		
Dumergue, A. F., 7 Montpellier-terrace	0	10	6	—		
Dwerryhouse, A. R., 8, Livingstone-av.	1	1	0	—		
Gair, H. W., Smithdown-rd., Wavertree	2	2	0	—		
Gamble, Col. C.B., Windlehurst, St. Helens	2	0	0	—		
Gamble, F.W., Owens College, Manchester	1	1	0	—		
Gaskell, Frank, Woolton Wood... ...	1	1	0	—		
Gaskell, Holbrook, J.P., Woolton Wood	1	1	0	—		
Gell, James S., High Bailiff of Castletown	1	1	0	—		
Forward ...£31	3	0	1	1	0	

Forward ...	31	3	0	1	1	0
Gibson, Prof. R. J. Harvey, 5, Adelaide-terrace, Waterloo ...	1	1	0	—		
Glynn, Dr., 62, Rodney-street ...	2	2	0	—		
Greening, Linnæus, 5, Wilson Patten-st., Warrington ...	1	1	0	—		
Gotch, Prof., Museum, Oxford ...	1	1	0	—		
Halls, W. J., 35, Lord-street ...	1	1	0	—		
Hanitsch, Dr., Museum, Singapore ...	1	1	0	—		
Henderson, W. G., Liverpool Union Bank	1	1	0	—		
Herdman, Prof., University College ...	2	2	0	—		
Holder, Thos. (the late), 1, Clarendon-buildings ...	1	1	0	—		
Holland, Walter, Mossley Hill-road	2	2	0	—		
Holt, Alfred, Crofton, Aigburth ...	2	2	0	—		
Holt, George, J.P., (the late), Sudley, Mossley Hill ...	1	0	0	—		
Hoyle, W. E., Museum, Owens College, Manchester ...	1	1	0	—		
Isle of Man Natural History and Anti-quarian Society ...	1	1	0	—		
Jones, C.W.,J.P., Field House,Wavertree	1	0	0	—		
Jones, C. E. E., Prenton-road, B'head	1	1	0	—		
Kermode, P. M. C., Hill-side, Ramsey...	1	1	0	—		
Lea, Rev. T. Simcox, 3, Wellington-fields	1	1	0	—		
Leicester, Alfred, Buckhurst Farm, Eden-bridge, Kent ...	1	1	0	—		
Macfie, Robert, Airds ...	1	0	0	—		
Meade-King, H. W.,J.P., Sandfield Park	1	1	0	—		
Meade-King, R. R., 4, Oldhall-street ...	0	10	0	—		
Melly, W. R., 90, Chatham-street ...	1	1	0	—		
Monks, F. W., Brooklands, Warrington	1	1	0	—		
Muspratt, E. K., Seaforth Hall ...	5	0	0	—		
Narramore, W., 5, Geneva-rd., Elm Park	1	1	0	—		
Forward ...	65	18	0	1	1	0

Forward ...	65	18	0	1	1	0
Newton, John, M.R.C.S., 44, Rodney-st.	0	10	6	—		
Poole, Sir James, Tower Buildings ...	2	2	0	—		
Rathbone, S.G., Croxteth-drive, Sefton-pk.	2	2	0	—		
Rathbone, Mrs. Theo., Backwood, Neston	1	1	0	—		
Rathbone, Miss May, Backwood, Neston	1	1	0	—		
Rathbone, W., Greenbank, Allerton ...	2	2	0	—		
Roberts, Isaac, F.R.S., Crowborough ...	1	1	0	—		
Simpson,J.Hope,Annandale,Aigburth-dr	2	2	0	—		
Smith, A. T., junr., 24, King-street ...	1	1	0	—		
Talbot, Rev. T. U., 4, Osborne-terrace, Douglas, Isle of Man	1	1	0	—		
Thompson, Isaac C., 53, Croxteth-road	2	2	0	—		
Thornely, James, Baycliff, Woolton ...	1	1	0	—		
Thornely, The Misses, Baycliff, Woolton	1	1	0	—		
Toll, J. M., Kirby Park, Kirby	1	1	0	—		
Walker, A. O., Nant-y-glyn, Colwyn Bay	3	3	0	—		
Walker, Horace, South Lodge, Princes-pk.	1	1	0	—		
Walters, Rev. Frank, B.A., King William College, Isle of Man...	1	1	0	—		
Watson, A. T., Tapton-crescent, Sheffield	1	1	0	—		
Weiss, Prof. F. E.,Owen's College, Man'tr.	1	1	0	—		
Westminster, Duke of, Eaton Hall ...	5	0	0	—		
Wiglesworth, Dr., Rainhill	1	1	0	—		
	£98	14	6	1	1	0

SUBSCRIPTIONS FOR THE HIRE OF " WORK-TABLES," OCCUPIED BY COLLEGES, &c.

Liverpool Museum Committee ...	£2	2	0
Owens College, Manchester	10	0	0
University College, Liverpool	10	0	0
	£22	2	0

THE LIVERPOOL MARINE BIOLOGY COMMITTEE.

Dr.   In Account with ISAAC C. THOMPSON, Hon. Treasurer.   **Cr.**

1896.	£	s.	d.	1896.	£	s.	d.
To Balance due Treasurer, Dec. 31st, 1895.........	5	13	5	By Subscriptions and Donations actually received......	107	1	6
,, Printing Reports, vol. iv. "Fauna," Plates, &c. ...	46	13	10	,, Amount received from Colleges, &c., for hire of "Work Tables"	22	2	0
,, Printing and Stationery	1	9	3	,, Dividend, British Workman's Public House Co., Ltd., Shares	5	18	9
,, Expenses of Dredging Expeditions.........	17	4	3	,, Sale of Reports and Volumes of Fauna	23	5	7
,, Boat Hire.........	8	16	2	,, Bank Interest.........	0	1	4
,, Reade's Portable Shellbend Boat.........	15	0	0	,, Admissions to Aquarium	3	17	0
,, Books and Apparatus at Port Erin Biological Station	20	6	7	,, Balance due Treasurer, Dec. 31st, 1896	4	10	9
,, Postage, Carriage of Specimens, &c.	4	9	1				
,, Salaries, Curator and Laboratory Boy.........	28	12	3				
,, Rent of Port Erin Biological Station	15	0	0				
,, Repairs, Bookcase, &c.	3	5	8				
,, Sundries	0	9	5				
	£166	19	11		£166	19	11

Endowment Investment Fund :—
British Workmans' Public House Co's. shares £173 1 0

ISAAC C. THOMPSON,
Hon. Treasurer.

LIVERPOOL, *December 31st* 1896.

Audited and found correct,

A. T. SMITH, JUNR.

FURTHER REPORT upon the FREE-SWIMMING COPEPODA of the WEST COAST OF IRELAND.

By Isaac C. Thompson, F.L.S.

[Read May 14th, 1897.]

The former Report (Trans. Biol. Soc., L'pool, Vol. X., p. 92) dealt with a collection of plankton, the result of a series of tow-nettings made by my friend Mr. Edward T. Browne, B.A., of London, off Valencia on the West of Ireland during the summer and autumn of 1895.

Mr. Browne again made Valencia his head-quarters in July 1896, remaining there until near the end of September, using the tow-net on most suitable days either inside or outside of the harbour. The results are contained in 41 bottles, numbered 1 to 41 in accompanying tabular resumé, the total number of species found being 29. During part of the time Mr. Browne had as companions Mr. A. O. Walker, F.L.S., of Colwyn Bay, and Mr. F. W. Gamble, of Owens College, Manchester. Both of these gentlemen have kindly sent me material they collected by tow-net. The conditions under which the latter were obtained being similar to those of Mr. Browne, they are included in the tabulated results, having no specially distinctive features. Through the kindness of the Misses Delap, daughters of the Vicar of Valencia, themselves trained naturalists, the work of tow-netting was continued throughout the autumn and winter up to March 1897, the bottles numbered 22 to 41 being contributed by them.

The previous Report had to do chiefly with tow-nettings taken during the months—April, May and June 1895, while those here referred to were collected from July 1896 to March 1897. We have thus a year's continuous series of material to report upon which should enable us to obtain a fair knowledge of the distribution of the free-swimming Copepoda of this district.

Mr. Browne informs me that the tide flows into Valencia Harbour from the ocean at a good pace. Two knots is the speed marked on the chart, being strong enough to keep the tow-net fully extended from a boat at anchor. The average depths to which the tow-net was lowered are given, but the results do not appear to show any great difference in the fauna at the bottom and at the top, possibly to be accounted for by the fact that the tide runs through a narrow entrance and the water is well stirred up.

All the material collected for examination was immediately preserved in Formaline. A 5% solution appears to be the strength best suited for these small delicate crustaceans, and is probably as efficient a preservative as is known.

NOTES ON THE SPECIES FOUND.

Calanus finmarchicus and *Clausia elongata* were found in all the bottles and were generally the prevailing forms. The much rarer *Calanus propinquus* occurred very sparingly during January and February. It differs from *C. finmarchicus* chiefly in the profusely plumose character of the setæ, and in the 5th swimming feet, and the diverging caudal segments of the male.

Pseudocalanus armatus was found but on one occasion (Feb. 27th) and then only one specimen.

Temora longicornis was generally abundant up to the end of October, but was not found during the mid-winter months, occurring again sparingly in February, and

becoming common in March. *Metridia armata* a species
not usually common, occurs in more than half the bottles and
on some occasions in fair numbers, especially during the
winter months. The gradually tapering setose antennæ
and the leaf-shaped terminations to the swimming feet,
distinguish it readily. One of the rarest species taken,
and previously unknown to the district is *Rhincalanus
cornutus*, easily distinguished by its long projecting
cruciform rostrum, and by its antennæ. On two occasions,
in August and October, one specimen only was found.
It has only I believe been previously reported in Britain
on one occasion, by Mr. Thomas Scott, off the Shetland
Isles.

Centropages hamatus occurs in half the bottles, but
usually sparingly, and *C. typicus* much less frequently.

In a tube sent to me by Mr. A. O. Walker were two
specimens belonging to the Dublin Museum collection,
and marked "West of Ireland," which on examination
proved to be *Euchæta marina* and *Pontella kroyeri* both
new to the district and very uncommon in our northern
seas, although I have previously taken one specimen of
each in the L.M.B.C. district.

Isias clavipes occurs on only three occasions. *Acartia
clausii* was generally found and fairly plentiful, except
during spring and early summer, and *Oithona spinifrons*
under much the same conditions. *Candace pectinata* a
species generally rare in our seas, has proved to be here not
uncommon, occurring, however, only during the summer
and early autumn. On only two occasions throughout
the year that very conspicuous species the beautifully
coloured *Anomalocera patersonii* was found, while as
alluded to in the last report it sometimes occurs in our
seas in dense shoals.

It was interesting again to find the southern species

alluded to in the previous report, *Corycæus speciosus*, though on only one occasion, and more frequently its very similar ally *C. anglicus*, the latter having been frequently reported from Plymouth.

Parapontella brevicornis occurred once in July and on several occasions in February. One specimen of *Monstrilla danæ* was taken on August 3rd. Great interest attaches to the family Monstrillidæ through the recent important discovery of Prof. Giard of Paris and confirmed by M. Malaquin that the early stages of one or more species of this group are spent parasitic in the body cavity of certain worms (see *Comptes rendus*, 16 novembre 1896, and 28 decembre 1896, and 11 janvier 1897).

Caligus rapax was collected in the tow-net on two occasions by the Misses Delap in December and January. The various species of the genus Caligus, though all fish parasites, are not uncommonly found as free-swimmers particularly at night.

The eight species, viz., *Harpacticus fulvus*, *Thalestris longimana*, *T. clausii*, *Ectinosoma spinipes*, *Longipedia coronata*, *Cyclopina littoralis*, *Porcellidium viride*, and *Laophonte hispida* are all fairly common littoral forms, generally found in rocky pools or near to shore. Their presence again suggests the regret that in addition to the valuable collection taken by tow-net, Mr. Browne and his co-workers did not preserve mud and sand dredged at various depths, and the washings of dredged material. For judging by results in Liverpool Bay it is to these sources we must chiefly look for forms new to science among the Copepoda. While we have doubtless still much to learn as to the causes which influence the distribution of the free-swimming forms, we can hardly now expect to add very many to the number of those already known. But with the vastly more numerous, mud loving, sedentary, and wholly or semi-parasitic species of Copepoda the case is different. Investigations as to the food of fishes have shown us that even scarce species, and but recently known Copepoda, as in the case of *Jonesiella hyænæ*, are found largely in the stomachs of young fishes, being evidently selected as their chief food and probably found in quantities near to the spawning grounds. From this aspect even apparently lifeless mud has under the microscope much to teach us. In addition to the examination of deposits it is evident that Sponges, Cœlenterates, Echinoderms, Worms, Molluscs, Nudibranchs, and Ascidians will all richly reward careful search; for judging from past results, they are the hosts of many highly organized species of Copepoda, each specially adapted to its particular host and that alone.

While sincerely thanking Mr. Browne and his friends for the amount of valuable material collected, I must again express the hope that we may at a future time be enabled to know as much of the Valencia sedentary species of Copepoda as we now do of the free-swimmers.

The accompanying table represents the distribution of free-swimming Copepoda in or about Valencia harbour from July, 1896 to March, 1897.

	1	2	3	4	5	6	7	8	9	10	11	12	13	14	15	16	17	18	19	20	21	22	23	24	25	26	27	28	29	30	31	32	33	34	35	36	37	38	39	40	41
Calanus finmarchicus	A	C	A		A	A	A	A		A	C	C	C	F	F	F	C	C	F	C	F	C	F	C	C	F	A	A	F	F	A	A	F	F	A		A	F	F	C	A	A
C. propinquus	A	A	A		F	A	A	A	A	A	C	C	A	A	C	C	A	C	F	C	C	A	C	A	C	C	A	A	A	A	A	A	C		C	A	A	A	A	A	A	A
Clausia elongata	A	C	C		A	A	A	A		C	C	C	F	C	C	C	C	C	F	C	C	F	C	A	C	A	A	A	A	A	A	F		F	C	C	C	C	C	A		C
Pseudocalanus armatus	C	F	A		C	F	F	C		C	C	F	F		F	P	P		C	F		C	F	F	F	F	A	F		A	F		F	C	F	C	C	C	C	C	A	C
Temora longicornis														S	P						C	F		S	C	C	C					A	F							S		
Metridia armata	F		F			F	F		F		P	C	F	C	F	C		P	C	F		S	C	C		C			F			F							S		F
Rhincalanus nasutus												S											S	C	C																
Centropages hamatus																	F						C	C	C		C														
C. typicus															P		P											F														
Euchaeta marina				F	F	F		F	F	F		F	F	P	P	C	C	F	A	F	C	F	F	P	F	C	A	A	A	C	A	A	C	F	A	A	A	A		C	F	
Isias clavipes					C	Y				C	F	F	O	P	P	F	A	F	C	C	C	F	C	C	C			F														
Acartia longiremis				F	F	Y		F	F	O		F	P	C	P	A	F	C	F	P	F	F	C	C	C	C	C	C	F				F	C	F	C	C	F	C	C	S	S
Candace pectinata										S																																
Anomalocera patersoni																																										
Pontella kroyeri		S		U		F	F			F	F		P	P	F		F	F		P	F	F	F		C	C	C	C	C	F	F				F	C	F		C	S	S	
Oithona spinifrons																																	F						F	F	F	
Paraponella brevicornis														S	P	P																								S		
Harpacticus fulvus														S	P	F																										
Thalestris longimana														P	F																											
T. clausii																																										
Peltionoma spinipes																																										
Longipedia coronata																																										
Cyclopina littoralis																S																							S			
Corycaeus speciosus																					S	P			F	F		F														
C. anglicus																					S	P			S																	
Porcellidium tenue																																										
Laophonte hispida																																										
Monstrilla danæ			S																																				S			
Caligus rapax																																										

*Date and locality not recorded.

A signifies Abundant.
C „ Common.
F „ Few.
S „ Scarce.

REVISED LIST OF HYDROMEDUSÆ of the L.M.B.C. DISTRICT.

By Edward T. Browne, B.A., F.Z.S.

[Read 12th January, 1897.]

Since the publication of my Report on the Medusæ of the L.M.B.C. District, in 1895, fresh researches in some cases have led to changes in the generic and specific names, and in other cases show that medusæ formerly regarded as distinct species are only early stages of other known species. I have given an account of these changes in nomenclature in a paper published in the Proceedings of the Zoological Society for 1896, upon which this Revision is based.

I append in tabular form a list of the Medusæ which I obtained at Port Erin in April, 1896, and notes on a few of the species.

HYDROMEDUSÆ.

I. Anthomedusæ.

Codonium pulchellum (Forbes).
Fauna,* IV., p. 374.
Corymorpha nutans, Sars.
P.Z.S., 1896, p. 463.
= *Steenstrupia rubra*, Forbes.
Fauna, IV., p. 375.
Cytaeandra areolata (sp. ?) (Alder).
Fauna, IV., p. 390.
Dipurena halterata (Forbes).
Fauna, IV., p. 375; P.Z.S., 1896, p. 473.
Euphysa aurata, Forbes.
Fauna, IV., p. 376; P.Z.S., 1896, p. 474.
Hybocodon prolifer, Agassiz.

* Fauna of Liverpool Bay, 1895, Vol. IV., pp. 371—414. Reprinted from Trans. Liverpool Biol. Soc., Vol. IX., pp. 243—286.

P.Z.S., 1896, p. 466.
= *Amphicodon fritillaria* (Steenstrup).
Fauna, IV., p. 379.
Lizzia blondina, Forbes.
P.Z.S., 1896, p. 475; Fauna, IV., p. 393.
= *Dysmorphosa minima*, Hæckel.
Fauna, IV., p. 388.
Margelis principis, Steenstrup.
Fauna, IV., p. 394.
Margelis ramosa, Hæckel.
= *Margelis britannica* (Forbes).
Fauna, IV., p. 395.
Margellium octopunctatum (Sars).
Fauna, IV., p. 398; P.Z.S., 1896, p. 477.
Podocoryne carnea, Sars.
P.Z.S., 1896, p. 463.
= *Dysmorphosa carnea* (Sars).
Fauna, IV., p. 388.
Sarsia tubulosa, Sars.
Fauna, IV., p. 375.
Tiara pileata (Forskal).
Fauna, IV., p. 386.
Turris neglecta, Lesson.
Fauna, IV., p. 388.

II. LEPTOMEDUSÆ.

Eutima insignis (Keferstein).
Fauna, IV., p. 410; P.Z.S., 1896, p. 492.
Laodice calcarata, Agassiz.
Fauna, IV., p. 404.
Melicertidium octocostatum (Sars).
Fauna, IV., p. 405.
Mitrocomella polydiadema (Romanes).
Fauna, IV., p. 407.
Obelia lucifera, Forbes.

Fauna, IV., p. 406.

Phialidium cymbaloideum (Van Beneden).

P.Z.S., 1896, p. 491.

Phialidium temporarium, Browne.

P.Z.S., 1896, p. 489.

= *Phialidium variabile*, Hæckel.

Fauna, IV., p. 409.

Saphenia mirabilis (Wright).

Fauna, IV., p. 410 ; P.Z.S., 1896, p. 493.

Tiaropsis multicirrata (Sars).

Fauna, IV., p. 406.

NOTES ON THE MEDUSÆ taken in April, 1896.

Margelis ramosa. A young stage with 3 to 4 tentacles in each of the four groups. The oral tentacles twice dichotomously divided.

Margellium octopunctatum. All young stages with medusa-buds upon the manubrium.

Phialidium cymbaloideum. Young stages usually having four perradial tentacles ; a few, however, showed the commencement of the four interradial tentacles.

Umbrella about 2 to 3 mm. in length and width, which is about twice the size of specimens belonging to a similar stage taken in Valencia Harbour, Kerry. The sudden increase in the number of specimens of the two species of *Phialidium* on 16th of April I am unable to account for.

Phialidium temporarium. All young stages, mostly with four tentacles. Umbrella about 1 mm. in diameter.

Podocoryne carnea. A young stage not long liberated from its hydroid. It is possible that the medusæ which I have described as *Cytæandra areolata* (Fauna, IV., p. 390) may be the later stages of a medusa budded off from one of the species of the hydroid *Podocoryne*.

Tiaropsis multicirrata. All the specimens belonged to young stages. Umbrella about 1 to 3 mm. in diameter.

LIST OF MEDUSÆ TAKEN AT PORT ERIN DURING APRIL, 1896.

APRIL	3	6	8	13	14	16	18	20	21	22	29
Phialidium temporarium, Browne	III	III	III	IV	I	VI	II	II	II	II	I
Obelia, sp.?	III	III	III	III	II	III	III	I	I	I	I
Tiaropsis multicirrata (Sars)	-	3	-	13	-	3	-	3	-	1	2
Margelis ramosa, Hæckel	-	1	1	-	-	-	-	-	-	-	-
Margellium octopunctatum (Sars)	-	-	1	-	1	-	-	-	-	1	1
Phialidium cymbaloideum (Van Beneden)	-	-	1	-	-	VI	II	I	-	-	2
Aurelia aurita (Ephyra-stage)	-	-	-	-	-	2	-	-	-	-	-
Sarsia, sp.?	-	-	-	-	-	2	1	1	-	-	-
Podocoryne carnea, Sars	-	-	-	-	-	-	-	1	-	-	-
Cyanaea (Ephyra-stage)	-	-	-	-	-	-	-	-	-	1	-

I Very Scarce.
II Scarce.
III Few.
IV Fairly Common.
V Common.
VI Abundant.
VII Very Abundant.

The "Arabic" figures show the actual number taken.

[WORK FROM THE PORT ERIN BIOLOGICAL STATION.]

ADDITIONAL NOTES on the TURBELLARIA of
the L.M.B.C. DISTRICT.

By H. Lyster Jameson, B.A.

With Plates V. and VI.

[Read May 14th, 1897.]

In this list, which is intended to supplement Mr. F. W.
Gamble's Report (Trans. L'pool Biol. Soc., Vol. VII.,
1893), I propose to record the Turbellaria which I found
in the neighbourhood of Port Erin, during a couple of
weeks I spent at the Laboratory of the Liverpool Marine
Biology Committee in April, 1897.

During my stay at Port Erin I confined my attention
to the marine species, the only fresh-water form that I
am able to record being *Polycelis nigra*, a number of
specimens of this Triclad having been found by Mr. A.
R. Jackson, B.Sc., Science Student at University College,
Liverpool, who kindly handed them over to me for identification.

Of the Marine species *Graffilla buccinicola* is new to
science, seven Rhabdocoeles and one Polyclad are not
recorded in Mr. Gamble's list; while two Polyclads and
eighteen Rhabdocoeles, already recorded in his report,
were also found by me.

I regret that the shortness of my stay at Port Erin
allowed me but few opportunities of exploring new collecting grounds on the Manx coast; but Port Erin Bay
and Port St. Mary supplied me with such a quantity of
material, that I devoted all my time to these two places.

I must here express my gratitude to Mr. Gamble for
his valuable advice as to the best localities for finding a

rich supply of material, and for many useful hints as to collecting and identifying my specimens.

As to the occurrence of the species: the commonest Rhabdocoele by far was *Macrorhynchus croceus; Monotus lineatus* and *M. fuscus* coming next in number of individuals. The two latter, taken together, hardly outnumbered the former, while to these three species belonged about half of the Turbellarians I examined. Of the remaining types the commonest were *Pseudorhynchus bifidus, Byrsophlebs intermedia, Promesostoma agile, Provortex balticus,* and *Macrorhynchus heligolandicus.*

My experience of the Parasitic Rhabdocoeles is limited to the new *Graffilla,* which occurred in numbers in the kidney of almost every *Buccinum undatum* and *Fusus antiquus* that I examined. The "Cocoons" of *Fecampia erythrocephala* were not uncommon between tide-marks, but I searched in vain for the worm itself.

TURBELLARIA.

I. TRICLADIDA.

Polycelis nigra, O. F. Müller.

New to Isle of Man fauna.

Nine examples of this very common species were found by Mr. A. R. Jackson in a small stream near Port Erin. They measured from 6 to 14 mm. in length and were of a particularly deep black, showing none of the variations so common in this species, which is frequently coloured brown or grey, and presents all gradations from these colours to black.

II. RHABDOCOELIDA.

A. ACOELA.

Aphanostoma diversicolor, Oersted.

Two examples, among seaweeds collected at low tide near Port Erin Breakwater.

Three specimens of an acoelous type, probably an *Aphanostoma*, in which no trace of gonads or genitalia was visible were taken by tow-netting in Port Erin Bay.

Convoluta paradoxa, Oersted.

One example only was found, in some seaweeds gathered at low-water opposite the Biological Station. This individual measured 2 mm. in length, and presented no trace of the transverse bars which sometimes occur.

B. RHABDOCOELA.

Family MESOSTOMIDÆ.

Promesostoma marmoratum (Schultze).

Two specimens taken in tidal pools near the biological station had the usual dark reticular pigment between the eyes. Five others found among seaweeds between tidemarks near the Breakwater were without reticular pigment, but provided with copious yellowish-red colouring matter in the epidermis, resembling the variety that v. Graff records from Naples (Monographie, p. 270).

Promesostoma solea (O. Schmidt).

Although Gamble does not record this species in his paper (4), he gives Port Erin as a locality in the list of British species which he has appended to his article on the Turbellaria in the Cambridge Natural History (vol. ii., p. 49). I procured two specimens at low water, among seaweeds, in front of the biological station.

This species can at once be distinguished from *P. ovoideum* (O. Schm.), by the presence of a curious process of the pigment-cup, which extends outwards over the lens of the eye.

My two specimens differed considerably in the density of their body pigment, in one it was as dense as in v.

Graff's figure of *P. ovoideum* (Monogr. Tab. vii., fig. 11), but not extending right to the anterior and posterior ends of the body, and thinning off between the eyes; in the other specimen it was altogether more diffuse. Length of both specimens ·6 mm.

Promesostoma agile (Levinsen).

New to L.M.B.C. District.

I obtained seventeen specimens of this worm among seaweeds collected at low tide on Port Erin Breakwater. They varied from ·35 to ·6 mm. in length. Colour, light red. The penis was in most of my specimens more strongly curved at the apex than it is in Gamble's figure (3, pl. xl., fig. 14), in some few it is even more so than in Levinsen's sketch (10, Tab. iii., fig. 5), but it is evident that this feature is one that varies.

I have in vain tried to make out the relations of the "receptaculum" in this species, the organ is visible behind the atrium in all specimens, and has a club-shaped appearance, but I can find no connexion between this structure and the other genitalia, nor am I quite clear as to the normal position of the genital atrium itself. These points can only be cleared up by the study of sections.

Very typical of this species are several very large granular salivary (?) glands, with ducts which converge just behind the pharynx, they do not seem to have been observed before.

Byrsophlebs intermedia, v. Graff.

Common on *Cladophora* collected in tide-pools near the biological station.

Proxenetes flabellifer, Jensen.

One specimen among *Cladophora*, in a rock-pool near the station.

Mesostoma neapolitanum, v. Graff (Pl. V., fig. 1).

New to L.M.B.C. District.

Under this name I introduce a single specimen, found among seaweeds collected between tide-marks at Port St. Mary, which conforms in all essential details with the descriptions of v. Graff and Gamble.

Length ·6 mm., white, gut slightly discoloured by yellowish food-stuff. Pharynx central (not in front of centre as described by v. Graff and Gamble, but this difference is perhaps due to unequal contraction in my specimen). Rhabdites very densely distributed in anterior end, forming two very well marked lines between the eyes. The eyes are small and reniform, provided with lenses, body pointed in front, rounded behind, genital pore close to posterior end of body. Testes lateral, elongated; seminal vesicle kidney shaped, the penis being attached to the concave side (fig. 1). Penis consists of a proximal muscular portion and a distal chitinous tube which is slightly more curved than in v. Graff's figure. The atrium is extremely large and conspicuous, copiously supplied with glands.

The female reproductive organs were not developed in my specimen. It is strange that in all the recorded occurrences (as far as I am aware) of this species, viz., von Graff's (Monographie, p. 310), Gamble (3), p. 26, and the present note, only single specimens have been found. I have preserved this specimen as a microscopic preparation, by Dr. M. Braun's method. ("Die Rhabdocoeliden Turbellarien Livlands"; Archiv. f. d. Naturkunde Livlands, &c., Bd. X., 1885.)

Family PROBOSCIDÆ.

Pseudorhynchus bifidus (M'Intosh).

Common between tide-marks at Port Erin and Port St. Mary. Some of those found on the Breakwater measured as much as 2·2 mm. in length, although v. Graff gives

1·7 mm. and Gamble 1—1·3 mm. as usual size. The number of turns in the spiral on the chitinous penis varied between six and seventeen. The spiral was right-handed in all the examples that I examined. The bursa, which has a delicate chitinous lining, was very obvious in some of the specimens I examined; it has been accurately figured by Jensen (7, Tab. iv., fig. 12).

When disturbed this worm retreats rapidly backwards in a very characteristic manner, attaching itself by its adhesive tail and drawing the body up, the movement being repeated in a leech-like manner.

Acrorhynchus caledonicus (Claparède).

Port Erin Breakwater, two specimens; Port St. Mary, between tide-marks, three specimens.

Macrorhynchus nägelii (Kölliker).

Among seaweeds at low-water opposite the biological station, six examples.

Macrorhynchus croceus (Fabricius).

New to L.M.B.C. District.

This species was the commonest Turbellarian during my stay at Port Erin; numerous specimens, from fully grown individuals with a ripe egg capsule in uterus, to young ones in which the gonads were only appearing, being found in every piece of seaweed examined, both from Port Erin and from Port St. Mary. I have been unable, although I had exceptional opportunities, to make out from pressure preparations the relations of the female gonads to the atrium, but I hope to be able to throw some light upon this question by the examination of sections of some specimens which I preserved for this purpose.

Macrorhynchus heligolandicus, Metschnikoff.

Occurred at Port St. Mary and at Port Erin among seaweeds, and in tide-pools.

Family VORTICIDÆ.

Provortex balticus (Schultze).

Common in tide-pools at Port Erin. A large proportion of those examined had ripe eggs in their uteri. A variety without the usual brown pigment was found along with normally coloured individuals among Cladophora.

Provortex affinis (Jensen).

New to L.M.B.C. District.

One example, among *Cladophora* in a tide-pool near the station. This species is easily distinguishable from *P. balticus* by the form of its penis, the distal part of the chitinous tube bending at an angle to the proximal part and bearing a leaf-like triangular plate.

Fecampia erythrocephala, Giard.

Gamble gives Port Erin as a locality for this parasitic form in the Cambridge Natural History, vol. ii., p. 50, although at the time of his publishing his paper in these " Transactions " he had not met with it.

I found numerous " cocoons " of this species under stones between tide-marks, but I failed to find the worm itself in any of the *Carcini* that I examined.

Graffilla buccinicola, n. sp. (Pl. V., figs. 3 to 6 ; Pl. VI., figs. 7 to 13).

The above name I propose to give to a parasite from the kidney of *Buccinum undatum* and *Fusus antiquus*, which I found infesting these two molluscs in considerable numbers. The genus *Graffilla*, von Jhering (8), was established to receive a species found in the kidney of *Murex*, which v. Jhering called *Graffilla muricicola*. To the same genus a parasite discovered by Lang (9) was referred by v. Graff (Monographie, p. 375) who called it *G. tethydicola*. V. Graff also refers the *Anoplodium mytili* of Levinsen (10) to this genus. Finally a fourth species was described by Dr. Ferdinand Schmidt (Archiv. fur

Naturgesch. v. lii., pt. 1, p. 305, 1886), from the liver of *Teredo*, under the name of *Graffilla braunii*. A valuable account of the anatomy of the first two species is given by Böhmig; (2). The genus *Graffilla* has not hitherto been found in British waters.

DESCRIPTION :—Length 1—2·5 mm. Breadth ·5—1 mm. Colour greyish yellow to reddish yellow, very opaque, in favourable pressure preparations groups of olive brown pigment granules (fig. 5) are visible evenly distributed over the body. Sections prove that they are situated in the parenchyma. Form, cylindrical in section ; outline of body varies according to condition of genital glands ; in small, 1—1·8 mm. long individuals, in which male organs alone are visible, it is elongate with the greatest breadth behind middle of body, and posterior third tapered off into a tail ; in larger specimens with ovaries and yolk glands highly developed the general outline is stouter, and the "tail" usually cannot be recognised (fig. 6).

Mouth on ventral aspect of anterior end of body, being in fact an excellent example of a terminal mouth. It leads into a pharyngeal sack, quite obvious in all sections of well preserved specimens, in which lies the small doliiform (v. Graff) pharynx, which can be protruded by the living animal to a certain extent. The pharynx measures $\frac{1}{12}$—$\frac{1}{15}$ of total length of body.

Nothing comparable to the "Haftapparat" of Böhmig is present in this species. The interior of the pharyngeal sack is not ciliated. Eyes, two ; small, reniform, provided each with three or four small lens cells, they are completely buried in the substance of the brain as will be seen in fig. 7. The body is uniformly ciliated, the cilia are short and comparatively thick, the cuticular layer of the epidermis is also thicker than is usual among the Turbellaria. Around the mouth I observed a few cilia longer than

the others and non-motile, they are presumably tactile.
No rhabdites were to be observed in the skin although in
living specimens pressed on the slide a considerable
amount of viscid mucous matter was extruded from the
epidermis. I have found no trace of the epidermal glands
so characteristic of *Graffilla tethydicola*.

There is an outer circular and an inner longitudinal
muscle layer. The pharynx passes off into a well marked
œsophagus (fig. 8). The œsophagus soon widens out
into the very extensive gut, which has histologically the
characters described by Böhmig. The extent of the lumen
of gut depends upon the amount of food recently absorbed
by the cells, and may be said to vary inversely as the
amount of granules and vacuoles in the gut cells themselves. The gut occupies the whole volume of the animal's
body posteriorly to the germ glands or testes, excepting
the small space left between it and the body wall, in
which the great yolk glands ramify.

The genital opening is a very short distance behind the
mouth, approximately on a level with the posterior opening
of the pharynx, so that in a section it is possible to get
the eyes, the genital opening and the posterior end of the
pharynx in the same plane, as was actually the case in the
specimen from which fig. 7 is copied. The whole relations
of the genital organs are so exceedingly variable according
to the age of the particular specimen examined that they
will require separate descriptions, just as if the worm was
of separate sexes.

This *Graffilla* presents one of the most extreme cases of
successive hermaphroditism with which I am acquainted
among the Turbellaria, specimens which present traces
of both male and female organs being very unusual. In
specimens measuring 1—1·8 mm. the male organs are
generally predominant, the most conspicuous organ is the

large seminal vesicle about twice as long as the pharynx, and lying just posterior to it. It completely conceals the atrium in pressure preparations, but the relations of the various parts have been drawn from a section in fig. 8. Here we see the penis projecting into the atrium ; and the diverticulum which will afterwards form the seminal receptaculum lying ventral and posterior to the seminal vesicle. The testes are lateral, they extend forward to the level of the seminal vesicle, with which they communicate by short ducts opening into the posterior end of the vesicle. Posteriorly they do not quite reach back to the middle of the body (fig. 3).

The penis is very difficult to observe, but my examinations of pressure preparations which I have been able to confirm by sections shew it to be a short thick tube, strengthened by several cuticular rings and capable of being retracted so as to appear as a rosette shaped organ on anterior end of seminal vesicle (fig. 10), or of being protruded as shown in figs. 11 and 13. The distal rings appear to be provided with fine serrations. I have not been able to make out the mechanism by which the penis is thus protruded.

In large individuals, in which the female organs are fully developed the testes and seminal vesicle together with the penis have atrophied and cannot be found either in pressure preparations or in sections. In one or two lucky pressure preparations I had the good fortune to find the seminal vesicle persistent, and the testes still visible ; in these specimens the germ glands were not mature, and the yolk glands were much less extensive than in the majority of cases.

In the typical "female" condition the receptaculum seminis is large, spherical, and lies dorsal and posterior to the atrium (figs. 4 and 12) which is provided with uni-

cellular shell (?) glands. The germ glands lie entirely in
the anterior third of the body, their proximal ends con-
verging forward to open into the posterior part of the
atrium in company with the yolk glands. Their distal
halves are bent at an acute angle to the proximal halves,
the apex pointing forwards and upwards, so that in a
pressure preparation the glands present the appearance
of two V's, one on either side of the body, occupying
much the position that the testes formerly did.

In minute structure they conform to the type which is
so distinctive of *Graffilla*, the individual ova being somewhat flattened and suggesting rouleaux of coins. The
ducts of the yolk glands lie internally (mesiad) to the
germaria and open into the posterior wall of the atrium.
A short distance behind the atrium they branch, and by
complex branching and anastomosing form a more or less
complete sheath around the gut extending from the brain
to the tail (fig. 4). Underneath the epidermis is the same
extraordinary system of supposed excretory tubes that
has been observed in other species of the same genus.
They ramify and anastomose all over the dorsal surface in
a most complex manner ; and, as in the other members
of this genus, no flame cells are visible. On either side
of the body, running from anterior to posterior end, is a
larger tubule, which presents a slight dilatation about the
middle of its length.

The chief characters which distinguish this species
from the other four members of the genus, are the position
of the genital pore, the form of the germ gland and the
penis, and the possession of pigment spots in the parenchyma. From *G. muricicola* it differs in the absence of
Böhmig's " Haftapparat " or " Bohrapparat," in the form
of the body, the position of the genital opening, the form
of the germ glands, the fact that the eyes are embedded

in the brain, &c. From *G. tethydicola* it is at once distinguishable by the possession of eyes, and absence of the very striking epidermal glands. *G. braunii* also possesses the "Haftapparat," while the "excretory" tubules are very distinctive, the testes are in front of the atrium, the germ glands also present quite different relations. With regard to *G. mytili*, as since Levinsen (10) published his original description no specimens have been found, its position is doubtful, but Levinsen's description supplies quite enough details to prove that it is distinct from the worm I am describing. "Ovaria longa, sacciforma intra margines laterales sita" is sufficient in itself to distinguish *G. mytili* from *G. buccinicola*, to say nothing of the further statement in the Danish text, where he describes it as extending itself along the side of the body from the eyes far backwards, as a pair of broad sinuous sacks. His sketch and description of the seminal vesicle and penis have also no resemblance to the same structures in my species.

Habitat :—In the kidney and renal duct of *Buccinum undatum* and *Fusus antiquus*. The greater number of specimens of both these molluscs were infected, the number of parasites in one host varied between four and several dozen. A few worms were generally present in the mantle cavity as well, into which they had probably escaped through the renal aperture. The movements of this species are sluggish, like the other members of the genus it frequently lies on its side with the body dorso-ventrally flexed and swims round in a circle. From my observations upon living examples, kept in sea-water, I find that it is extremely sensitive to light, hiding under any object that is put into the vessel with it.

As the specimens of *Buccinum* which I examined were procured from the fishermen (who use them as bait), and had been out of the water for a couple of days before I received them, I cannot say whether the host suffers serious consequences from the attacks of the parasites ; the kidney, in many cases, was distended, and contained a large quantity of mucus, and its walls seemed to be locally attenuated, being almost transparent in places.

LOCALITY :—Port Erin, Isle of Man.

C. ALLOIOCŒLA.

Family PLAGIOSTOMIDÆ.

Plagiostoma koreni, Jensen.

New to L.M.B.C. District.

Four specimens occurred among seaweeds collected in tide-pools near the biological station.

Body oval, pointed behind, rounded in front, broadest about the middle.   Colour, under a low power or pocket lens, greyish white with a broad dark band across the body about the middle.   The colour is due to two kinds of pigment, a dark brown reticular pigment in the paren-chyma, which corresponds to the dark zone ; and a lighter brown granular pigment in the epidermis, which, although denser in the middle of the body, extends to the head and tail as well.

The eyes are very variable in form, carmine red, and as far as I can make out have no lens.   One of my examples presented a particularly beautiful variation in the form of the eyes.   Here they were triangular, carmine red, with a violet spot in the centre which sent processes of the same colour to the three angles.   I am inclined to believe that in this species we have an eye of the *Monotus* type, *i.e.*, a simple mass of pigment without lens.

Plagiostoma vittatum (Frey and Leuckart).

One specimen in a pool near the biological station,

colour distribution as in v. Graff, Monographie, Tab. xxii., fig. 6, *f*.

Vorticeros auriculatum (O. F. Müller).

Three specimens among seaweeds, opposite the biological station.

An individual 1·8 mm. long was taken in the tow-net by Mr. A. R. Jackson, at sunset on the 26th April in Port Erin Bay.

Cylindrostoma quadrioculatum (Leuckart).

A few specimens among *Cladophora*, collected in shallow rock pools near high-water mark, beside the biological station. This species seems particularly sensitive to light, and only appeared at the surface of the vessels, in which I put the *Cladophora*, after dark.

Family MONOTIDÆ.

Monotus lineatus (O. F. Müller).

Port Erin and Port St. Mary, very common among seaweeds collected between tidemarks.

In the great majority of the specimens that I examined the eye spot had the form of an irregular semilunar patch of pigment, such as is figured by Boeck (7), Pl. F., figs. 2, 6 and 9, or by von Graff (6), Tab. xx., fig. 18. One individual, however, taken at Port St. Mary presented such an extreme reduction of the "eye" that I have figured it (Pl. V., fig. 2). Here it is merely represented by a faint row of granules, forming a semicircle in front of the otolith.

Monotus fuscus (Oersted).

Common in tide-pools, and among tidal refuse, at Port Erin. Also found at Port St. Mary, but nowhere was it quite so numerous as *M. lineatus*.

III.　POLYCLADIDA.

A. ACOTYLEA.

Family PLANOCERIDÆ.

Stylochoplana maculata, Quatrefages.

New to L.M.B.C. District.

Two specimens, under stones, at low-water, in front of the biological station.

Length of my two specimens 11 and 13 mm. respectively, body elongated; broad and leaf-like in front, tapering behind. Tentacles in the first fifth of body, the base of each is surrounded by a circle, which in examination of the living animal suggests a thickening of the epidermis, and gives one the idea of a depression into which the tentacle can be partly retracted. Five to seven eyes on base of each tentacle; between tentacles, and running toward anterior margin are two groups of eight or nine eyes, besides which, one of my specimens showed several minute black specks intermixed with these intertentacular eyes.

Colour, to naked eye, greyish brown; under microscope the ground colour is yellowish-brown, with darker spots, and a number of large white blotches, very well developed in one specimen, but smaller and less clearly marked in the other. Two genital openings are present.

I kept both specimens under observation for some hours, and noted that they swim with dorso-ventral flexion of the whole body, much as the medicinal leech does; and if disturbed, when creeping on the wall of the vessel, they retreat with a curious wriggling movement, which is produced by the alternate use of the two sides of the anterior end of the body; the left anterior corner being extended, and the body drawn up, when immediately the right side is pushed forwards and the process repeated alternately by the two sides, the worm literally dragging itself along "hand over hand."

Verrill has suggested the generic name *Heterostylochus* for this species, considering that by the possession of two genital openings it has a right to generic distinction. (Trans. Connecticut Acad., vol. viii., 1892, p. 467.)

Family LEPTOPLANIDÆ.

Leptoplana tremellaris (O. F. Müller).

Several specimens, under stones, at Port St. Mary and Port Erin.

B. COTYLEA.

Family EURYLEPTIDÆ.

Cycloporus papillosus, Lang, var. *lævigatus*, Lang.

Two specimens at Port St. Mary under the same stone one of which measured 13·5 the other 11 mm.

Two found by Prof. Herdman, adhering to a colony of *Botrylloides* outside Port Erin Harbour, length 9 mm. and 6 mm. respectively. I have not had sufficient experience of this species to say whether this marked difference in size between the two individuals which usually consort together is or is not constant; further observations upon this point will be of interest, as it is possible that these couples which are so frequently found together represent stages in which the male and female elements are respectively at the maximum of development.

The relation between *Cycloporus* and the Ascidians with which it is so frequently associated is also a mystery, whether *Cycloporus* eats the Ascidian or merely derives moisture from it during the period it is exposed by the tide remains to be determined.

POSTCRIPT :—Since the above was written I have found

a previous record of *Stylochoplana maculata* and of
Fecampia erythrocephala for the district. Report Brit.
Assoc., 1894. Proc. Sect. D, p. 318. *Fecampia* has also
been recently observed at Plymouth by Mr. Garstang
(Journal Marine Biol. Assoc., vol. III., p. 217).

PAPERS REFERRED TO.

1. BOECK; in Gaimard "Voyages de la commission
   scientifique du Nord en Scandinavie, &c., pendant
   les années, 1838—40, Atlas de Zoologie, imp. fol.,
   Paris 1842—1845.
2, BÖHMIG, L. "Untersuchungen über rhabdocöle Turbellarien; I, das genus Graffilla." Zeitschrift fur
   wissenschaftliche Zoologie, Bd. XLIII.
3. GAMBLE, F. W. "British Marine Turbellaria"
   (Quarterly Journ. Microscop. Science; April, 1893).
4. GAMBLE. "Report on the Turbellaria of the
   L.M.B.C. District" (Trans. L'pool. Biol. Soc.,
   Vol. VII., 1893).
5. GAMBLE. Article "Turbellaria" in Cambridge
   Natural History, vol. II., London, 1896.
6. VON GRAFF, L. "Monographie der Turbellarien,"
   I. Rhabdocoelida, Leipzig, 1882.
7. JENSEN. "Turbellaria ad litora Norvegiae occiden-
   talis." Bergen, 1878.
8. VON JHERING. "Graffilla muricicola, eine parasit-
   ische Rhabdocoele." (Zeitschrift fur wiss. Zool.
   Bd. XXXIV., pp. 147—174, Tab. VII., 1880.)
9. LANG. "Notiz über einen neuen Parasiten der
   Tethys, aus der Abtheilung der Rhabdocoelen Turbellarien." (Mittheilungen aur der Zoologischen
   Station zu Neapel; Bd. II., 1881.)
10. LEVINSEN, G. M. R. "Bidrag til Kundskab om
    Grönlands Turbellarie Fauna." (Vidensk. Meddel.

fra den naturh. Foren. i Kjöbenhavn, 1879, pp. 1—
42, Tab. III.)

11. SCHMIDT, FERDINAND. *Graffilla braunii*, n. sp. Archiv. fur Naturgesch., 1886. Bd. I.

12. VERRILL. Trans. Connecticut Acad., vol. VIII., p.
467, 1892.

EXPLANATION OF PLATES.

PLATE V.

Fig. 1. Penis of *Mesostoma neapolitanum*, v. Graff.
v.s., seminal vesicle ; *ch.*, chitinous tube which
projects into the spacious atrium.

Fig. 2. Anterior end of variety of *Monotus leneatus* (O.
F. M.), with greatly reduced eye spot. *e.*,
eye ; *ot.*, otolith ; *g.*, gut ; *t.c.*, tactile cilia on
anterior margin.

Fig. 3. *Graffilla buccincola*, n. sp., from a pressure
preparation, the male organs are developed in
this individual. *ep.*, epidermis ; *ph.*, pharynx ;
e., eye ; *s.gl.*, salivary (?) glands ; *g.*, gut ; *te.*,
testis ; *v.s.*, seminal vesicle ; *pe.*, penis.

Fig. 4. Anterior end of an individual in which the
female elements are developed. *gcr.*, germ
gland ; *y.*, yolk gland, which is not figured on
the right side of the sketch ; *at.*, atrium ; *r.s.*,
receptaculum seminis ; other letters as in
fig. 3.

Fig. 5. Pigment spots, composed of groups of granules,
from the subcutaneous parenchyma of *G.
buccinicola.*

Fig. 6. Four sketches, to illustrate the variations in form,
of *G. buccinicola.* *a.* and *b.*, examples in the

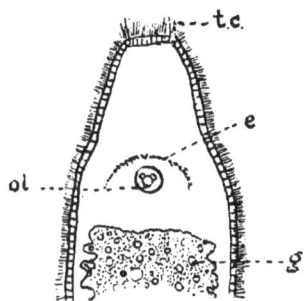

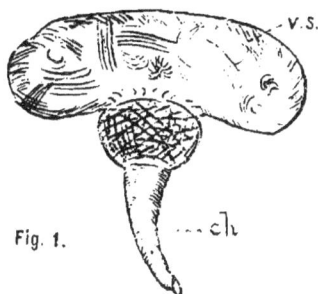

Fig. 1.

Fig. 2.

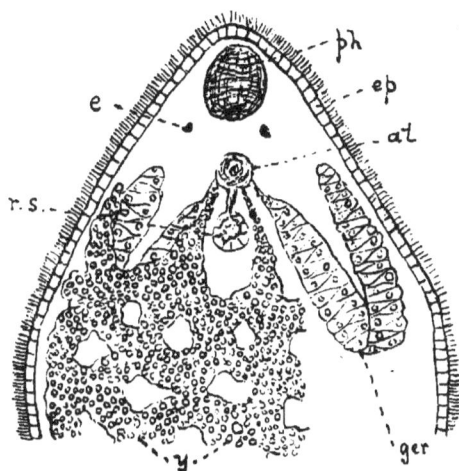

Fig. 4.

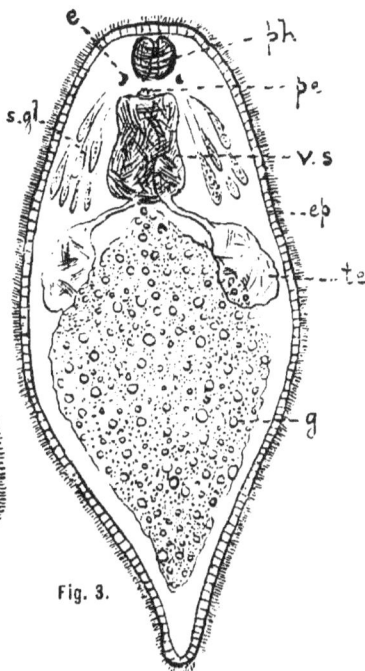

Fig. 3.

Fig. 5.

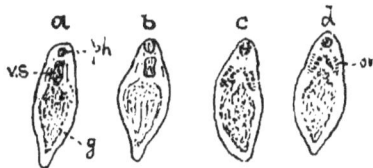

Fig. 6.

H. L. J., del.

L.M.B.C. TURBELLARIA.

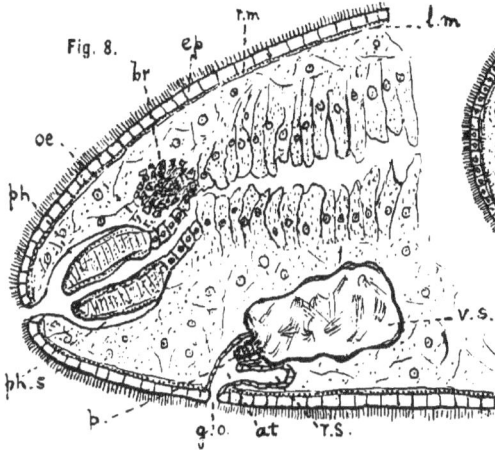

Fig. 8.

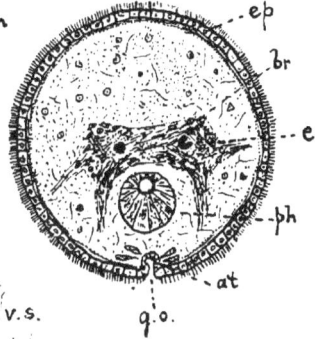

Fig. 7.

Fig. 9.

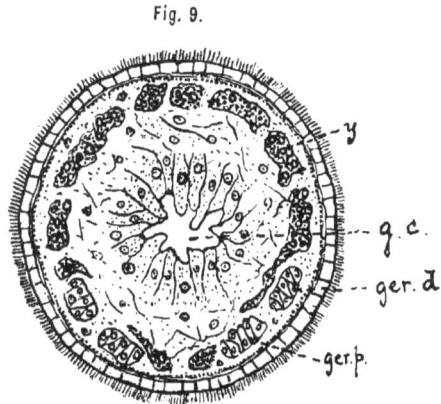

Fig. 10.

Fig. 11.

Fig. 12

Fig. 13.

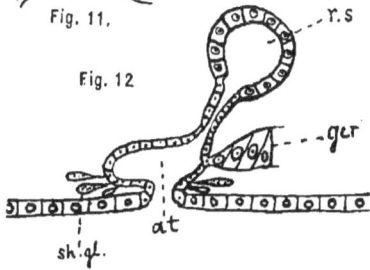

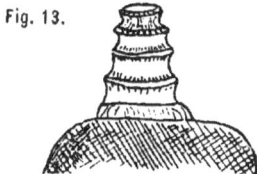

H. L. J., del.

L.M.B.C. TURBELLARIA.

male stage; *c.* and *d.*, do. in female stage; *ph.*, pharynx; *v.s.*, seminal vesicle; *g.*, gut; *ov.*, germ gland.

PLATE VI.

Fig. 7. Transverse section through region of brain of *Graffilla buccinicola*; *ep.*, ciliated epithelium; *br.*, brain; *ph.*, pharynx (posterior end); *at.*, genital atrium; *g.o.*, genital pore; *e.*, eye.

Fig. 8. Longitudinal section of specimen with male organs developed. *ep.*, epidermis; *r.m.*, circular muscles; *l.m.*, longitudinal muscles; *br.*, brain; *ph.*, pharynx; *oe.*, oesophagus; *ph.s.*, pharyngeal sack; *g.o.*, genital pore; *r.s.*, rudiment of receptaculum seminis; *p.*, penis; *v.s.*, seminal vesicle.

Fig. 9. Tranverse section of *G. buccinicola* in the female stage, passing through body near the apex of the loop formed by the germ glands. *g.c.*, cavity of gut; *ger.p.*, proximal limb of germ gland; *ger.d.*, distal do.; *y.*, yolk gland.

Fig. 10. Seminal vesicle with penis retracted.

Fig. 11. Do. penis protruded, from the same specimen under higher pressure.

Fig. 12. Scheme of relationships of the female apparatus constructed from sections. *at.*, atrium; *r.s.*, receptaculum seminis; *sh. gl.*, shell glands; *ger.*, germ glands.

Fig. 13. Penis, strongly magnified, from a pressure preparation.

ELEVENTH ANNUAL REPORT of the LIVERPOOL MARINE BIOLOGY COMMITTEE and their BIOLOGICAL STATION at PORT ERIN.

By Professor W. A. Herdman, D.Sc., F.R.S.

[Read December 10th, 1897.]

The past year, though comparatively uneventful, has been marked, as the following pages will show, by much solid biological work carried on at Port Erin and elsewhere on our littoral; and several of our local workers have opened up interesting lines of investigation of both scientific and economic importance. There was experimental fish hatching at Easter, several meetings have been held at Port Erin during the year and lectures and practical demonstrations given, the College tables have been well occupied, some additions have been made to our faunistic lists, and several notable papers published in scientific journals upon the results of work done at the Biological Station.

The fact that we have had fewer dredging expeditions than in some previous years is probably due to the absence of the Hon. Treasurer in Scotland and of the Hon. Director in America during a considerable part of the summer. But the new season will see renewed activity in this direction. Plans are being laid for a more detailed survey of our submarine area, for a systematic exploration of the problems of distribution and environment. The changes of the plankton, or floating life of the sea, a knowledge of which is so important in fishery questions, has already engaged our special attention, and, as will appear further on in this Report, a scheme is in operation for the simultaneous observation and record of the organisms on the surface at a number of stations in our district.

It is these general problems, sometimes extending over neighbouring sciences and requiring the co-operation of several specialists, which are now of the greatest interest and practical importance. Our specialists in marine biology are becoming more minute in the details of their work, but, at the same time, wider in their knowledge, in their outlook, and in the applications of their research. Biology—which has given not only to science but to all departments of knowledge the educational method of laboratory work and the great fundamental principle, Evolution, which underlies all advance—is ever ready to adopt methods and results from other sciences as an aid in the investigation of her special problems on land and sea. And in this age, pre-eminently that of Biology—the age of Darwin, Pasteur, and Lister—it is coming to be recognised equally over Europe and America that nowhere more than in Marine Biological Stations has the work of the great masters been followed up and extended, and that nowhere else can be found a more natural and happy union of the philosophy of science and of the industrial applications. It is that that gives to marine laboratories their first-rate importance both in pure science and in the work of Sea-Fisheries Committees, and which is causing universities all over the world to establish and maintain Biological Stations as a necessary condition for the advance of natural knowledge. Thus the University of Paris has Roscoff and Banyuls, Vienna has Trieste, St. Andrew's has just opened the Gatty Marine Laboratory, and Glasgow the Millport Station. We have our modest workshop at Port Erin, and our more extensive Fisheries Institution at Piel, in Lancashire, but we may well hope for and claim a larger and better equipped laboratory at

Port Erin, or at Hilbre, one more worthy of our University and of this great seaport. Liverpool owes much to the sea, it is asking but little that she should take her place in supporting oceanographic research.

STATION RECORD.

The following naturalists have worked at the Port Erin Laboratory during the past year :—

DATE.	NAME.	WORK.
January.	Mr. I. C. Thompson, Liverpool	Collecting.
—	Prof. W. A. Herdman, Liverpool	
March.	Mr. H. Murray, Manchester	Algæ.
—	Prof. F. E. Weiss, Manchester	
—	Mr. I. C. Thompson, Liverpool ...	Copepoda.
—	Mr. F. W. Gamble, Manchester ...	Annelids.
—	Mr. Cole, Liverpool	Compound Ascidians.
April.	Prof. W. A. Herdman, Liverpool	
—	Mr. I. C. Thompson, Liverpool ...	Copepoda.
—	Mr. Cole, Liverpool ...	Ascidians.
—	Mr. H. Murray, Manchester	Algæ.
—	Prof. F. E. Weiss, Manchester	
--	Mr. F. W. Gamble, Manchester	Annelids.
—	Mr. J. H. Ashworth, Manchester	
—	Mr. Mundy, Manchester ...	General.
—	Mr. Claxton, Liverpool ...	General.
—	Mr. Wadsworth, Manchester	General.
—	Miss Hiles, Manchester	General.
—	Miss Pratt, Manchester	General.
—	Mr. Jameson, London	Turbellaria.
—	Mr. Jackson, Liverpool	General.
—	Mr. Gunn, Liverpool ...	Collecting.
—	Mr. A. Watson, Sheffield ..	Annelids.
—	Dr. Hurst, Dublin	Ascidians.
—	Mr. R. A. Dawson, Preston	Fish Hatching.
—	Mr. R. L. Ascroft, Lytham	
May.	Mr. F. W. Gamble, Manchester ...	Annelids.
June.	Mr. F. W. Gamble, Manchester	Annelids.
—	Mr. R. H. Yapp, St. John's Coll., Camb.	General.
—	Rev. T. S. Lea, Liverpool... ...	Photography.
July.	Rev. T. S. Lea, Liverpool...	Photography.

July.	Prof. W. A. Herdman, Liverpool		General.
—	Mr. I. C. Thompson, Liverpool ...	...	Copepoda.
—	Mr. Cole, Liverpool ...	...	Ascidians.
—	Mr. R. H. Yapp, Cambridge ...	...	General.
—	Mr. J. A. Clubb, Liverpool ...	...	Actinians.
—	Mr. Keeble, Manchester ...	...	Algæ.
—	Mr. Jackson, Liverpool ...	...	Arachnida.
August.	Mr. Keeble, Manchester ..	...	Algæ.
—	Mr. Jackson, Liverpool ...		Arachnida.
September.	Mr. Keeble, Manchester ...	...	Algæ.
—	Mr. I. C. Thompson, Liverpool ⎫		
October.	Mr. I. C. Thompson, Liverpool ⎬		Copepoda.
—	Prof. W. A. Herdman, Liverpool		Ascidians.
—	Prof. R. Boyce, Liverpool ⎫		
—	Dr. Warrington, Liverpool ⎬ ···		General.
November.	Prof. W. A. Herdman, Liverpool ...		General.
—	Mr. I. C. Thompson, Liverpool ...		Copepoda.
—	Dr. R. T. Herdman, Edinburgh ⎫		
.–	Mr. P. M. C. Kermode, Ramsey ⎬		General.
—	Mr. R. Okell, Douglas ⎭		
December.	Mr. H. C. Chadwick, Bootle ...		Collecting.
—	Prof. R. J. Harvey Gibson, Liverpool ...		General.

This is about the same total number as in the previous year. We have not this time the distinguished foreign Biologists who visited us after the British Association meeting in Liverpool; but our numbers of students and of ordinary workers are steadily increasing.

Amongst those in the above list are several students from Owens College, Manchester, and University College, Liverpool, who, along with members of the staff of the biological departments, have made use of the work places rented at the Station by the two Colleges. All of the students who took up Zoology as one of the subjects for their final B.Sc. examination in Victoria University, and of whom three passed with Honours, took advantage of the Port Erin Station during some part of their final year of study. During the year the Owens College table has been used by one professor, three demonstrators, two junior assistants, and three students. The University College table has been used by one professor, two demonstrators, one assistant, one former assistant, and three students.

In addition to workers and students, we had many visitors, and on July 9th, the members of the Isle of Man Natural History Society spent a day at the Station. A public meeting of the Society and others was held in the Laboratory, and your Director gave them an Address upon "The Study of Marine Biology." On that occasion about thirty boys with some of the masters from King William's College, Castletown, also visited the Station, and took a lively interest in the Aquarium tanks and the specimens under microscopes.

Later in July Mr. T. S. Lea, who was working at the time at Port Erin, organised the visit of about thirty Liverpool Board School boys to the Biological Station, along with Mr. H. Edwards, one of their masters. They were taken for a zoological ramble round by Spanish Head and the Calf Sound, came back to Port Erin for tea, and afterwards examined the Aquarium tanks, and were taken to hunt the rock pools at low tide. Nothing is better calculated than marine biology—with its endless variations of form and colour, interesting habits, and instructive adaptations to environment and circumstance —to impress the youthful mind with a love of nature, to encourage powers of observation, to excite curiosity as to the causes of things, and to open up to those accustomed only to a town life some glimpses of the beautiful world of nature.

It has become evident to the Committee that, in the interests of the College students who are now attending the Biological Station, it is necessary to obtain a more highly qualified Curator than the Laboratory lad who has looked after the place for the last couple of years. They feel also that the presence of a scientific man constantly at Port Erin will result in an improvement in the Aquarium and in the experimental fish hatching. The collecting and recording of specimens and physical observations which has depended so much in the past upon the chance visits of members of the Committee and other investigators ought, under a resident naturalist, to become systematised, and yield valuable results. The Committee consider they are fortunate in having been able to arrange with Mr. H. C. Chadwick—formerly of Owens College, and for some time Assistant Curator in the Bootle Museum— that he shall go into residence at Port Erin at the beginning of the new year, and shall devote his attention, in addition to the routine duties of the post, to a series of observations and investigations upon lines drawn up by the Committee.

It is becoming more evident year by year that both for the purposes of scientific Biology and also in the interests of fishery questions we must endeavour to gain a more intimate and detailed knowledge of the statistics of communities or assemblages of animals on the sea-floor, and of their habits and inter-relations.

A couple of years ago we published* some statistics of dredging on different grounds. This work should be continued and extended. Mr. A. O. Walker has lately† made comparison between the fauna on shallow and that on deep mud, in our area, with the result that the shallow mud shows by far the greater number of genera, species, and specimens of Crustacea. That is a valuable opinion, but refers to one group of organisms only. The individual members of our Committee are specialists—each with his

* Ninth Annual Report, p. 25, 1896.
† Liverpool Biological Society, November, 1897.

own absorbing interest—and though thoroughly alive to the importance of these general questions, they have rarely time or opportunity for sufficiently extended or continuous observations.

In this, as in other departments of work, we hope for much help from our new Curator. Regularity of investigation, observation, and record, and the accumulation of statistics as to modes of occurrence will soon give us a body of evidence from which to draw definite conclusions.

THE AQUARIUM.

Over 350 persons paid for admission to the Aquarium during last summer. The Committee do not consider this a large number. The result of several years' experience is that when naturalists are at the Station— especially responsible members of the Committee — it is easy to attract numbers to a demonstration in the Aquarium. The visitors are interested and anxious to learn when there is anyone to show them what and how to observe, and to explain wherein the importance of the observation lies. In the absence of a scientific zoologist, the Aquarium has languished. Our new Curator will meet this want. He proposes to fill the shelves with collections of local animals, to re-stock the tanks and vessels, to lay out some microscopic preparations, and otherwise to make that part of the institution open to the public more attractive and more efficient.

THE BOAT.

The " Shellbend " folding dinghy, the acquisition of which was recorded in our last Report, has proved a very serviceable boat, and keeps in excellent condition. She holds three comfortably for tow-netting work about the bay, can be expanded for use or folded up again by one

person in about twelve seconds, and is light to carry up
and down the shore. We are indebted to Mr. M. Treleaven
Reade, of Liverpool, the inventor of the " Shellbend "
patent, for the use of the accompanying cuts showing
(fig. 1) the bottom of the boat when folded up, and (fig. 2)
the interior when half expanded.

THE EASTER PARTY.

Notwithstanding rather boisterous weather the usual
L.M.B.C. Easter Dredging Expedition was carried out
with success, and the Port Erin Biological Station was
never before so full of workers as it was during April.
In the actual Easter week the rather limited accommodation was more than fully occupied, and the Committee are
in hope that an extension may be provided, which will
give several additional working places, before next Easter.

The Colleges of Liverpool and Manchester, it will be
remembered, last year acquired the right to send members
of their staff or science students to occupy certain work
places for specified periods at the Port Erin Laboratory.
On this occasion the Owens College was represented by
Professor Weiss, Mr. F. W. Gamble, Mr. Ashworth,
Mr. H. Murray, Mr. Wadsworth, Mr. Mundy, Miss Hiles,
and Miss Pratt; University College, Liverpool, by Professor Herdman, Mr. F. J. Cole, Mr. Jackson, Mr. Claxton,
and Mr. W. Gunn. Amongst other workers at the Station
were—Mr. Arnold Watson (Sheffield), Mr. Isaac Thompson
(Liverpool), Dr. Hurst (Dublin), and Mr. Lyster Jameson
(Royal College of Science, London). The Lancashire
Sea-Fisheries steamer " John Fell " (with Mr. Dawson,
the Superintendent, and Mr. Ascroft, a member of the
Fisheries Committee), was at Port Erin during the Easter
week carrying on trawling investigations, and several
general dredging expeditions were made with her. Spawn-

Fig. 1.   10 ft. Shellbond Dinghy.   Fig. 2.

ing fish were procured to the west of the Isle of Man, and the tanks in the Biological Station were supplied with developing Lemon Soles and "Witches" (White Soles), and with a cross between the "Megrim" and the Cod. Elsewhere in this Report will be found the accounts given by several of the above-named naturalists of the work which they were engaged in at the Laboratory during the Easter vacation.

Fig. 3. Eastern half of the Port Erin Biological Station, from the steps, showing the apparatus—trawl, dredges, tow-nets, sieves, aquaria, pails, collecting jars and baskets—brought back from a dredging expedition.

SEA-FISH HATCHING.

In continuation of the experiments carried out at Easter, 1896, last April the Lancashire Sea-Fisheries steamer, "John Fell," came to Port Erin under the direction of Mr. R. A. Dawson, the Superintendent, for the purpose of searching for spawning fish. On Saturday,

April 17th, we trawled mature fish of various kinds both flat and round, but did not succeed in getting both males and females of the same species in the ripe condition, and consequently no eggs were fertilized. On Monday, April 19th, we were more fortunate, and obtained to the north-west of Port Erin Lemon Soles and "Witches" spawning, and were able to fertilize the eggs. We also found spawning Megrims *(Arnoglossus laterna)*, and, as an experiment, we fertilized the eggs with the milt of a ripe Cod. As a result large numbers of the following embryos were started on their development in the Aquarium on the afternoon of the 19th :—

In tank I.—Lemon Soles *(Pleuronectes microcephalus)*.

In tank II.—"Witches" *(Pleuronectes cynoglossus)*.

In tank III.—Ova of Megrim fertilized by milt of Cod.

The arrangement of the hatching tanks, and the apparatus for the circulation of the water was described and illustrated in last year's Report.* The water during the hatching kept at a specific gravity of from ·26 to ·27 and at a temperature of from 46° to 47° Fahr. The cross between the Megrim (a flat fish) and the Cod only developed for from three to four days, and then all the embryos became abnormal and distorted, and died.

On April 26th the Witches began to move inside the egg covering, on the 27th the Lemon Soles were wriggling, and on the 28th both hatched out, eight and a half days after fertilisation of the eggs.

We have now shown conclusively that sea-fish hatching can be carried on readily at Port Erin, and beyond this experimental stage, dealing with only a few hundreds of eggs at a time, we cannot with our present accommodation and appliances pretend to go. If hatching on an industrial

* See Trans. Biol. Soc., vol XI., p. 67 and Pls. I—IV

scale is to be carried on at Port Erin, it will be necessary
to erect a separate building—the hatching house—with an
adjacent concreted pond and a boat jetty, alongside the
Biological Station. The hatchery house, made two storeys
high and placed in the gap to the west of our Aquarium,
so that the lower floor would open on the beach and the
upper floor from the Aquarium room, could be put up
for a comparatively small sum by our local builder at
Port Erin. The necessary concreted pond, to be used
sometimes for spawning fish and sometimes for rearing
young, could be readily made on the beach below, using
the cliff as one side, while the opposite wall of the pond
could be run out as a boat jetty. Such a hatchery would
be available both for sea-fish eggs and also for hatching
young lobsters, and its connection with the Biological
Station should be an advantage to both institutions, and
should especially conduce to the efficiency of the hatchery.

"Plankton" Observations.

From an early period in the L.M.B.C. work attention
has been directed to the importance of careful observations
on the periodic variations in the amount and nature of
the plankton or assemblage of drifting organisms on or
near the surface of the sea.

In 1888, during our first year of work at Puffin Island,
we started our Curator of the Station taking weekly
gatherings of surface organisms, which were sent to
Liverpool and examined by Mr. Thompson. This was
kept up intermittently during the five years of our occupation of the Puffin Island Station. During the first
year of the Committee's work (1885) we noticed (see
Report 1, p. 21, and Report 3, p. 8) in some of the
gatherings the presence of those extraordinary numbers
of *Halosphæra*, *Tetraspora*, and other minute gelatinous

Algæ which periodically cause, round our coast, what has been called "foul water." This was again noticed in 1886, and in subsequent years (Report 3, p. 8). In 1889 (Report 2) we noted the occasional occurrence of phenomenal numbers of *Anomalocera patersonii* over certain tracts of sea, and its subsequent complete disappearance. We also in that year made our first observations upon the effect of "baiting" the tow-net with an electric light, for use after dark both at the surface and at the bottom of the sea (for details see Report 2, p. 17). Further observations of this nature were made in 1889 (Report 3, p. 27), and in the same year's Report we published a summary of the observations throughout the year upon the temperature of the sea and the condition of the organisms upon its surface. A further observation of surface organisms in connection with "foul water" will be found in the next year's Report, for 1890. Other odd notes on the subject occur scattered throughout our ten previous Reports, and in Mr. Thompson's various papers reprinted in the volumes of the "Fauna" (see also "Fauna," vol. I., p. 324, for lists of surface organisms taken at Port Erin in the Summer of 1886).*

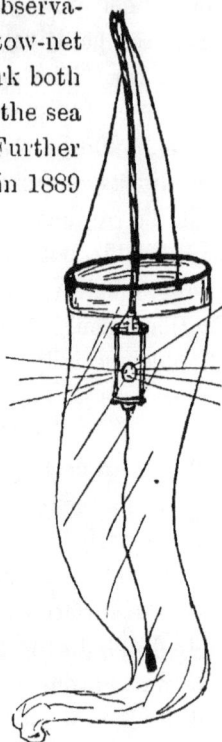

Fig. 4. Tow-net with electric light.

Last year we went a step further, and, with the help

---

* Prof. M'Intosh had carried out similar investigations for the Scottish Fishery Board in 1888 (see Seventh Ann. Rep. Fish. Bd., Scot., p. 259, 1889). More recently Messrs. Bourne, Bles, Garstang, and others at Plymouth have recorded the variations in the plankton at different times of the year.

of Mr. Andrew Scott, I organised a scheme for the weekly collection of surface plankton throughout 1897 at six stations in our district. The localities were Port Erin (I. of Man), New Brighton (near Liverpool), Lytham and mouth of Ribble (coast of Lancashire), Piel (Barrow Channel), and from the Fisheries steamer, at sea, wherever she happened to be. The collections were taken, preserved, and sent to my Laboratory at Liverpool, where they were

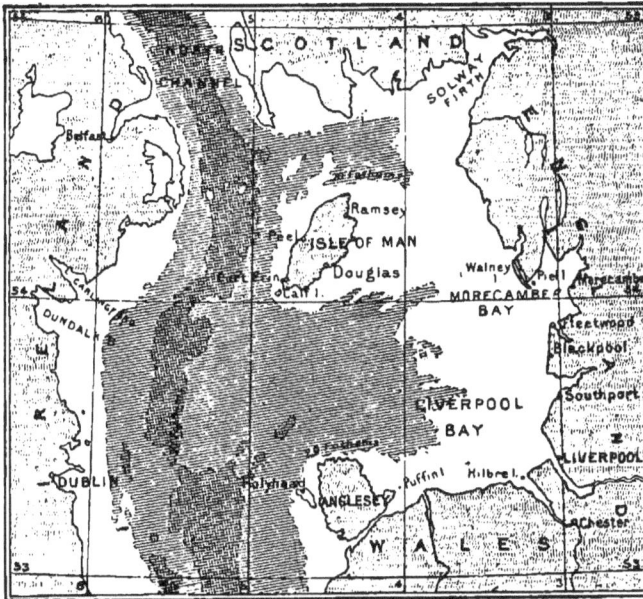

Fig. 5. Plan of the L.M.B.C. District.

measured by Mr. Scott and then examined in detail. The scheme was started towards the end of January, and was kept up as regularly as possible—perfect regularity is not possible, first, on account of the weather, and secondly, because the bailiffs who take the gatherings are liable to be called off occasionally to other duties. During the first fourteen weeks the number of gatherings received out of the possible six were—5, 6, 4, 3, 4, 2, 3, 4, 4, 3, 3, 5, 3, 4.

These gatherings, which have been worked up fully, bring the record up to the end of April. The rest of the collection, which is now in process of being examined, consists of some sixty tubes, giving an average of nearly two a week for the remainder of the year. The results will be given in the Report on the work of the Sea-Fisheries Laboratory, to be published in a few weeks. Taking these statistics, along with the many previous less complete records that we have, extending back for ten or twelve years, there are some prominent features of the collections taken week by week and month by month that arrest attention—the abundance of *Sagitta* in January and February; the comparative scarcity of Copepoda early in the year; the abundance of diatoms, such as *Biddulphia, Coscinodiscus, Rhizosolenia*, and *Chœtoceros*, in February and early spring; the appearance of Nauplei and then other larval forms in February and March; the comparative scarcity of plankton all round in February and March (except when gelatinous Algæ sometimes swarm in the latter month, and even later); the increase in April, and especially the increased abundance of pelagic Coelenterates and of Copepoda in early summer; the appearance of fish eggs and embryos and larval fish in abundance about Easter; the disappearance of Nauplei and other larvæ as summer goes on, and the great increase in Medusæ and Ctenophora; the quantities of *Oikopleura* which appear in the height of the summer; the abundance of Dinoflagellates in late summer and autumn ; the great relative abundance of life in general during July, August, and September, and finally the rapid diminution in the amount and variety of plankton during the last few months of the year.

There are, on the other hand, some organisms, such as the Algæ *Halosphæra* and *Tetraspora*, the Infusorian *Noctiluca*, and the Copepod *Anomalocera*, which seem to vary greatly in their abundance from year to year; but probably when we have a more complete knowledge of the plankton of the North Atlantic, and of the relations existing between physical conditions and the distribution of organisms, we shall be able to assign rational causes for these curious irregularities in the floating population of our seas.

We have arranged that in 1898 gatherings will be taken weekly at the same six stations, but rather further out at sea, so as to avoid the disturbing influence of the shore.

It is interesting, in this connection, to note that of all the tow-net gatherings which I took this summer in crossing the Atlantic twice, between Liverpool and Quebec, once at the beginning of August, and again at the end of September, those from the sea around Port Erin, between Liverpool bar and the north of Ireland, were the richest in species. The following are the lists of organisms observed in the gatherings in question, quoted from the paper recently published by Herdman, Thompson, and Scott[*] :—

[OUTWARD JOURNEY—LIVERPOOL to QUEBEC.]

"1. August 5th, 9 p.m., to August 6th, 8 a.m. From 50 miles off Liverpool to about Rathlin Island, Ireland. Water slightly phosphorescent; no *Ceratium* observed; an ordinary Irish sea gathering, containing :—Fish eggs, Gastropod larvæ, *Limacina retroversa*, *Nyctiphanes norvegica* (small), Nauplei, and Zoeas (Crab), Amphipoda (fragments), *Philomedes interpuncta*, *Sagitta bipunctata*, *Mitraria* larva, Medusoids (small), Campanularians

* Trans. L'pool. Biol. Soc., vol. XII., p. 33.

(broken), *Oithona spinifrons* (common), *Calanus finmarchicus* (common), *Paracalanus parvus* (few), *Temora longicornis* (few), *Acartia clausii* (few), *Metridia armata* (rare), *Pseudocalanus elongatus* (few), *Centropages hamatus* (few), *C. typicus* (few), *Anomalocera patersonii* (few), *Longipedia coronata* (rare), *Ectinosoma atlanticum* (rare), *Thalestris serrulata* (rare), *Alteutha interrupta* (rare)."

[RETURN JOURNEY—QUEBEC TO LIVERPOOL.]

" **P,** on bath tap, October 3rd, 5 p.m., to October 4th, 10 a.m. (off Port Erin). *Ceratium tripos, C. fusus, Peridinium divergens, Dictyocysta elegans*, Diatoms, *Globigerina bulloides, Calanus finmarchicus* (common), *Centropages hamatus* (few), *Metridia armata* (few), *Pseudocalanus elongatus* (common), *Oithona spinifrons* (few), *Acartia longiremis* (few), *Isias clavipes* (scarce).

"**XXI.** and **XXII.** October 3rd, 7 p.m., to October 4th, 9 a.m.; from 20 miles north of Tory Island to 15 miles west of Peel. Great deal of material. Part of it probably from entrance to Lough Foyle, where we stopped for an hour at midnight. Nets at overflow pipe had Sagitta, Medusæ, Amphipoda, and the larger Copepoda; nets at tap had much finer stuff—the Protozoa and the smaller Copepoda. *Ceratium tripos* (few), *C. fusus* (few), *Navicula* sp., *Dictyocysta elegans, Globigerina bulloides*, Unicellular Algæ, Medusæ (small), *Sagitta* sp. (abundant), Megalopa, *Gastrosaccus spinifer, Hyperia galba, Euthemisto compressa* (common), *Calanus finmarchicus* (common), *Acartia longiremis* (common), *Metridia armata* (common), *Pseudocalanus elongatus* (common), *Centropages hamatus* (few), *C. typicus* (few), *Anomalocera patersonii* (few).

"**XXIII.** October 4th, 9 a.m. to 2 p.m.; from 15 miles west of Peel to near Liverpool Bar. Nets at overflow pipe. *Sagitta bipunctata* (abundant), *Ceratium tripos* (common), *C. fusus* (common), and *C. furca* (common), *Dictyocha speculum*, *Coscinodiscus radiatus* (common), *Globigerina bulloides* (common), *Biddulphia* sp. (common), *Calanus finmarchicus* (common), *Acartia longiremis* (abundant), *Metridia armata* (few), *Pseudocalanus elongatus* (abundant), *Centropages hamatus* (few), *Isias clavipes* (common), *Parapontella brevicornis* (few), *Labidocera wollastoni* (common).

"**XXIV.** October 4th, same time and locality as **XXIII.** Nets at tap (fine stuff). *Ceratium tripos* (abundant), *C. furca*, and *C. fusus*, *Coscinodiscus radiatus*, *Peridinium divergens*, *Dictyocha speculum*, *Tintinnus acuminatus*, *Codonella campanula*, *Halosphœra viridis*, *Rotalia beccarii*, *Calanus finmarchicus* (common), *Acartia longiremis* (common), *Metridia armata* (few), *Pseudocalanus elongatus* (few), *Temora longicornis* (few), *Centropages typicus* (few), *Labidocera wollastoni* (common), *Oithona spinifrons* (common)."

FAUNISTIC AND OTHER SPECIAL INVESTIGATIONS.

During the year a good deal of collecting has been done on the shores of the south end of the Isle of Man, and the Laboratory Assistant has made many gatherings of Copepoda from the sand and mud at low tide, and has sent them to Mr. Thompson. The tow-net has also occasionally been attached to the buoy at the entrance to Port Erin bay, and left out all night. One such gathering taken in November was sent to Mr. A. O. Walker, and he reports to me that it contained :—

Siriella armata (M. Edw.), several young.
Cuma scorpioides (Mont.), one only.
Iphinoe trispinosa (Goodsir), one only.
Bathyporeia pelagica, Bate, one only.
Apherusa bispinosa (Bate).

Atylus swammerdamii, M. Edw.

Dexamine spinosa (Mont.).

Mr. I. C. Thompson reports as follows in regard to COPEPODA :—

"During the 'John Fell' expedition of April 19th, a quantity of fish were trawled. A search for fish parasites led to two additions to the L.M.B.C. fauna, viz. : *Trebius caudatus*, Kroyer, and *Caligus diaphanus*, Müller, of which the former were taken in numbers, both males and females, from the Hake, and of the latter a few specimens were found on the Cod.

"On a specimen of *Calanus finmarchicus*, I found the small parasitic Isopod *Microniscus calani*, belonging to the Epicaridæ. Although not infrequently met with on Calani taken in more northern regions where the latter are found in profusion, it appears not to be common about our shores, and has hitherto escaped record in our district.

"W. Bridson, the Laboratory boy, has occasionally, by means of the very convenient " Shellbend " boat, fixed a tow-net to the buoy, near the breakwater, where it has remained all night. Although no forms of Copepoda new to the district have hitherto been met with through this method, the results in material have been such as to warrant our continuing it when practicable throughout the coming year.

"Washings of mud and sand taken at low water from various parts of Port Erin bay, have resulted in our obtaining a large variety of sedentary Copepoda, and among them Mr. A. Scott has recently found some very minute Harpacticidæ of one or more species new to science, but which we are not yet prepared to describe. Some of the rarer sedentary species are occasionally found by this means in very large numbers, as was the case in

October last, when some mud taken near the breakwater was found to swarm with *Laophonte lamellifera*. The rocky pools about Port St. Mary, too, have yielded good results."

Dr. George W. Chaster, of Southport, who has undertaken to report upon the FORAMINIFERA of the district, writes to me :—

" There is one aspect of the work in this group I should like to enter upon. Foraminifera are so enormously abundant, and multiply so rapidly, that they must afford an important food supply to some other animals. I only know of one group of Mollusca—the Scaphopoda—which subsist solely, or almost so, on Foraminifera. There must be many other consumers, and I propose to examine stomachs of various other animals that seem likely. I shall be pleased to examine consignments of stomach-contents which have a gritty feel, if sent to me, with full data. Fish stomachs should yield good results. The reproduction of Foraminifera is also a matter requiring further investigation."

Professors Boyce and Herdman still continue their work on the bacteriology and other diseased conditions of Oysters. In the report of the " Oyster " Committee to the British Association Meeting at Toronto last September, it was shown that there are considerable quantities of copper in certain green leucocytes found in a diseased condition of the American Oyster. The Oysters in this state are always more or less green, and the colour is due to the presence of a compound of copper. The amount of copper is far in excess of what can be accounted for as due to hæmocyanin, and it may be explained as due to a disturbed metabolism, whereby the normal copper of the body becomes stored up in certain cells. The *cause* of this diseased condition is still undetermined.

Mr. F. W. Gamble and Mr. J. H. Ashworth, of Owens College, Manchester, spent three weeks of the Easter vacation, 1897, at the Port Erin Biological Station, working at the anatomy of the species of *Arenicola*, and Mr. Gamble again visited the Station for a few days at Whitsuntide for further research into the same subject. The following is a summary of the results obtained :—

" The shores of Port Erin and Port St. Mary are inhabited by three well-marked species of *Arenicola, A. marina, A. ecaudata,* and *A. grubii.* The best known of these, *A. marina,* the common 'lug-worm,' is represented on the coarse sandy beaches by the small littoral variety characterised best by the delicate gill plumes being each composed of four or five lateral branches on either side of the main vascular axis, and by constructing U-shaped galleries in the sand. In the gravel and amongst the debris of decomposing rock on the shore of Bay-ny-Carrickey a large, dark variety occurs with gills of the same type, but strongly pigmented. The blood-vessels are very large and turgid. This variety, which may attain a length of one foot, is distinguished by the form of its gills from a somewhat similar, dark, bulky variety, occurring at extreme low water on the Lancashire coast (and known to the fishermen as 'worms,' in contradistinction to the littoral form or 'lugs'), in which the gills have ten or eleven lateral branchlets, and hence a pinnate form. This 'Laminarian' variety has not hitherto been found at Port Erin, but this is probably due to the great depth (up to four feet) to which it is known to burrow.

" The two species *A. grubii* and *A. ecaudata* belong to a division of the genus characterised by the continuation of the gills and parapodia to the hinder end of the body. They are distinguished from one another by important anatomical peculiarities, which have hitherto escaped notice, with the result that no record of *A. ecaudata*, on British shores, excepting only that made by Dr. G. Johnson, can be accepted as trustworthy. Even the record in the Tenth Annual Report of the L.M.B.C., p. 26, should be *A. grubii*, and not *A. ecaudata*, as printed.

"To establish the differences between these two forms was the chief result obtained by Messrs. Gamble and Ashworth. They have found that *A. ecaudata* has the first gill implanted immediately behind the 16th notopodium on each side, whereas, in *A. grubii*, the first pair of gills occurs behind the 12th notopodia; that the nephridia open on segments 5—17 inclusive, in *A. ecaudata*, while in *A. grubii* there are only five pairs of these organs opening on segments 5—9; that, in the former species, the gonads are massive and compact; in the latter, diffuse and floating freely during maturation.

"The anatomy of Johnston's *A. ecaudata* has hitherto remained unknown. Our researches have shewn it to be by far the most interesting member of the genus, in that it possesses *thirteen* pairs of nephridia, the first of which opens on the 5th chætigerous segment, and the last on the 17th, or the second of the gill-bearing segments. Hitherto it has been supposed that the six pairs of nephridia of *A. marina* were typical for the genus both in number and in form. It now appears, however, that the nephridia of the true lug-worms are reduced in number, as compared with the similar organs found in *A. ecaudata*, and that the relation of these nephridia to the gonads, is of a more intimate and less transient character than is usually supposed to be the case, and though different in detail, is essentially similar in plan, in these two species. The characters by which *A. ecaudata* may be recognised are—the presence of the first pair of gills on the 16th chætigerous segment, and of well-marked notopodia on the first segment. The corresponding neuropodia are not closely approximated in the mid-dorsal line.

" The third species. of *Arenicola* found in the district occurred along with *A. ecaudata* in fine gravel on the south side of Bay-ny-Carrickey, and also, though not so abundantly, on the south side of Port Erin bay. It is certainly the *A. grubii* of Claparède, which has been hitherto described from Naples, and agrees with his description (incomplete though this is) in all essentials as in the number and position of the gills, the number (five pairs) and relations of the nephridia, and the form of the chætæ.

" In our specimens, the first pair of notopodia are absent or extremely reduced, so as to be very rarely visible externally, while the corresponding neuropodia are continued upwards and inwards, so that they almost meet in the median dorsal line. The first pair of gills is attached to the 12th chætigerous segment. Of the five pairs of nephridia, the first opens on the 5th segment (as in *A. ecaudata*), and corresponds in position with the 2nd nephridium of *A. marina*. In addition to these observations (which introduce *A. grubii*, hitherto only known from the Mediterranean, into the British Fauna), some further work was done upon the anatomy of *A. marina*, and the result of this investigation will appear shortly in the ' Quarterly Journal of Microscopical Science.'

" Of these three species, *A. marina* contained little or no traces of reproductive organs. Ripe females of *A. ecaudata* and of *A. grubii* were abundant, though males were scarce and immature (March 26th to April 16th, and the same appeared to hold good at Whitsuntide). It appears that the shore lugs are ripe in the middle of summer, specimens from Port Erin examined last August proving to be as completely so as the large 'Laminarian worms' were at Blackpool from the end of January to the middle of May. Meanwhile, Mr. Gamble would be very greatly obliged if his fellow workers, during the coming spring, could send to Owens College, Manchester, any specimens of what are suspected to be larvæ or post-larval stages of the common lug-worm, the eggs and development of which are still quite unknown."

Amongst the workers at the Biological Station at Easter was Dr. C. H. Hurst, of Dublin. Dr. Hurst kindly offered to do some work for me, so I suggested that he should make a series of observations and experiments in regard to the currents of water entering and leaving the two apertures of *Polycarpa glomerata*, a gregarious red-coloured simple Ascidian found in great abundance under the limestone masses at Perwick Bay, and clothing the sides of the caves at the sugar-loaf rock. Dr. Hurst has sent me the following account of his results along with some diagrams and an exact record of all his experiments :—

" At the request of Prof. Herdman, I made some simple experiments at the Port Erin Laboratory of the Liverpool Biological Committee, at Easter, 1897, with the object of discovering whether or not the direction of flow into and out of the branchial and atrial cavities of Polycarpa was constant.

" To test the direction of the currents, the water in the immediate neighbourhood of the branchial and atrial apertures was made streaky by causing small vortex rings of water-colour paint (lamp black) to pass close to these apertures. This was easily effected by means of a small glass cannula with an india-rubber cap. To eliminate the effect of gravitation, the paint was mixed in the first instance with sea-water, and fresh-water was then added drop by drop till the specific gravity, as tested by the behaviour of small vortex rings, was identical with that of the sea-water in which the Polycarpæ were living.

" Results :—(1) At the moment of contraction of the body water was driven violently out by the branchial and less violently out by the atrial aperture.

" (2) When undisturbed a slow steady current enters by the branchial aperture.

" (3) When undisturbed the current at the atrial aperture is slow, and is sometimes inwards and sometimes outwards. It was observed to flow thus in opposite directions, while the inward current at the branchial aperture remained constant and steady.

" (4) When a moderately strong mixture of the paint was used, the entrance of it by either aperture was followed (sometimes immediately, but at other times only after a considerable quantity had been inhaled) by sudden contraction of the body and expulsion of water by both apertures, but chiefly by the branchial. In some individuals, at least, this contraction followed more rapidly upon entrance of the paint by the atrial aperture than by the branchial.

" These results would seem to verify the suggestion of Prof. Herdman* that the occurrence of tentacles about the atrial aperture is connected with an occasional inhalation by that aperture, the tentacles serving to detect impurities in the water so inhaled, the stimulation being followed by contraction of the body and expulsion of the impure water."

* British Association Report, Edinburgh, 1892, p. 788 ; and Bulletin Scientifique de la France et de la Belgique, t. xxv., p. 56, 1893.

Mr. Arnold Watson, of Sheffield, sends me the following report upon his work :—

" My object in going to the Port Erin Biological Station in April last was chiefly to obtain fresh material for studying the spinning gland of *Panthalis oerstedi*, and so far as the dredging operations were concerned, the object was attained, since we captured the only complete specimen we have ever got, our previous captures being (with the exception of a very young one, which lived in my tank a year) fragments, chiefly the anterior portions, though once a posterior part was taken. The complete specimen above-mentioned was 3¼ inches long when at rest. When moving it would be rather longer. I was unable to get it to settle in my tank, though I provided it with mud, &c., and at the end of ten days I killed it, as it was evidently ailing. This specimen was lent to Prof. M'Intosh, who desired to make a drawing from it for his forthcoming Monograph on British Annelids, and he has since returned it.

" We also got an anterior portion of another *Panthalis*, which lived in my tank until the end of June. It had commenced renewing its hinder quarters, in fact had grown two small anal cirri, when I found it necessary to kill and preserve it. We got a considerable number of mud tubes belonging to *Panthalis*, nearly all of which were empty.

" I am sorry to say that ill-health has prevented my making use of the material in the way I intended, and so far I have no further progress to report, though I hope in the near future to again take up the study of the spinning gland, and give a further description of *Panthalis* and its mode of working.

" I think there are only the following two additions to make to your list of species. They are worms which were

dredged up with *Panthalis*, from deep water between Port
Erin and Ireland:—*Glycera alba* (Rathke), and *Praxilla
gracilis* (Sars). We have previously got them from the
Panthalis ground, but have not recorded them.

"I got a very fine specimen of *Owenia filiformis* from
low water mark opposite Port Erin. I notice that
Hornell has not recorded it as found at the Isle of Man,
though I was under the impression we have got it by
dredging off Bradda Head. I also got *Magelona papillicornis* from low water mark at Port Erin. It also is not
recorded in Hornell's list, except in the Lancashire and
Cheshire column. So these are additions to the Manx
fauna.

"You will be glad to hear that my observations of
Owenia have been very successful. I have induced the
beast to show me his building operations, and by means
of experiments, have ascertained how the imbrication of
the sand on the tubes is produced. The animal has
been good enough to exhibit the action of the labial
organ (metastomium), and I have seen it licking the
grains of sand, building with them (for which purpose
it selects the flat ones), and burrowing or digging down
into the sand. That organ is not used for burrowing
upwards. The surface of the sand in which the worm
lives is reached by a peculiar screwing and undulatory
motion of the worm inside its tube. The latter, held by
the innumerable uncini, travels with the worm inside it,
and is made to stretch until the surface of the sand is
reached, though, very occasionally, the hinder part of the
tube breaks off (I have only seen this happen once out of
a considerable number of trials). It is very much easier
for the worm to work its way to the surface than to go
down into the sand (if by chance stranded), although, in
the first case, the sand on the tube is so arranged as to

cause much more friction, *i.e.*, all the edges of the flat particles are exposed to the sand, but they will, I suppose, act to some extent like the cutting edge of a drill. In going down the worm and tube seem always to travel the other way, so there is less friction, but evidently much more work. I have not yet quite finished my observations."

Professor Weiss and his Museum Assistant, Mr. H. Murray, collected Algæ at Port Erin during the Easter vacation. Prof. Weiss gives me the following list of species which have not previously been recorded, and are, therefore, new to the locality :—

Dermocarpa prasina, Born.   On *Catenella opuntia* and *Laurencia pinnatifida*.

Spirulina tenuissima, Kütz. Sparingly amongst *Rhizoclonium riparium*.

Rivularia atra, Roth.   In shallow pools of fresh water, near high water line, which would be over-run by the sea at high tide.

Rhizoclonium riparium, Harv.

Urospora speciosa, Holm. et Batt.

Enteromorpha ramulosa, Hook.

Ralfsia verrucosa, Aresch.

Callophyllis laciniata, Kütz.

Melobesia corallinæ.

Lithothamnion colliculosum, f. rosea, Fosl. Pt. St. Mary.

Mr. F. W. Keeble, Assistant Lecturer in Botany in the Owens College, Manchester, occupied the Owens College work table during the summer vacation, and was engaged in work on some physiological problems in the nutrition of red sea-weeds. Mr. Keeble reports as follows :—

"During my stay of about two months at Port Erin I was engaged in investigating the significance of the red pigment of the Floridcæ. The research, although incomplete, seems to point to several interesting conclusions as to the nature of the so-called Floridean starch, the destruction of the red pigment, and the starch-depletion of the cells in intense sunlight. Cases of apparent etiolation-phenomena among the red sea-weeds were recorded, and stages in germination of their spores under various conditions observed."

Prof. Herdman and Mr. F. J. Cole have commenced an investigation on the process of budding and the formation of colonies in various genera of the Compound Ascidians in the hope of being able to throw some light upon the curiously contradictory accounts which have been given by different writers (such as Pizon, Caullery, Ritter, Hjort, and others) of late years, and some of which, if established, would have an important influence on our views as to certain current biological theories. It is also proposed to include in the investigation the comparison and correlation (if any correlation is possible) of the development of the bud with the development of the embryo. Large numbers of colonies have now been collected, preserved, and sectioned at different times of the year from last April onwards. The state of affairs in *Botrylloides rubrum*, and in a species of *Amaroucium*, of which the most complete series have been obtained, will be investigated first.

So far as these observations have yet gone, the process seems to be in agreement with that described by Ritter, and by Lefevre in his recent paper, which has appeared since the present investigation commenced.

Mr. H. Lyster Jameson, B.A., from Dublin, spent two weeks in April at the Laboratory, in work on the TURBELLARIA in continuation of the researches of Mr. F. W. Gamble on that group in a former year. Mr. Jameson gives me the following summary of his results so far :—

"The species recorded are the following, those new to the district being marked with an asterisk(*):—

"*Polycelis nigra* (O.F.M.) [in fresh-water]; *Aphanastoma diversicolor*, Oe.; *Convoluta paradoxa*, Oe.; *Promesostoma marmoratum* (Schultze); *P. solea* (O. Schm.); *P. agile* (Levinsen); *Byrsophlebs intermedia*, v. Grff.; *Proxenetes flabellifer*, Jensen; *Mesostoma neapolitana*, v. Gr.; *Pseudorhynchus bifidus* (M'Intosh); *Acrorhynchus caledonicus*, Claparéde; *Macrorhynchus nägelii* (Köll.); *M. croceus*, Fab.; *M. heligolandicus*, Metschnikoff; *Provortex balticus* (Schultz); *P. affinis* (Jensen); *Fecampia erythrocephala*, Giard; *Graffilla buccinicola*, n. sp.; *Plagiostoma koreni*, Jensen; *P. vittatum*, Frey and Leuck.; *Vorticeros auriculatum* (O.F.M.); *Cylindrostoma quadrioculatum* (Leuck.); *Monotus fuscus*, Oe.; *M. lineatus* (O.F.M.); *Stylochoplana maculata* (Quatref.); *Leptoplana tremellaris*, O.F.M.; *Cycloporus papillosus*, Lang, var. *lævigatus*, Lang.

"The occurrence of *Mesostoma neapolitana* is of interest, it has hitherto only been found at Plymouth and at Naples. Only the cocoons of *Fecampia erythrocephala* were found. I have since found this worm in more than one locality on the Irish Coast.

"Besides identifying these species I have commented in my report upon certain anatomical characters in various of them, and have recorded one or two interesting varieties. I have also endeavoured to determine the relative abundance of the different species that came under my notice.

"I also describe a new species, which I refer to the genus *Graffilla*, v. Jhering, under the name of *Graffilla buccinicola*. It was found in the kidney of *Buccinum undatum* and *Fusus antiquus*. This species differs from the other members of the genus in the form of the body, in the form and relations of the gonads, in the position of the reproductive aperture, and in the possession of pigment spots in the superficial parts of the body parenchyma. It comes nearer to *G. muricicola* than to either *G. tethydicola*, v. Graff, or *G. brauni*, Ferdinand Schmidt. *G. mytili* (Levinson) has been described in such a way that its position in the genus cannot be determined until the species is rediscovered. The genus *Graffilla* is a new record for British seas."

Mr. J. A. Clubb, M.Sc. (Vict.), and the Rev. T. S. Lea, M.A., were both at work in different ways on Sea-Anemones during a part of the summer. Mr. Clubb is investigating the animals from an anatomical and histological point of view, dealing with variations in the mesenteries, and studying the structure of the species according to the lines laid down by Prof. Haddon ; while Mr. Lea is observing the habits and photographing different conditions and positions—feeding them under the eye of the camera, and taking pictures of them at home in their rock-pools. We shall probably have papers from both these workers, giving their results, at the Biological Society during the winter.

Mr. A. Randall Jackson, B.Sc. (Vict.), spent the Easter vacation at Port Erin, and occupied the University College table in the Laboratory. Besides taking part in the dredging expeditions, in tow-netting on the bay, in general out-door zoology, and in preparing for his Victoria University Examination—which he afterwards passed with first-class Honours—Mr. Jackson commenced to collect and to study the spiders of the neighbourhood. After taking his degree, Mr. Jackson returned to Port Erin for part of the summer vacation, and continued to collect and work out the spiders. He has already identified about forty species of the group, and he proposes to return to the

Laboratory next year and complete a " Report upon the Araneida of the District."

The remarkable new green Gephyrean worm referred to at p. 24 of last year's Report, has since been fully described, discussed, and illustrated, in a recent part of the "Quarterly Journal of Microscopical Science,"* under the name of *Thalassema Lankesteri*. It is in some respects intermediate in its characters between *Hamingia arctica* from Norwegian seas and *Thalassema gigas* from Trieste. It might have been described as a new genus lying between *Hamingia* and *Thalassema*, and forming a term in the series—*Echiurus, Thalassema*, the Port Erin form, *Hamingia, Bonellia ;* but I preferred to enlarge rather the genus *Thalassema* for its reception. The remarkable green colour (which I have called " Thalassemin ") has been discussed by Prof. Lankester in a paper in the same number of the Quarterly Journal, and has been contrasted with " Chætopterin," " Bonellin," and some other green tegumentary pigments.

In connection with this new British worm *(Thalassema Lankesteri)*, Mr. Lyster Jameson writes to me :—" I have found in the Dublin Museum Collection a specimen of the *T. gigas* group (with two nephridia) from the west coast of Ireland that is probably the same species as your new one. I thought at first it was *T. gigas*, but have recently procured Müller's paper, and find it is quite different. It was taken from the stomach of a Cod during the Dublin Royal Society Survey. There is no trace of green colour, but in a partly digested specimen that may well have gone."

We must endeavour to obtain some more specimens of

* On a new British Echiuroid Gephyrean, with remarks on the genera *Thalassema* and *Hamingia*, by W. A. Herdman, F.R.S.; "Quart. Journ. Mic. Sci.," December, 1897, p. 367 (with two plates).

this interesting form by trawling in the deep channel (see fig. 6) between Port Erin and Ireland.

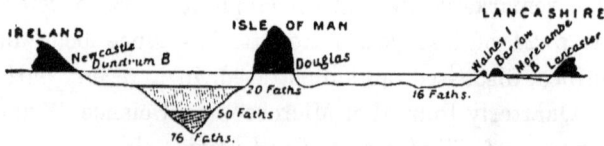

Fig. 6—Section across the Irish Sea through Douglas.

L.M.B.C. PUBLICATIONS.

Since the last Annual Report the following L.M.B.C. papers have appeared in the Transactions of the Biological Society. As usual, extra copies of the sheets have been struck off and stored for publication in volume V. of the "Fauna," which will probably be ready to be issued in a year or so :—

1. Revised list of Hydromedusæ of the L.M.B.C. district, by Edward T. Browne, B.A., F.Z.S.

2. Additional notes on the Turbellaria of the L.M.B.C. district, by H. Lyster Jameson, B.A.

The following may also be noted under this head :— (1) In the Sea-Fisheries Laboratory Report for 1896, there have been published some observations on the development of the Gurnard, the Lemon Sole, and the Witch at Port Erin ; (2) the Catalogue of the Fisheries Collection, in the Zoological Museum at University College, contains a list of Fishes of the neighbourhood—a group upon which we have not yet had a full report ; and (3) Prof. Herdman's paper in the "Quarterly Journal of Microscopical Science" gives an account of the large green *Thalassema* trawled in the deep water off Port Erin.

CONCLUDING REMARKS.

As we have recorded, in the earlier part of this Report, science students from our colleges are beginning to attend the Biological Station for purposes of work. That is very satisfactory; but we shall not be content with science students alone. We desire to interest and educate the general public in Natural History, and to give all university students opportunities of studying living nature. Students of science study, to some slight extent at least, Arts subjects—Literature, History, Languages, and, it may be, Philosophy—but how very few of the ordinary Arts students have even the most elementary acquaintance with any experimental or natural science. Fortunately, it is now becoming rare to hear an educated person boasting of ignorance or indifference to science, but it is still very unusual to find anyone who has received a non-scientific education and who understands and appreciates the natural phenomena by which he is surrounded. The elements of nature-knowledge should surely always form part of a liberal education ; and a most instructive portion of the course on nature-knowledge would be a couple of weeks spent amongst the researchers at a biological station. It is a revelation and an inspiration to the young student, or the inexperienced, to spend a forenoon on the rocks exploring and collecting with specialists who can point out at every turn the working of cause and effect, adaptation to environment, and the results of Evolution. It is equally instructive and inspiring to have a day at the microscope with, say, our authority on Copepoda, studying the nature and ways of animals which are probably of greater economic importance to the world than the wheat plains of Manitoba or the gold of Klondike.

Biology is the department of human knowledge which more than any other has been at the foundation of those advances of civilization which are most characteristic of the last half century. It has done much in its applications to lengthen life, to relieve its burdens and ameliorate conditions, to improve, purify, and advance the world. Biological Stations have now been adopted by the universities, they will in time be established also in all sea-port towns as municipal institutions. The great French biologist and educationalist, Paul Bert, has spoken of a similar institution to ours, the Station Zoologique d'Arcachon, as "un établissement d'utilité publique de l'ordre de ceux dont, dans d'autres branches, la création incombe à l' Etat."

Nowhere in all the broad field of knowledge will a man learn better to think exactly than in the natural sciences. Nowhere will he be more impressed with the importance of truth, for truth's sake, than when trained in accurate observation and impartial record at a Biological Station.

Appendices on (A), the Constitution and Regulations of the Committee, and (B), the Hon. Treasurer's Statement and List of Subscribers to the Funds, follow as usual.

APPENDIX A.

THE LIVERPOOL MARINE BIOLOGY COMMITTEE (1897).

At the meeting of the Committee held in December, 1896, Lord Henniker and Mr. Hoyle were elected in place of Sir Spencer Walpole and Mr. Vicars, resigned.

R. D. DARBISHIRE, Esq., B.A., F.G.S., Manchester.

PROF. R. J. HARVEY GIBSON, M.A., F.L.S., Liverpool.

HIS EXCELLENCY LORD HENNIKER, Governor of the Isle of Man.

PROF. W. A. HERDMAN, D.Sc., F.R.S., F.L.S., Liverpool, Chairman of the L.M.B.C., and Hon. Director of the Biological Station.

W. E. HOYLE, Esq., M.A., Manchester.

A. LEICESTER, Esq., formerly of Liverpool.

SIR JAMES POOLE, J.P., Liverpool.

DR. ISAAC ROBERTS, F.R.S., formerly of Liverpool.

I. C. THOMPSON, Esq., F.L.S., Liverpool, Hon. Treasurer.

A. O. WALKER, Esq., F.L.S., J.P., Colwyn Bay.

CONSTITUTION OF THE L.M.B.C.

(Established March, 1885.)

I.—The OBJECT of the L.M.B.C. is to investigate the Marine Fauna and Flora (and any related subjects such as submarine geology and the physical condition of the water) of Liverpool Bay and the neighbouring parts of the Irish Sea; and if practicable to establish and maintain a Biological Station on some convenient part of the coast.

II.—The COMMITTEE shall consist of not more than 12 and not less than 10 members, of whom 3 shall form a quorum; and a meeting shall be called at least once a year for the purpose of arranging the Annual Report, passing the Treasurer's accounts, and transacting any other necessary business.

III.—During the year the AFFAIRS of the Committee shall be conducted by an HON. DIRECTOR, who shall be Chairman of the Committee, and an HON. TREASURER, both of whom shall be appointed at the Annual Meeting and shall be eligible for re-election.

IV.—Any VACANCIES on the Committee, caused by death or resignation, shall be filled by the election, at the Annual Meeting, of those who, by their work on the Marine Biology of the district, or by their sympathy with science, seem best fitted to help in advancing the work of the Committee.

V.—The EXPENSES of the investigations, of the publication of results, and of the maintenance of the Biological Station shall be defrayed by the Committee, who for this purpose shall ask for subscriptions or donations from the public, and for grants from scientific funds.

VI.—The BIOLOGICAL STATION shall be used primarily for the Exploring work of the Committee, and the SPECIMENS collected shall, so far as is necessary, be placed in the first instance at the disposal of the members of the Committee and other specialists who are reporting upon groups of organisms; work places in the Biological Station may, however, be rented by the week, month, or year to students and others, and duplicate specimens which, in the opinion of the Committee, can be spared may be sold to museums and laboratories.

LIVERPOOL MARINE BIOLOGICAL STATION
at PORT ERIN.

REGULATIONS.

I.—This Biological Station is under the control of the Liverpool Marine Biological Committee, the executive of which consists of the Hon.Director(Prof.Herdman,F.R.S.) and the Hon. Treasurer (Mr. I. C. Thompson, F.L.S.).

II.—In the absence of the Director, and of all other members of the Committee, the Station is under the temporary control of the Resident Curator or Laboratory Assistant, who will keep the keys, and will decide, in the event of any difficulty, which places are to be occupied by workers, and how the tanks, boat, collecting apparatus, &c., are to be employed.

III.—The Resident Curator will be ready at all reasonable hours and within reasonable limits to give assistance to workers at the Station, and to do his best to supply them with material for their investigations.

IV.—Visitors will be admitted, on payment of a small specified charge, to see the Aquarium and the Station, so long as it is found not to interfere with the scientific work. Occasional lectures are given by members of the Committee.

V.—Those who are entitled to work in the Station, when there is room, and after formal application to the Director, are :—(1) Annual Subscribers of one guinea or upwards to the funds (each guinea subscribed entitling to the use of a work place for four weeks), and (2) others who are not annual subscribers, but who pay the Treasurer 10s. per week for the accommodation and privileges. Institutions, such as Colleges and Museums, may become subscribers in order that a work place may be at the disposal of their staff for a certain period annually; a subscription of two guineas will secure a work place for six weeks in the year, a subscription of five guineas for four months, and a subscription of £10 for the whole year.

VI.—Each worker* is entitled to a work place opposite a window in the Laboratory, and may make use of the microscopes, reagents, and other apparatus, and of the boats, dredges, tow-nets, &c., so far as is compatible with the claims of other workers and with the routine work of the Station.

VII.—Each worker will be allowed to use one pint of methylated spirit per week, free. Any further amount required must be paid for. All dishes, jars, bottles, tubes, and other glass may be used freely, but must not be taken away from the Laboratory. Workers desirous of making, preserving, and taking away collections of marine animals and plants, can make special arrangements with the Director or Treasurer in regard to bottles and preservatives. Although workers in the Station are free to make their own collections at Port Erin, it must be clearly understood that (as in other Biological Stations) no specimens must be taken for such purposes from the Laboratory stock, nor from the Aquarium tanks, nor from the steam-boat dredging expeditions, as these specimens are the property of the Committee. The specimens in the Laboratory stock are preserved for sale, the animals in the tanks are for the instruction of visitors to the Aquarium, and as all the expenses of steam-boat dredging expeditions are defrayed by the Committee the specimens obtained on these occasions must be retained by the

---

* Workers at the Station can always find comfortable and convenient quarters at the closely adjacent Bellevue Hotel; but lodgings can readily be had by those who prefer them.

Committee (*a*) for the use of the specialists working at the Fauna of Liverpool Bay, (*b*) to replenish the tanks, and (*c*) to add to the stock of duplicate animals for sale from the Laboratory.

VIII.—Each worker at the Station is expected to lay a paper on some of his results—or at least a short report upon his work—before the Biological Society of Liverpool during the current or the following session.

IX.—All subscriptions, payments, and other communications relating to finance, should be sent to the Hon. Treasurer, Mr. I. C. Thompson, F.L.S., 53, Croxteth Road, Liverpool. Applications for permission to work at the Station, or for specimens, or any communication in regard to the scientific work should be made to Professor Herdman, F.R.S., University College, Liverpool.

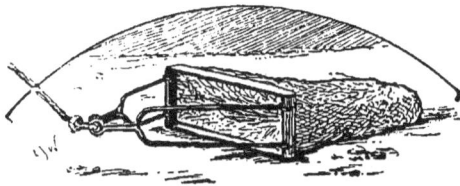

APPENDIX B.

SUBSCRIPTIONS and DONATIONS.

	Subscriptions.			Donations		
	£	s.	d.	£	s.	d.
Anon per Prof. Herdman ...	—			1	1	0
Ayre, John W., Ripponden, Halifax ...	1	1	0	—		
Banks, Prof. W. Mitchell, 28, Rodney-st.	2	2	0	—		
Bateson, Alfred, Harrop-road Bowdon ...	1	1	0	—		
Beaumont, W. I., Cambridge	1	1	0	—		
Bickersteth, Dr., 2, Rodney-st.	2	2	0	—		
Brown, Prof. J. Campbell, Univ. Coll....	1	1	0	—		
Browne, Edward T., B.A., 141, Uxbridge-road, Shepherd's Bush, London ...	1	1	0	—		
Brunner, Sir J. T., Bart., M.P., Druids Cross	5	0	0	—		
Boyce, Prof., University College ...	1	1	0	—		
Caton, Dr., 86, Rodney-street ...	—			1	1	0
Clague, Dr., Castletown, Isle of Man ...	1	1	0	—		
Clague, Thomas, Bellevue Hotel, Port Erin	1	1	0	—		
Claxton, E. J.	1	0	0	—		
Comber, Thomas, J.P., Leighton, Parkgate	1	1	0	—		
Crellin, John C., J.P., Ballachurry, Andreas, Isle of Man	0	10	6	—		
Darbishire, R.D., Victoria-pk., Manchester	1	1	0			
Gair, H. W., Smithdown-rd., Wavertree	2	2	0	—		
Gamble, Col. C.B., Windlehurst, St. Helens	2	0	0	—		
Gamble, F.W., Owens College, Manchester	1	1	0	—		
Gaskell, Frank, Woolton Wood	1	1	0	—		
Gaskell, Holbrook, J.P., Woolton Wood	1	1	0	—		
Gibson, Prof. R. J. Harvey, 5, Adelaide-terrace, Waterloo ...	1	1	0	—		

Forward ...£29 10 6   2 2 0

	Subscriptions.			Donations.		
	£	s.	d.	£	s.	d.
Forward ...	29	10	6	2	2	0
Glynn, Dr., 62, Rodney-street	2	2	0	—		
Gotch, Prof., Museum, Oxford	1	1	0	—		
Halls, W. J., 35, Lord-street	1	1	0	—		
Hanitsch, Dr., Museum, Singapore ...	1	1	0	—		
Henderson, W. G., Liverpool Union Bank	1	1	0	—		
Herdman, Prof., University College	2	2	0	—		
Holland, Walter, Mossley Hill-road	2	2	0	—		
Holt, Alfred, Crofton, Aigburth	2	2	0	—		
Holt, Mrs. George, Sudley, Mossley Hill	1	0	0	—		
Hoyle, W. E., Museum, Owens College, Manchester	1	1	0	—		
Isle of Man Natural History and Antiquarian Society	1	1	0	—		
Jameson, H. Lyster, Dublin	1	0	0	—		
Jones, C.W.,J.P., Field House,Wavertree	1	0	0	—		
Kermode, P. M. C., Hill-side, Ramsey...	1	1	0	—		
Lea, Rev. T. Simcox, 3, Wellington-fields	1	1	0	—		
Leicester, Alfred, Buckhurst Farm, Edenbridge, Kent ...	1	1	0	—		
Mactie, Robert, Airds	1	0	0	—		
Meade-King, H. W., J.P., Sandfield Park	1	1	0	—		
Meade-King, R. R., 4, Oldhall-street	0	10	0	—		
Melly, W. R., 90, Chatham-street ...	1	1	0	—		
Monks, F. W., Brooklands, Warrington	1	1	0	—		
Mundy, Randal, Manchester ...	0	10	0	- -		
Muspratt, E. K., Seaforth Hall	5	0	0	—		
Newton, John, M.R.C.S., 44, Rodney-st.	0	10	6	—		
Poole, Sir James, Tower Buildings	2	2	0	—		
Pratt, Miss, Manchester	1	0	0	—		
Rathbone, S.G.,Croxteth-drive, Sefton-pk.	2	2	0	—		
Rathbone, Mrs. Theo., Backwood, Neston	1	1	0	—		
Forward ...	£67	6	0	2	2	0

	Subscriptions.			Donations.		
	£	s.	d.	£	s.	d.
Forward ...	67	6	0	2	2	0
Rathbone, Miss May, Backwood, Neston	1	1	0		—	
Rathbone, W., Greenbank, Allerton	2	2	0		—	
Reade, M. T., Blundellsands ...		—		3	0	0
Roberts, Isaac, F.R.S., Crowborough ...	1	1	0		—	
Simpson, J. Hope, Annandale, Aigburth-dr	1	1	0		—	
Smith, A. T., junr., 24, King-street ...	1	1	0		—	
Talbot, Rev. T. U., 4, Osborne-terrace,						
Douglas, Isle of Man	1	0	0		—	
Thompson, Isaac C., 53, Croxteth-road	2	2	0		—	
Thornely, James, Baycliff, Woolton ...	1	1	0		—	
Thornely, The Misses, Baycliff, Woolton	1	1	0		—	
Toll, J. M., Kirby Park, Kirby	1	1	0		—	
Walker, A. O., Nant-y-glyn, Colwyn Bay	3	3	0		—	
Walker, Horace, South Lodge, Princes-pk.	1	1	0		—	
Walters, Rev. Frank, B.A., King William						
College, Isle of Man...	1	1	0		—	
Watson, A. T., Tapton-crescent, Sheffield	1	1	0		—	
Weiss, Prof. F. E., Owen's College, Man'tr.	1	1	0		—	
Westminster, Duke of, Eaton Hall ...	5	0	0		—	
Wiglesworth, Dr., Rainhill	1	1	0		—	
Yapp, R. H., Cambridge...	0	10	0		—	
	£93	15	0	5	2	0

SUBSCRIPTIONS FOR THE HIRE OF " WORK-TABLES," OCCUPIED
BY COLLEGES, &c.

Owens College, Manchester		£10	0	0
University College, Liverpool	...	10	0	0
		£20	0	0

THE LIVERPOOL MARINE BIOLOGY COMMITTEE.

In Account with ISAAC C. THOMPSON, Hon. Treasurer.

Dr. 1897.	£	s.	d.	Cr. 1897.	£	s.	d.
To Balance due the Treasurer, Dec. 31st, 1896	4	10	9	By Subscriptions and Donations actually received	92	11	0
,, Printing Reports	16	6	3	,, Amount received from Colleges, &c., for hire of "Work Tables"	20	0	0
,, Printing and Stationery	0	12	0	,, Dividend, British Workman's Public House Co., Ltd., Shares	5	18	9
,, Expenses of Dredging Expeditions	35	1	6	,, Sale of Reports and Volumes of Fauna	3	6	0
,, Boat Hire	2	1	6	,, Interest on British Association Fund	2	12	5
,, Books and Apparatus at Port Erin Biological Station	17	2	7	,, Bank Interest	0	8	0
,, Postage, Carriage of Specimens, &c.	1	10	6	,, Admissions to Aquarium	2	18	6
,, Salaries, Curator and Laboratory boy	35	1	1	,, Balance due Treasurer, Dec. 31st, 1897	2	15	4
,, Rent of Port Erin Biological Station	15	0	0				
,, Repairs	0	16	8				
,, Sundries	2	7	2				
	£130	10	0		£130	10	0

Endowment Investment Fund :—
British Workmans' Public House Co's. shares £173 1 0

ISAAC C. THOMPSON,
Hon. Treasurer.

Audited and found correct,

A. T. SMITH, Junr.

Liverpool, *December 31st* 1897

L.M.B.C. NOTICES.

The public are admitted by ticket to inspect the Aquarium at suitable hours daily, when the Assistant will be, as far as possible, in attendance to give information. Tickets of admission, price threepence each, to be obtained at the Biological Station or at the Bellevue Hotel. The various tanks are intended to be illustrative of the marine life of the Isle of Man. It is intended also that short lectures on the subject should be given from time to time by Prof. Herdman, or by others of the Committee.

Applications to be allowed to work at the Biological Station, or for specimens (living or preserved) for Museums, Laboratory work, and Aquaria, should be addressed to Professor Herdman, F.R.S.; University College, Liverpool.

Subscriptions and donations should be sent to Mr. I. C. Thompson, F.L.S., 53, Croxteth Road, Liverpool.

The surplus copies of the five Annual Reports upon the Marine Biological Station formerly on Puffin Island (1888 to 1892, the complete set) have been collated and bound up to form an 8vo. volume of about 180 pages, illustrated with cuts and plates, and containing the original lithographed covers. There are 20 copies of this vol., which are now offered by the Committee at 3/- each nett.

Copies of the Annual Reports for 1893, 1894, and 1895 can also be had, price one shilling each (all post free).

The L.M.B. Committee are publishing their Reports upon the Fauna and Flora of Liverpool Bay in a series of 8vo. volumes at intervals of about three years. Of these there have appeared :—

Vol. I.   (372 pp., 12 plates), 1886, price 8/6.
Vol. II. (240 pp., 12 plates), 1889, price 7/6.
Vol. III. (400 pp , 24 plates), 1892, price 10/6.
Vol. IV. (475 pp., 53 plates), 1895, price 10/6.

Copies of any of the above publications may be ordered from the Liverpool Marine Biology Committee, University College, Liverpool, or from the Hon. Treas., 4, Lord Street, Liverpool.

ISAAC C. THOMPSON,

Hon. Treas.

[WORK FROM THE PORT ERIN BIOLOGICAL STATION.]

NOTE ON A TETRAMEROUS SPECIMEN
OF ECHINUS ESCULENTUS.

By H. C. CHADWICK.

Curator of the Biological Station, Port Erin.

With Plate XVII.

[Read May 13th, 1898.]

THE subject of this note was dredged by Professor Herdman in Port Erin Bay, during Easter week, 1898, and lay with several other normal specimens in the Laboratory of the Biological Station until the following day, when its abnormal character was fortunately noticed, and it was laid aside for careful examination.

The test measured 5 c.m. in diameter, and was composed of four ambulacra alternating with four interambulacra, one of which was slightly wider than the rest. The apical system (Pl. XVII., fig. 5) consisted of four genital and four ocular plates. One of the latter was much larger than its fellows, and apparently occupied the position of a genital (*o'*, fig. 5) in the system, though it abutted externally upon the summit of one of the ambulacra, and its pore was minute, like those of the other oculars. The normal number of five pairs of peristomial plates with their tube-feet,* and five pairs of tegumentary gills were borne by the peristome (Pl. XVII., fig. 1), and in addition to these, one pair of tube-feet (*tf.'*, fig. 1) which, as will be presently shown, probably

---

* In "Forms of Animal Life," second edition, p. 560, it is stated that "The feet belonging to the buccal plates of the peristome end not in a disc but in two or more processes." This is not correct if applied to *E. esculentus*, in which the peristomial tube-feet do end in a disc supported by a calcareous rosette similar to those of the ordinary tube-feet of the test.

represented those of the absent ambulacrum. Each of these tube-feet was seated upon a minute ossicle (*os.*, fig. 3), through the centre of which ran a perforation. Upon the distal margin of the ossicle, and encircling the tube-foot, spines and pedicellaria were seated (*sp*, *ped.*, fig. 2). In the semi-contracted condition in which it was examined the ossicle and its tube-foot together measured 3 mm. in length.

The test being opened by a meridional incision, the intestine was found to follow the normal course, and four well-developed ovaries occupied the interambulacral areas. The jaw apparatus, or "Aristotle's lantern," was quite normal, all its parts numbering five. But, there being only four ambulacra, two of the five pairs of adductor muscles of the teeth were attached to the margin of one interambulacrum (*amt.*, fig. 4), and, between the two pairs, a feebly developed pair of opening muscles of the teeth were also attached. Each ambulacrum bore a well-developed auricula. The circum-oral water vessel bore five Polian vesicles, and five radial vessels proceeded from it. Of these latter four traversed the corresponding ambulacra to the apex of the test in the normal way, while the fifth passed through the peristome near its margin, and ended in the pair of tube-feet described above. This anatomical relationship points to the view suggested above, *i.e.*, that the tube-feet represented those of the absent ambulacrum, upon which it follows that the perforated ossicles upon which they were seated were the feeble and displaced representatives of the plates of that ambulacrum.

Assuming that the madreporite occupied its normal position on the right anterior genital plate (genital 2 according to Lovèn's formula), it is evident that the genital marked * in fig. 5, occupied the position of the

Fig. 1.

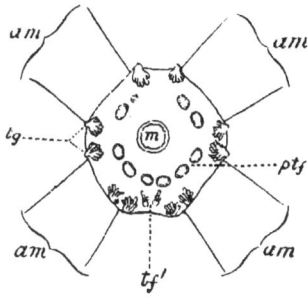

Fig. 2.

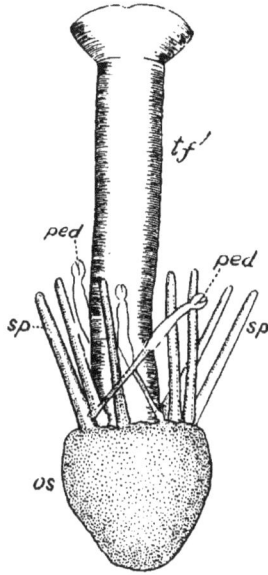

Fig. 3

Fig. 4.

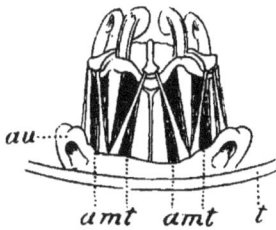

Fig. 5.

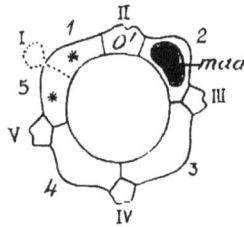

H. C. C. del.

TETRAMEROUS ECHINUS.

normal genitals 1 and 5, ocular I. being entirely absent. The absent ambulacrum is therefore I., the left one of the three which normally form the trivium.

EXPLANATION OF PLATE XVII.

Fig. 1. Diagrammatic figure of the peristome and ambulacra, showing the exact positions of the tegumentary gills, peristomial tube-feet and the pair of tube-feet representing those of the absent ambulacrum.

Fig. 2. One of the pair of tube-feet marked *tf'* in fig. 1, magnified.

Fig. 3. The ossicle *os* in fig. 2, viewed from above to show the perforation through which the radial water vessel opened into the tube-foot.

Fig. 4. The "Lantern of Aristotle," viewed from the side, to show the abnormal arrangement of the two pairs of adductor muscles of the teeth.

Fig. 5. Diagrammatic figure of the apical system of plates × 3. The ocular plates are indicated by Roman numerals, the genital plates by Arabic numerals. The dotted outline indicates the position in the system of the absent ocular plate.

LIST OF REFERENCE LETTERS.

am., ambulacra; *amt.*, adductor muscles of the teeth; *au.*, auricula; *m.*, mouth; *mad.*, madreporite; *ped.*, pedicellaria; *ptf.*, peristomial tube-feet; *sp.*, spines; *t.*, test; *tf'.*, abnormal tube-feet.

[WORK FROM THE PORT ERIN BIOLOGICAL STATION.]

ACTINOLOGICAL STUDIES.

I.—The MESENTERIES and ŒSOPHAGEAL GROOVES of ACTINIA EQUINA, Linn.

By JOSEPH A. CLUBB, M.Sc. (VICT.),

Victoria University Scholar in Zoology;
Assistant Curator of the Derby Museum, Liverpool.

With Plate XX.

[Read May 13th, 1898.]

UP to comparatively recently external characters alone served as a means of identification and as a basis for the classification of the Anemones, and thanks to the work of Prof. E. Forbes, P. H. Gosse and others, we have excellent accounts of the external appearance and habits of our British Actinians. But latterly, the attention of investigators has been directed more and more to the details of the internal anatomy, as forming a sounder and more scientific method of arriving at a true understanding of the natural relations of this group of animals. The brothers Hertwig, by the publication of their paper on the Anatomy of Actinians (1879), have done much to bring this about, and later, Dr. Richard Hertwig, in his Report on the "Challenger" Actinians (1882), laid down lines of classification based entirely on anatomical characteristics. In 1889 Prof. Haddon published Part I. of a "Revision of the British Actiniæ," in the introduction to which he makes the statement that "apart from external characters, we are unable to assign to most of them (the British species) a position in the groups proposed by Prof. R. Hertwig, on account of the absence of any knowledge of their anatomy." The great importance and necessity of a work of this character is then apparent, and it is the unfeigned regret of all workers in this group of animals that Prof. Haddon has not completed this revision.

The principal morphological characters of the Hexactiniæ are the hexamerous arrangement of the mesenteries in pairs and the presence of two " œsophageal grooves "—the " siphonoglyphs " of some authors—associated with two pairs of " directive " mesenteries. But a large amount of variation from this typical condition has been shown to exist in a number of species, both in the arrangement of the mesenteries and in the number of œsophageal grooves and directive mesenteries. It is, therefore, of the greatest importance that not only should there be a thorough knowledge gained of the anatomy of these animals, but that that knowledge should be founded on the examination, where possible, of a large number of specimens, so as to prevent the possibility of mistaking individual variations for characteristics of the species. It is only by this means, in view of the amount of variation shown to exist, that a true and faithful estimate can be formed of the morphological and phylogenetic value of the various anatomical characters.

Acting on the kind suggestion of Prof. Herdman, to whom I wish to record my sincere thanks for much valuable advice and active interest in the work, I have commenced a series of observations on the various species of Actinians found in this neighbourhood. It is my intention to take them in rotation, and examine a large number of specimens of each species, in order to determine the constancy or otherwise of the different points of their anatomy. This, my first contribution, deals with the well-known species, *Actinia equina*, Linn. — the

A. mesembryanthemum of Gosse, popularly known as the Beadlet, and a very common form around our shores. The specimens examined were all collected in the neighbourhood of Port St. Mary, Isle of Man—the major portion by myself during a short holiday spent there in June of last year, and a second consignment was obtained later from the Port Erin Biological Station, collected by the Curator, Mr. H. C. Chadwick. This species varies greatly in its external colouration, and, although most of my specimens were of the bright-red or crimson variety, I obtained a good many of the various shades of green and liver-brown.

The killing, fixing and preserving of Anemones is by no means easy to do satisfactorily, for it is of great importance in view of future work, especially of an anatomical character, that not only should they be thoroughly hardened and preserved, but also, that they should be fixed in an expanded condition. The method adopted in the present case, with a fair amount of success, was as follows :—The animals were allowed to expand in shallow dishes of fresh sea - water, when magnesic sulphate was cautiously added in sufficiently large quantity, so as to make a fairly strong solution. This is the stupefying method recommended by Tullberg (1891). On the expiration of several hours, the tentacles become totally irresponsive to stimuli, and then, and not till then, concentrated formalin was added to the sea-water in sufficient quantity to make about a two per cent solution. This fixes the tissues, and the Anemones were then arranged in a single layer, and a second solution of formalin of 10 per cent strength poured over them. It is of the greatest importance in preserving Anemones that the preserving fluid should have free access to every part, and for purposes of histological research it is advisable to syringe the internal cavity through the mouth opening. Otherwise, and if packed too soon into bottles, where they lie one on top of another, a certain amount of maceration takes place, and disappointment is the result.

The specimens were obtained as large as possible, and usually possessed five or six circles of tentacles, arranged in the usual manner. The "marginal spherules," or "bourses chromatophores," were of the usual azure blue colour, which varied considerably, however, in intensity. They also varied much in number and size. The following is a brief record of the essential features of the internal anatomy of *A. equina*, in its normal condition :— The œsophagus is usually much corrugated on its inner surface, and produced at its lower end into two processes, the œsophageal lappets. In the living condition the two œsophageal grooves can be easily distinguished by the colour, being of a greenish cast, quite different from the other parts of the wall of the œsophagus. This distinction was observable in the specimens preserved in formalin, when I afterwards commenced work on them a month or so later. But I was compelled to transfer all my preserved material into spirit, owing to the extremely irritating action of the fumes of the formalin on the mucous membrane of the nose and throat, and the painful effect on the eyes. The spirit effectually destroyed this colour distinction, and it became necessary to examine with greater care in order to determine the œsophageal grooves. For, although in most cases they were distinguishable by their greater depth, in many instances this was not so, and it was only by careful examination of the surface epithelium, confirmed by a transverse section, that their existence could be with certainty demonstrated. Oftentimes, from a cursory examination only of the œsophagus, I formed erroneous conclusions,—

in some cases the siphonoglyphs were so indistinct as not to be distinguishable from the ordinary longitudinal grooves of the œsophagus, and in other cases some of the ordinary longitudinal grooves were so enlarged and exaggerated as to lead me to think they were true siphonoglyphs.

The mesenteries are attached lengthwise to the sides of the column, the "complete" mesenteries stretching across and uniting with the œsophagus, the "incomplete" mesenteries falling short and hanging freely in the cavity. Most of my specimens exhibited twelve pairs of complete mesenteries representing the first and second cycles. There were usually three other cycles of "incomplete" mesenteries present. This arrangement is represented by the diagrammatic fig. 1, Pl. XX. Of the complete mesenteries, two kinds are distinguished by the arrangement of the mesenterial muscles. Each mesentery has longitudinal muscle fibres on the one face, and transverse muscle fibres on the other, forming in each case a slightly raised layer. Each pair of mesenteries has similar muscles on the sides facing each other, that is, projecting into the endocoele or space enclosed between each pair of mesenteries. In all cases except two, it is the longitudinal muscle series which are so arranged, and these two exceptions, which are known as the "directive" mesenteries (Pl. XX., fig. 1, *d.*), have the transverse muscles facing the endocoele or intra-mesenteric space, and the longitudinal muscles facing the exocoele or inter-mesenteric space. The remaining complete mesenteries are known as the "non-directives" (Pl. XX., fig. 1, *n. d.*).

I have examined 165 specimens of *Actinia equina*, and out of these, 158 had the arrangement of mesenteries, the number of œsophageal grooves and directive mesenteries

in accordance with the details sketched above ; seven, or
4·24 per cent only showing variation from this typical
condition. Many of the specimens were sufficiently large
to dissect with knife and scissors, by the aid of a simple
dissecting microscope ; the others were cut transversely
with a common razor and the details of the arrangement
of œsophageal grooves and mesenteries made out from
transverse sections.

In discussing the variation in anatomical characteristics
which I have found in *Actinia equina*, I will briefly refer,
for purposes of comparison, to the observations made by
other investigators on variations in other species.

The Hexamerous arrangement of Mesenteries.—A number of species have been described in specimens of which
heptamerism, octamerism, or decamerism has obtained,
instead of the normal hexamerism, or arrangement of
mesenteries in six or multiples of six.

Last year Prof. McMurrich (1897) described seven
specimens of *Sagartia spongicola*, in which in two the
mesenteries were arranged on a heptamerous plan, and
in three were arranged on an octamerous plan. In the
whole of the 165 specimens of *Actinia equina* examined
by me, there was not one which departed from the
hexamerous arrangement. This question is usually
decided on the evidence of the "complete" mesenteries
only. All of these are of the "incomplete" series at
first, and it is merely a question of age as to when they
become of the "complete" series. In several of the
specimens of *Actinia equina*, I found some of the mesenteries of the same circle complete and some incomplete,
sometimes even the same pair has had one mesentery
complete and the other incomplete. Again, the attachment of the mesenteries to the œsophagus gradually
extends downwards from above, and if the arrangement

is being determined by transverse section only, and the
section not being quite horizontal passes lower on the
one side than on the other, it may show the mesenteries
of the same circle incomplete on the one side and
complete on the other. It is of the greatest possible
importance to bear these points in mind, in deciding
the question as to whether the hexamerous arrangement
is retained or not, and special care should be taken to
distinguish the primary and secondary circles of mesenteries from each other.

Œsophageal Grooves or Siphonoglyphs.—The instances
of variation found by me in *Actinia equina* are all in
the number and position of the œsophageal grooves, and
it is in these particulars that other investigators have
found most variation in other species. Out of 131 specimens of *Metridium marginatum*, a common American
species, Parker (1897) records that 59 per cent had only
one siphonoglyph, a single specimen had three and the
remaining 41 per cent *only* had two. G. Y. Dixon (1888,
p. 120) observed that in *Sagartia venusta*, *S. nivea*, and
S. miniata, one or two siphonoglyphs may occur; the
brothers Dixon (1891, p. 19), confirmed by the observations of several others, show the presence of either one or
two in *Metridium dianthus*; while Haddon and Shackleton (1893) record that from two to seven occur in different
specimens of *Condylactis ramsayi*.

In the case of *Sagartia spongicola*, previously referred
to, McMurrich (1897) records variations in this respect
in all the seven specimens examined. He speaks of
the " directive " mesenteries only, but in the absence
of any statement to the contrary, it may be concluded
that each pair of directives had an " œsophageal groove "
associated with it. He describes one with two pairs of
directives, but not arranged opposite each other, five with

three pairs and one with four pairs. Hence it is seen that in many species uniformity in the number of œsophageal grooves is by no means characteristic.

As before mentioned, in *Actinia equina* all the seven variations which occurred among the 165 specimens examined were in the arrangement and number of the œsophageal grooves. In four specimens a single œsophageal groove only existed, in one specimen three existed, and in two specimens two œsophageal grooves were present, but were not arranged opposite each other as in the normal condition. It will be noticed that a decrease in the number is more common than an increase, and this is in accordance with the records of other investigators, and to which McMurrich draws attention in his paper on *Sagartia spongicola* (1897, p. 116).

Of the four specimens with the single œsophageal groove, two had two cycles of mesenterics complete (1 pair of directives, 11 pairs of non-directives); one had the primaries alone complete (1 pair directives, 5 pairs non-directives); and the remaining one had the primaries and four of the secondaries complete (1 pair directives, 9 pairs non-directives), the two remaining pairs of mesenteries of the second cycle being incomplete. This last condition is shown diagrammatically in Pl. XX., fig. 2. (In the diagrammatic figs. 2, 3, 4, and 5, the primary and secondary mesenteries only are shown).

The specimens possessing three œsophageal grooves exhibited perfect hexamerism like all the others, and possessed twelve complete mesenteries (3 pairs directives (*d.*), 9 pairs non-directives (*n. d.*), see Pl. XX., fig. 3). Of the three œsophageal grooves (*œ. g.*), two occupied the normal position opposite each other.

The third type of variation, found in the two specimens in which the two œsophageal grooves present were not arranged opposite each other, is shown in Pl. XX., figs. 4 and 5. The one differs from the other in the relative position of the œsophageal grooves (*œ. g.*). In one (fig. 4) they are close together and are associated with adjacent pairs of mesenteries of the first order; and in the other (fig. 5) they are separated from each other by three pairs of complete mesenteries of the first order on the one side, and by one pair on the other. In both cases the first cycle of mesenteries only is complete.

Directive Mesenteries.—In every instance of the 165, I found each œsophageal groove in relation with a pair of directive mesenteries, and I never found directive mesenteries otherwise than in relation with œsophageal grooves. Of the 131 specimens of *Metridium marginatum* examined by Parker (1897), he records that in every case, whether possessing one, two or three œsophageal grooves, there was always associated with each a pair of directive mesenteries. This exact correlation of the two structures is, however, not the case in all Actinians. McMurrich (1891) and others have shown that in certain forms *(Peachia, Oractis, Ptychodactis)* directive mesenteries occur, and no œsophageal grooves are discernable. But this want of correlation between the two sets of structures is very exceptional, and the great majority of the species investigated show the same exact correlation as is seen in *Metridium marginatum* and *Actinia equina*.

My observations on the whole tend to show that *Actinia equina* has a constancy of character not possessed by many other Actinians, and it is of importance to find a dominant species, as *A. equina* undoubtedly is, so free from variation as this record shows. That certain species are not so liable to variation as others, and are much more constant in their essential characters, has been shown by other investigators to be the case. Even species of the same genus show marked differences in this respect. The brothers Dixon record that in four specimens of *Bunodes thallia* (1889, p. 318), one, two, three, and four œsophageal grooves were found; while in *Bunodes verrucosa*, on the other hand, the whole of twenty-three specimens examined had the normal number. I consider, then, that the constant hexamerism which I find obtains without exception in the specimens of *Actinia equina* examined by me, even where variation in other directions has crept in, is an interesting feature of the species.

———

PAPERS REFERRED TO.

1879. HERTWIG, O. and R. Die Actinien anatomisch und histologisch mit besonderer Berücksichtigung des Nervensystems untersucht. Jena. Zeitschr., Bd. XIII., pp. 457-640, Taf. XVII.-XXVI.

1882. HERTWIG, R. Report on the Actiniaria dredged by H. M. S. "Challenger" during the years 1873-76. The Voyage of H.M.S. "Challenger," Zoology. Vol. VI., pp. 1-136, Pls. I.-XIV.

1888. DIXON, G. Y. Remarks on *Sagartia venusta* and *Sagartia nivea*. Sci. Proc. Roy. Dublin Soc., N.S. Vol. VI., pp. 111-127.

1889. DIXON, G. Y., and DIXON, A. F. Notes on *Bunodes thallia, Bunodes verrucosa*, and *Tealia crassicornis*. Sci. Proc. Roy. Dublin Soc., N.S. Vol. VI., pp. 310-326, Pls. IV. and V.

1889. HADDON, A. C. A Revision of the British Actiniæ. Part I. Sci. Trans. Roy. Dublin Soc. 2nd Series. Vol. IV., pp. 297-361, Pls. XXXI.-XXXVII.

1891. DIXON, G. Y., and DIXON, A. F. Report on the Marine Invertebrate Fauna near Dublin. Proc. Roy. Irish Acad. Third Series. II.

1891. McMURRICH, J. P. Contributions on the Morphology of the Actinozoa. III. The Phylogeny of the Actinozoa. Jour. Morph. Vol. V., pp. 125-164, Pl. IX.

1891. TULLBERG, T. Ueber Konservierung von Evertebraten in ausgedehntem Zustand. Biol. Fören, Förhdlgr. Bd. IV., pp. 4-9.

1893. HADDON, A. C., and SHACKLETON, A. A Revision of the British Actiniæ. Part II. Sci. Trans. Roy. Dublin Soc. 2nd Series. Vol. IV., pp. 609-672, Pls. LVIII.-LX.

1893. HADDON, A. C., and SHACKLETON, A. Descriptions of some new species of Actiniaria from Torres Straits. Sci. Proc. Roy. Dublin. Soc. N.S., VIII.

1897. McMURRICH, J. P. Contributions on the Morphology of the Actinozoa. IV. On some irregularities in the number of the Directive Mesenteries in the Hexactiniæ. The Zoological Bulletin. Vol I., No. 3, pp. 115-122. Boston, October, 1897.

1897. PARKER, G. H. The Mesenteries and Siphonoglyphs in *Metridium marginatum*, M. Eds. Bull. Mus. Comp. Zool., Harv. Coll. Vol. XXX., No. 5, pp. 259-273, Pl. I.

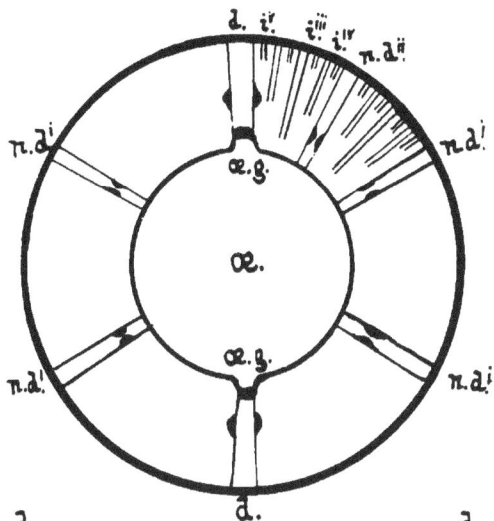

Fig. 1.

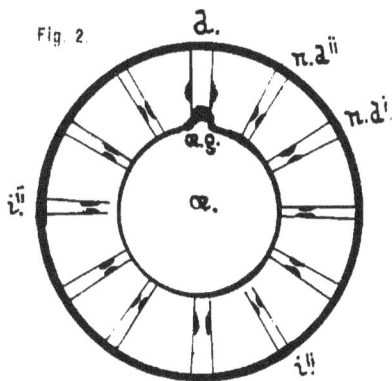

Fig. 2.

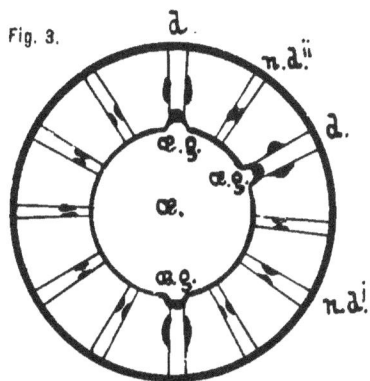

Fig. 3.

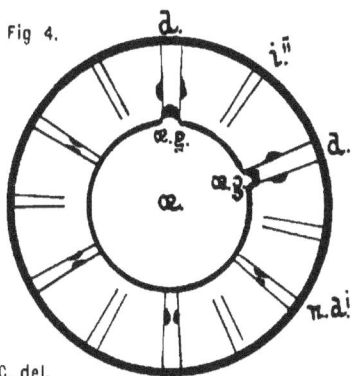

Fig 4.

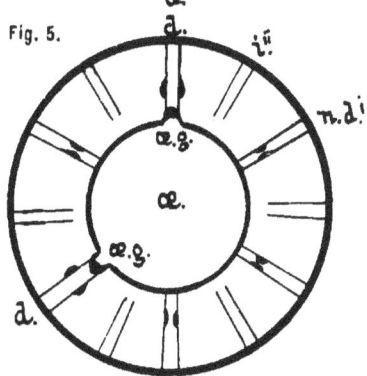

Fig. 5.

J. A. C. del.

ACTINIA EQUINA, Linn.

EXPLANATION OF PLATE XX.

REFERENCE LETTERS :— $d.$, directive mesenteries; $n.d.^{i,}$ non-directive mesenteries of the first cycle; $n.d.^{ii,}$ non-directive mesenteries of the second cycle; $i^{ii,}$ $i^{iii,}$ $i^{iv,}$ and $i^{v,}$ incomplete mesenteries of the second, third, fourth and fifth cycle respectively; $œ.$, œsophagus; $œ. g.$, œsophageal grooves.

Fig. 1. Diagrammatic T. S. showing the arrangement of mesenteries in the typical specimen. The first cycle of mesenteries is shown complete, and the arrangement of the remaining mesenteries is represented in one-sixth part only.

Fig. 2. Diagrammatic T. S. illustrating the condition found in the specimen with a single œsophageal groove, and four pairs of mesenteries of the second cycle complete, and two incomplete.

Fig. 3. Diagrammatic T. S. illustrating the arrangement of mesenteries in the specimen with three œsophageal grooves.

Figs. 4 and 5. Diagrammatic T. Ss. showing the arrangement in the two specimens where the relative situations of the œsophageal grooves were abnormal.

TWELFTH ANNUAL REPORT of the LIVERPOOL MARINE BIOLOGY COMMITTEE and their BIOLOGICAL STATION at PORT ERIN.

By Professor W. A. HERDMAN, D.Sc., F.R.S.

THE Report affords an opportunity of laying annually before the Biological Society, our subscribers, and the larger public of Liverpool and the neighbourhood, not only a statement of the statistics of the Biological Station during the year, and a record of the work done, but also an outline of such plans for the future, extensions of the work, and promising investigations as seem to the Director worthy of being published, in the interests of Biological science, in order that some possibly who are able may follow up the suggestions, and that others who can afford it may, if they care, provide funds to enable special pieces of work to be undertaken.

The year's work will, as usual, be reported upon first, and then certain proposed investigations and more general matters will be discussed (see p. 242).

THE STATION RECORD.

During the past year the following naturalists have worked at the Station, in addition to the Curator, who has been constantly in attendance :—

DATE.	NAME.	WORK.
January	Mr. I. C. Thompson, Liverpool	Copepoda.
—	Professor W. A. Herdman, Liverpool	General.
—	Dr. Christophers, Univ. College*... ...	General.

* Univ. College after a name indicates that that worker made use of the " table " secured by the College Council

DATE.	NAME.	WORK.
March.	Mr. I. C. Thompson	Copepoda.
—	Prof. W. A. Herdman	Comp. Ascidians.
—	Mr. F. J. Cole, Univ. College	Comp. Ascidians.
—	Mr. R. L. Ascroft, Lytham	Fishes.
April.	Mr. F. J. Cole, Univ. College ...	Comp. Ascidians
—	Mr. A. R. Jackson, Univ. College	Arachnida.
—	Mr. J. C. Mann, Univ. College ...	General.
—	Prof. W. A. Herdman	Comp. Ascidians.
—	Mr. Mundy, Owens College ...	General.
—	Mr. I. C. Thompson ...	Copepoda.
—	Mr. A. O. Walker, Colwyn Bay ...	Amphipoda.
May.	Rev. L. J. Shackleford, Clitheroe	Mollusca.
June.	Rev. T. S. Lea, Liverpool ...	Algæ & photography
July.	Mr. F. J. Cole, Univ. College ...	Comp. Ascidians.
—	Rev. T. S. Lea	Algæ, &c.
—	Prof. W. A. Herdman	Comp. Ascidians.
—	Mr. I. C. Thompson	Copepoda.
—	Mr. A. O. Walker	Amphipoda.
—	Mr. J. A. Chubb, Liverpool Museum	Actinia.
—	Mr. R. L. Ascroft	Fishes.
August.	Mr. C. Crossland, Clare College, Camb. ...	General.
—	Mr. R. L. Ascroft	Fishes.
—	Rev. L. J. Shackleford ...	Mollusca.
September.	Mr. H. Yates, Manchester...	...Polychæta & Plankton.
—	Rev. L. J. Shackleford ...	Mollusca.
October.	Mr. I. C. Thompson	Copepoda.
—	Prof. W. A. Herdman ...	General.
—	Mr. F. J. Cole, Univ. College ...	Comp. Ascidians.
December.	Mr. I. C. Thompson ...	General.
—	Prof. W. A. Herdman ...	General.

The work of a number of these Naturalists will be
found discussed further on in this report.

THE CURATOR'S REPORT.

Mr. H. C. Chadwick entered upon his duties early in
January, 1898. Since then he has kept the Station in
excellent condition, has added greatly to the attractiveness
of the Aquarium, has carefully attended to the wants of

workers in the laboratory, and has done a considerable amount of collecting on the shores and in the bay. A most effective addition to the collecting apparatus which can be worked from a rowing boat in the bay is a small iron trawl of 4½ feet beam, and fitted with the cod-end of a shrimp net. This is lighter to pull than a dredge, sweeps a greater area, and gives better results.

The Lords Commissioners of Her Majesty's Treasury have kindly presented to the library of the Station the three large quarto volumes containing the well-known elaborate Report upon the "Challenger" Amphipoda by the Rev. T. R. R. Stebbing, F.R.S., a classic work in marine biology.

Mr. Chadwick reports as follows :—

" The Port Erin Biological Station has been open every week day from January 24th to the present. The Laboratory, together with the instruments, dredges, and other implements has been kept in a clean and efficient condition. A few additions have been made to the stock of reagents, and the supply of all has been kept up to the requirements of workers. A few additions, in the shape of reprints and one volume (purchased) have been made to the library. To all donors an acknowledgement was promptly sent.

" Meteorological observations have been taken and carefully recorded twice almost every day, the very occasional omissions being caused by the Curator's absence on a dredging excursion.

" In January, the temperature of the sea was almost uniformly 2° higher than that of the air, while in February it averaged 4° to 5° higher. In March, it averaged about 3° higher, and in April, 1° higher. From the beginning of May to the end of September, the temperatures of air and sea averaged the same. The greatest disparities occurred in May, when, during the afternoon of the 13th,

the temperature of the sea was 11° higher than that of the
air, and during the afternoon of September 5th, when the
temperature of the sea was 12° less than that of the air.
The density of the sea-water has shown very little
variation, and only twice have I noticed a slight loss of
density to occur *immediately* after a few hours rain.   The
average may be stated as 1·026.

Fig. 1.   The "Shellbend" punt on the shore—starting for a
"plankton" trip.   [From a photo by Mr. Edwin Thompson.]

"The 'Shellbend' folding boat has been frequently used,
and is still in a thoroughly seaworthy condition, though
beginning to show signs of wear.   In launching and
housing the boat residents and many visitors have
rendered willing assistance.

"In spite of much rough weather, over fifty tow-nettings
have been taken and carefully preserved.   The Curator's

time has been so fully occupied with other duties that he has not yet been able to carry out the projected careful examination of these week by week. *Sagitta* appeared in great abundance on June 21st, but has occurred only sparingly since. *Pleurobrachia* has occurred in all the gatherings from early spring onwards, and the same remark applies to *Appendicularia*.

"The Scyphomedusæ have been very sparingly represented this year. Several trips were taken in the 'Shellbend' in quest of *Aurelia aurita*, but none were obtained. A large number were, however, cast ashore on one day early in July. About twenty large specimens of *Rhizostoma pulmo* were cast ashore early in March, in the gastric canals of each of which were a number of specimens of *Hyperia galba*, and one living specimen was seen in the bay on October 12th. *Cyanea capillata* and *Chrysaora isosceles* have not come under the Curator's notice.

"Eleven dredging excursions have been undertaken in a small boat, and hauls taken in various parts of the bay, outside as well as inside the break-water. In one such excursion the following species were taken :—*Clytia johnstoni, Obelia geniculata, Asterias rubens, Astropecten irregularis, Ophiura ciliaris, O. albida, Echinus esculentus, E. miliaris, Nereis* sp., *Portunus pusillus, P. puber, Stenorhyncus rostratus, Eupagurus bernhardus, E. prideauxii* with *Adamsia palliata, Crangon vulgaris, Virbius varians, Galvina farrani, Facelina drummondi, Aplysia punctata, Rissoa parva,* var. *interrupta, Phasianella pullus, Sepiola* sp., egg capsules of *Loligo forbesi,* with developing embryos.

"The frequent occurrence of rough weather has prevented the Curator from gaining more than a general idea of the nature of the bottom of the bay beyond the lowest tide mark. The shoreward border of the great sea-weed bed, which trends in a south-westerly direction from the spot where the drainage of Port Erin is discharged, has, however, been traced.

"Shore collecting has been carried on at intervals all through the year, and much knowledge gained of the precise localities in which many species may be found.

"On August 15th the buoy at the end of the breakwater was brought ashore for its annual cleaning, and with the assistance of Mr. C. Crossland of Clare College, Cambridge, the Curator took from its flat bottom the following species :—

" *Sycon compressum, Bougainvillia ramosa, Coryne* sp., *Tubularia larynx, Eudendrium ramosum, Obelia geniculata, Clytia johnstoni, Sertularella rugosa, Nereis* sp., *Terebella* sp., *Sabella* sp., a Nemertine, *Bicellaria ciliata, Scrupocellaria* sp., *Membranipora*, several Amphipods and Isopods, *Nymphon gracile, Doto fragilis, Ciona intestinalis* (very abundant), *Ascidia mentula, Styela grossularia* (abundant), *Corella parallelogramma*, and at least half-a-dozen as yet undetermined species. Such a large assemblage of animals, over thirty species, formed a striking commentary on the efficacy of the anti-fouling paint with which the buoy had been covered, and which, the harbour master said, had been guaranteed by the makers to poison everything.

On November 2nd there was another incursion of the Amphipod *Orchestia gammarellus* from the shore. As on previous occasions they swarmed all over the laboratory.

"The number of visitors to the Aquarium again shows a substantial increase, over 500 having paid for admission. The constant care which its successful maintenance has entailed has been repaid by the interest shown by the visitors, and by the favourable comments of many who had visited it in former years. A considerable proportion of the visitors were families of children who were brought by their parents to learn something about the structure and habits of the animals seen in their rambles amongst the rocks on the shore. The fish-hatching tanks have attracted a great deal of attention, and the many questions asked by the visitors—questions based on information gained from newspapers and magazines—showed that the interest in fish-hatching is becoming wide-spread.

"We also had in July a visit from a number of the pupils of the Arnott Street Board School in Liverpool, the party being under the guidance of Rev. T. S. Lea. The Curator hopes before next season to arrange a series of tanks, for workers at the Station, through which sea-water will constantly circulate.

"The publication of a price list of specimens has not as yet had very much result, comparatively few orders having been received. The list, thoroughly revised and with prices lowered as much as possible, is now appended to this report."

ADDITION TO THE LABORATORY.

During the past year we have added very considerably to the accommodation for students at the Biological Station by placing a wooden staircase and a new upper floor in the laboratory. This upper chamber is sub-divided by partial wooden partitions into five alcoves, each having a small window in the roof, and each containing a fixed work table for one student, measuring over 8 feet by 3 feet. This increase of accommodation permits the Committee now to carry out a plan they have long had in view, viz., to arrange an

EASTER STUDENTS' PARTY.

It is proposed that any persons who desire to join as students should pay (in addition to their own expenses)

Fig. 2. Port Erin Biological Station and its surroundings, at low tides, from the sea. [From a photo by Rev. T. S. Lea.]

a small fee to the funds of the institution. * Dredging, tow-netting, and shore collecting expeditions will be organised on certain of the days, according to tides and weather; and at other times there will be occasional lectures and demonstrations in the Biological Station, and opportunities for laboratory work. Several members of the Committee have promised to join the party, and the scheme offers a very unusual opportunity for collecting and studying in the company of specialists which will probably be of great service to junior students of marine biology and to some amateurs.

DREDGING EXPEDITIONS.

In addition to frequent dredging and trawling from small boats in the bay, the following expeditions in steam trawlers have been organised :—

   I. April 9th, 1898. In Fisheries steamer "John Fell"—

     1. Two miles off Fleshwick, 20 faths. (2 hauls of shank net).

     2. Two miles off Dalby, 35 faths., bottom mud (with shank net).

     3. Two-and-a-half miles off Dalby, 40 faths., bottom mud (with fish trawl).

  II. April 11th, 1898. In Fisheries steamer "John Fell "—

     1. One to two miles off Dalby, 20-35 fathoms.

     2. Between Bradda Head and Calf Sound, 17 faths.

 III. April 20th, 1898. In steam trawler "Tudor Prince "—

     From three to five miles off Entrance to Glen Meay, 22 to 30 faths.

* A circular, giving all particulars, will be issued. Apply to the Curator.

IV. July 16th, 1898. In steam trawler "Rose Ann," when we dredged and trawled at the following localities :—

Fig. 3. Plan of the L.M.B.C. District.

1. Two miles south of Spanish Head, 17 faths., a neritic bottom.

2. Four miles east of Port St. Mary, 18 faths., many Ophiuroids.

3. Four miles east of St. Ann's Head, 20 faths., gravel bottom.

4. Three-quarters of a mile east of St. Ann's Head, 13 faths., melobesia bottom, dragging inwards to Algæ and Polyzoa, with enormous masses of *Perophora* (chiefly on *Vesicularia spinosa*) and other Ascidians. In this last haul, 14 species of Tunicata were

obtained, as follows:—*Perophora listeri,
Clavelina lepadiformis, Ascidia mentula,
A. scabra, Corella parallelogramma, Molgula* sp., *Molg.* sp., *Polycarpa pomaria,
P. comata, Styelopsis grossularia, Amaroucium proliferum, Aplidium* sp., *Diplosoma*
sp., and *Botrylloides* sp.

We still have living healthily in the tanks specimens of
the following moderately deep-water forms obtained at
one of the dredging expeditions last summer :—*Stichaster
roseus, Porania pulvillus, Venus verrucosa, Pectunculus
glycimeris,* and *Aporrhais pes-pelicani.*

The following rare or interesting animals have been
kept in the tanks of the Aquarium house during the
year :—

Zoothamnion dichotomum.

Ephyræ of *Aurelia aurita.*

Lineus gesserensis.

Carinella annulata.

Phyllodoce laminata.

Cucumaria hyndmanni.

Hyperia galba.

Eggs of *Loligo forbesi.*

The sea slugs *Galvina cingulata* and *Eolis farrani
Lepadogaster decandollii.*

The Amphipod *Tritæta gibbosa,* infesting colonies of *Amaroucium argus.*

A specimen of *Eupagurus prideauxii* (the hermit crab),
which had *Adamsia palliata* (the cloak sea-anemone) on its
shell, after being kept for some time, adopted a new shell—
the *Adamsia* accompanying it.

Early in July great shoals of young *Gadus virens* (the
saithe) and *Pleuronectes platessa* (the plaice) were seen
close to the shore, in front of the laboratory. They were

feeding upon the "fairy shrimp," *Mysis inermis*, which swarmed at the time in countless thousands. Mr. Andrew Scott finds, on the Lancashire coast, that young Herrings, 2 to 3 inches long, feed largely upon *Mysis*. There can be no doubt that this Crustacean is an important fish food.

NOTES ON WORK DONE AT THE STATION.

Mr. J. A. CLUBB, M.Sc. (Vict.), is now carrying out his investigations on the structure and variations of sea anemones upon material from Port Erin. He spent some time at the station in summer collecting material for his work, the first part of which has just been published in our twelfth volume.

A quotation from Mr. Clubb's report to me runs :—

" I had a very good collecting time during the remainder of my stay at Port Erin. We dredged *Adamsia palliata*, obtaining about half-a-dozen good specimens, and I succeeded in fixing three of them well expanded. I did not, however, get *Corynactis viridis*, although out dredging for it two or three times. I got a number of the littoral forms (five different species), and have fixed them all for dissection and histological work."

Mr. J. NEWTON COOMBE, late Chairman of the Sheffield School Board, has been supplied frequently with gatherings of diatoms from the surface during the spring and early summer. From this material he has been making observations upon the life history and reproduction. and his results, he informs me, are nearly ready for publication.

Mr. A. O. WALKER sends me the following " Report on Malacostraca in 1898 " as the result of his work this year:—

" There has been a remarkable scarcity of Amphipoda this summer on the coasts of Liverpool Bay within a mile of the shore. At Bull Bay, Anglesey, in June ; Port Erin in July; and Colwyn Bay in September, very few individuals were taken in the bottom tow-net. It is possible that this may partly account for the small takes of fish at Rhos Weir this summer.

"The following are the more noteworthy Crustacea taken during the past season. :—

"*Mysis longicornis*, M. Edw. [G. O. Sars, Middelhavets Mysider, p. 22 (separate copy) pls. ix and x.]

"This species, hitherto recorded only from the Bay of Naples, may be at once distinguished from all the other British species of the old genus *Mysis*, except *M. (Neomysis) vulgaris*, J. V. Thompson, by the absence of a notch, or cleft, at the end of the telson, and from that species by its telson having the end rounded, and the margins furnished with closely set setæ of unequal length. On the other hand, the resemblance of the telson and antennal scales to the genus *Leptomysis* is remarkable. The female of this species is practically indistinguishable from the same sex in *Leptomysis apiops*, G. O. Sars, except by the length of the spine on each side of the two small central spines at the apex of the telson. These in *M. longicornis* are but little longer than the other long spines on the margin, while in *L. apiops* they are quite twice as long. The resemblance extends to the curious trigonate points of the spines round the end of the telson, which I have not observed in any Crustacean except these two. A new genus is required for this species, the characters of which do not agree with any of the existing genera of Mysidæ.

"One female was taken in the bottom tow-net at Bull Bay, June 10th, 1898, in 20 fath. Length 6 mm.

"*Cumella pygmæa*, G. O. Sars. Several females and three males taken at the same time and place as the last. Not previously recorded in Liverpool Bay.

"*Argissa hamatipes* (Norman); = *Syrrhoë hamatipes*,

Norman, Brit. Assoc. Report on Shetland Dredgings, 1868, p. 279; = *Argissa typica*, Boeck, Crust. Amph. Bor. et Arct. 1870.

"A single male specimen was taken in the bottom townet at Colwyn Bay in 2½ fath., Sept. 13th, 1898. I have Dr. A. M. Norman's authority for saying that his species is identical with Boeck's, and I have seen the type specimen. Previously recorded from the Isle of Man coast in 12 fath.

"*Gammarus duebeni*, Lilljeborg.

"A single specimen, dredged off Glen Meay, about four miles from land in 22—30 fath., April 20th, 1898.

"There is, perhaps, no aquatic animal which appears to be able to exist under such different conditions as this species. I have specimens collected by Dr. R. F. Scharff, in Lough Doon, Co. Kerry, "about 1,000 ft. above sea-level;" Mr. Thos. Scott records it from Loch Ruan, near Campbeltown, "several hundred feet above sea level" (Fifteenth Ann. Rep. of the Fishery Board for Scotland, Part III., p. 322); and Prof. G. O. Sars says that it occurs "in brackish pools among shore rocks, considerably above high-water mark," and in "warm springs of south Greenland." (Amphipoda of Norway, p. 503).

"*Pleonexes gammaroides*, Sp. Bate.

"Several specimens taken at Port Erin by Mr. Chadwick, in April, 1898, probably in rock pools. Not previously recorded in the Isle of Man."

Mr. ARNOLD T. WATSON, F.L.S., reports to me :—

"During the past year I have continued the observations on the habits of *Owenia filiformis*, to which reference was made in the last Report, and I am now working at one or two points which still remain unsettled. In due course I hope to publish a full account, but mean-

time it may perhaps be of interest to state that, by
establishing a colony of these annelids in my aquarium, I
have been enabled to ascertain their method of reproduction, and to obtain the larval form. This proves to be
Mitraria, the parentage of which was, I believe, previously
unknown."

The Rev. T. S. LEA has continued his interesting work
on the preparation of photographic records of the habitats
and appearance of the littoral plants and animals under
natural conditions, either when exposed at low tide or
when actually under water. He has produced a most
beautiful series of plates showing sea-anemones expanded
in shore pools, in some cases catching and swallowing
food. Last year Mr. Lea photographed (half-plate size)
a marked area of rock covered with adhering animals.
This year he has taken the same area, on the same scale,
and finds (see fig. 4) that the population has changed
almost entirely. All last year's limpets are gone, leaving
their scars on the rock; a few of the barnacles seem to
be the same individuals. Many thousands of new animals
have appeared.

Miss L. R. THORNELY has found on material from
Port Erin the following interesting Polyzoa new to our
district :—*Mucronella peachii*, var. *labiosa*, previously
known from Belfast and Guernsey ; and *Mucronella
abyssicola*, on a shell, previously known only from
Shetland.

Dr. G. W. Chaster informs me that from material
dredged in our district by the Fisheries steamer he has
obtained the mollusca *Eulima intermedia* (alive and very

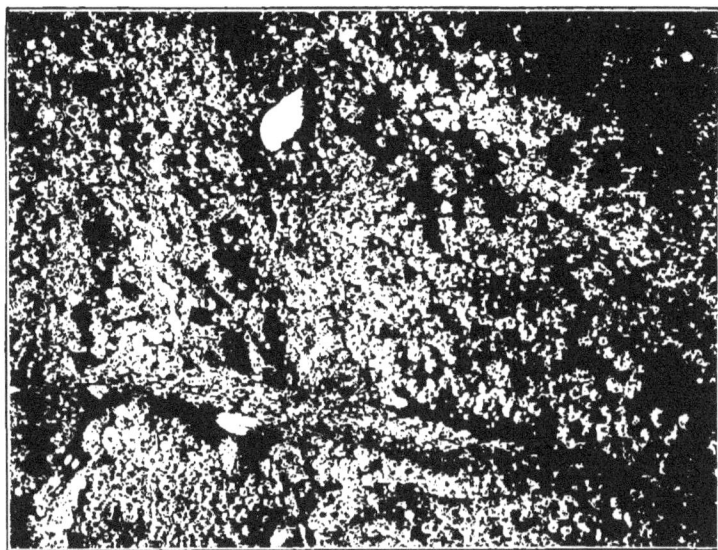

June, 1897.

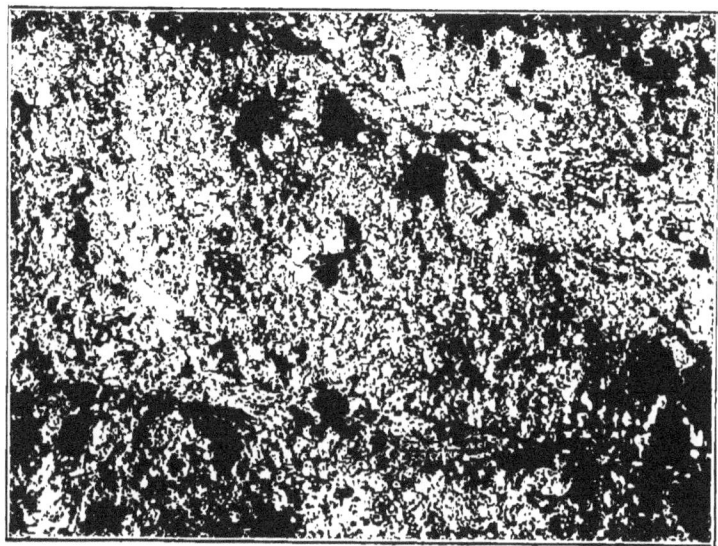

June, 1898.

Fig. 4. Vertical rock-face, Spaldrick Bay, Port Erin, about 6 feet above low water mark.   [From photos by Rev. T. S. Lea.]

fine) and *Neolepton obliquatum*, the latter a species which he added to the British fauna from Irish dredgings.

It will be remembered that in our Sixth Annual Report (December, 1892), I gave an account of the colour varieties of the little shore prawn, *Hippolyte* (or *Virbius*) *varians*, and of their "protective" nature, as found on different kinds of sea bottoms at Port Erin. A coloured plate (Pl. VI.) showed the green form on the green plant *Zostera;* a red form on the sea-weeds *Delesseria* and *Rhodymenia;* and a dark olive brown form on the bushes of the brown Alga, *Halidrys siliquosa.* Four possible explanations of the facts were given, and it was stated that an experimental enquiry into the matter was in progress.

The following year, in the Seventh Report (1893, p. 35), the results of the experiments, so far as then obtained, were given, leading to the conclusion "that the adult animal *can* change its colouring very thoroughly, although not in a very short space of time." A short description of the conditions of the differently coloured chromatophores or pigment cells of the skin during these changes was given, and their microscopic characters briefly discussed. It was stated "it would be interesting to determine whether . . . . the modification of the chromatophores is due to nerve action and is dependent upon sight, or is the result of the direct action of light upon the integument," but no further experiments bearing upon this point were then made.

During the following year the experiments were repeated at Port Erin, with practically the same results. Glass jars, painted on one side, or covered with coloured papers, as well as those containing coloured weeds, were used; and the prawns, while changing colour, were occasionally examined under the microscope, and some coloured drawings were made of the elaborately branched chromatophores, showing the distribution of the blue, yellow, red, brown, and chocolate pigment. The whole subject, so far as then known, was discussed in a lecture at Port Erin, to the Isle of Man Natural History Society, in the summer of 1894, and a passing reference to it was made in the Tenth Annual Report, in connection with other cases of colour change, such as those of some fishes. There I left the matter, and I am glad to say that Mr. James Hornell, in Jersey, and Messrs. F. W. Gamble and F. W. Keeble, in our own district, have now taken up the investigation, with the intention of carrying it further.

Messrs. GAMBLE and KEEBLE, both Demonstrators of Biology in the Owens College, have been making a number of interesting observations and experiments in regard to the physiology of the colour changes of *Hippolyte* and other prawns during this past summer. Most of their work has been carried out at the Piel Sea-Fish Hatchery, but they have in part made use of specimens obtained at Port Erin. Their results are not yet ready for publication, but I may remark that the chief novelty in their methods is the use of water-tight glass chambers, through which fresh aërated sea-water can constantly pass, and in which the prawns can be kept under constant observation while different colours of light are applied. A fuller account of this method, by Messrs. Gamble and Keeble, will be given in the forthcoming Annual Report to the Lancashire Sea-Fisheries Committee.

Professors BOYCE and HERDMAN, with the help of Dr. KOHN, have continued during the year their work on Oysters and Disease, and presented their third, and final, report upon the subject to Section D of the British Association at the Bristol meeting in September. Amongst the conclusions at which they have arrived may be quoted the following :—

"1. There are several distinct kinds of greenness in oysters. Some of these, such as the green Marennes oysters and those of some rivers on the Essex coast, are healthy; while others, such as some Falmouth oysters containing copper, and some American oysters re-bedded on our coast, and which have the pale green leucocytosis we described in the last report, are not in a healthy state.

"2. Some forms of greenness (e.g., the leucocytosis) are certainly associated with the presence of a greatly-increased amount of copper in the oyster, while other forms of greenness (e.g., the Marennes) have no connection with copper, but depend upon the presence of a special pigment Marennin, which may be associated with a certain amount of iron.

"3. We see no reason to think that the iron in the latter case is taken in through the surface epithelium of the gills and palps; but regard it, like the rest of the iron in the body, as a product of ordinary digestion and absorption in the alimentary canal and liver.

"4. We do not find that there is any excessive amount of iron in the green Marennes oyster compared with the colourless oyster; nor do the green parts (gills, palps, &c.) of the Marennes oyster contain either absolutely or relatively to the colourless parts (mantle, &c.) more iron than colourless oysters. We therefore conclude that there is no connection between the green colour of the 'Huitres de Marennes' and the iron they may contain.

"5. On the other hand, we do find by quantitative analysis that there is more copper in the green American oyster than in the colourless one; and more proportionately in the greener parts than in those that are less green. We therefore conclude that their green colour is due to copper. We also find a greater quantity of iron in these green American oysters than in the colourless; but this excess is, proportionately, considerably less than that of the copper.

"6. In the Falmouth oysters containing an excessive amount of copper, we find that much of the copper is certainly mechanically attached to the surface of the body, and is in a form insoluble in water, probably as a basic carbonate. In addition to this, however, the Falmouth oyster may contain a much larger amount of copper in its tissues than does the normal colourless oyster. In these Falmouth oysters the cause of the green colour may be the same as in the green American oysters.

"7. The Colon group of bacilli is frequently found in shellfish, as sold in towns, and especially in the oyster; but we have no evidence that it occurs in Mollusca living in pure sea-water. The natural inference that the presence of the Colon bacillus invariably indicates sewage contamination must, however, not be considered established without further investigation.

"8. The Colon group may be separated into two divisions —(1) those giving the typical reactions of the Colon bacillus, and (2) those giving corresponding negative reactions, and so approaching the typhoid type; but in no case was an organism giving all the reactions of the *B. typhosus* isolated. It ought to be remembered, however, that our samples of oysters, although of various kinds and from different sources, were in no case, so far as we are aware, derived from a bed known to be contaminated or suspected of typhoid.

"9. Consequently, as the result of our investigations, and the consideration of much evidence, both from the oysters growers' and the public health officers' point of view, we beg to recommend : —

"(a) That the necessary steps should be taken to induce the oyster trade to remove any possible suspicion of sewage contamination from the beds and layings from which oysters are supplied to the market. This could obviously be effected in one of two ways, either (1) by restrictive legislation and the licensing of beds only after due inspection by the officials of a Government department, or (2) by the formation of an association amongst the oyster-growers and dealers themselves, which should provide for the due periodic examination of the grounds, stores and stock, by independent properly-qualified inspectors. Scientific assistance and advice given by such independent inspectors would go far to improve the condition of the oyster beds and layings, to re-assure the public, and to elevate the oyster industry to the important position which it deserves to occupy.

"(b) Oysters imported from abroad (Holland, France, or America) should be consigned to a member of the 'Oyster Association,' who should be compelled by the regulations to have his foreign oysters as carefully inspected and certificated as those from his home layings. A large proportion of the imported oysters are, however, deposited in our waters for such a period before going to market that the fact of their having originally come from abroad may be ignored. If this period of quarantine were imposed upon all foreign oysters a great part of the difficulty as to inspection and certification would be removed.

"(c) The grounds from which mussels, cockles and periwinkles are gathered should be periodically examined by scientific inspectors in the same manner as the oyster beds. The duty of providing for this inspection might well, we should suggest, be assumed by the various Sea-Fisheries Committees around the coast."

Mr. A. R. Jackson, B.Sc., has continued his work on the Spiders of the neighbourhood of Port Erin, and has now drawn up a list of species which has been communicated to the Biological Society.

Mr. Andrew Scott, working chiefly at Piel Island, has made some interesting observations upon the food of young fishes. He finds that young Mullet up to an inch and a little more in length are still feeding entirely upon Diatoms (chiefly *Navicula*). At a length of one-and-a-half inches they feed on both Diatoms and Copepoda (*Tachidius*). Young Herring up to three inches long were found feeding on *Mysis*.

Mr. F. J. Cole has continued throughout the year his preparation of material for his joint research with Prof. Herdman on the process of budding and the formation of colonies in compound Ascidians. He has visited Port Erin about four times during 1898 for the purpose of collecting and preserving old and young colonies of the species commonly known as *Amaroucium argus*, which, however, is seen by its anatomical character to belong to the genus *Morchellium*. Mr. Cole and Prof. Herdman have also kept various kinds of compound Ascidians alive under observation in the tanks, so as to record the mode of growth and the positions of the new buds in the colony. Before regarding the collection of material as complete, it has been considered advisable to examine colonies at different times of the year so that no stage of importance might be omitted. Colonies of Ascidians have therefore been collected from month to month from Easter, 1897, and this will be continued until next April. Owing to the difficulty experienced in satisfactorily staining the preparations, the microscopic part of the work is somewhat laborious and takes considerable time. Any detailed statement of results of work would, at the present stage of the enquiry, be premature, and perhaps lacking in accuracy, but it may be remarked that what is seen in the first series of sections of the young stages of buds is more in accord with the conclusions of Ritter and of Lefevre than of other investigators.

EXTENSION OF THE WORK.

THERE are several matters, in addition to the routine work of the Station, which call for special attention this year.

One of these is the necessity for further exploration in the North Atlantic. Attention has repeatedly been drawn of late years to the importance, both from the purely scientific and the industrial points of view, of the problems involved. The Scandinavians (Petterssen, Ekman, Hjort, and others) have succeeded in unravelling some of the interlacing belts of water from Arctic, Baltic, North Sea, and Atlantic sources which sweep past their coast, and affect the movements of migratory fish. It is only by such work that we can hope to explain rationally the mysterious movements of the Herring—perhaps the most important food fish on our coast.

It was formerly supposed* that when the Herrings left our shores in autumn they retired to the far north, and next season started from the Arctic regions on their annual migration, led by one large old fish—the "King of the Herrings." We now believe that breeding and feeding are the two impulses that govern the movements of a fish. The Herring comes into shallow water on our coast to spawn, and when it migrates in search of food from the Atlantic to the North Sea, or from our West Coast out into deep water, we have reason to believe that it is following those minute organisms which

* *E.g.*, see Pennants' British Zoology, and article Ichthyology in Ency. Brit. for 1857.

form the plankton carried along in particular currents of water, characterised by the temperature, the salinity, and the microscopic fauna.   It is possible by these characters to recognise the currents, to trace their variations from year to year, and so to some extent to determine and predict the movements of the shoals of fish.   We owe this view chiefly to Professor G. O. Sars, of Christiania.   It is, then, to the physical conditions and the biological contents of the water that the movements of the Herring are due, and these are matters within the scope of man's investigation, but outside his power to regulate by local bye-laws, imperial legislation, or even international treaty.   This should be recognised by the fishermen and by Sea-Fishery authorities.

It follows, then, that one of the most important things the Biologist can do to add to our knowledge of life in the sea is to make a survey of the microscopic floating and drifting life of the sea, and its relation on the one hand to the physical conditions at the time (especially the temperature and salinity of the water), and on the other to the food materials found in the stomachs of the fish.

The "pump plankton" method, which I described fully before this Society last session, might do much if systematically worked, but it seems difficult to induce anyone on a ship, except a naturalist, to undertake such work, and, moreover, the organisms collected may, unless great care be taken, suffer so much damage in the process, as to be difficult of identification.   How, then, are we to sample the plankton of our oceans?   The attempt has been made lately to arrange for the stoppage of a steamer at certain fixed points in the ocean long enough to permit of surface and deep-sea gatherings being taken.   This method is difficult to arrange, liable to failure, and very expensive considering the very small number of observations that could be taken—such as eight in a proposed traverse from England to the West Indies.

A much more satisfactory plan, from the scientific point of view, would be to have a series of special cruises, in a vessel of the type of a fisheries steamer, a steam trawler, or a steam yacht able to keep at sea, carrying lines of closely-placed observations across those areas that influence our British Marine Fauna. I would suggest:—

1. A line from Land's End to Porto Rico and Jamaica in the West Indies crossing Rennel's Current, and the mid-Atlantic south of the Gulf Stream area, and tracing back those elements of our floating fauna that we owe to the S.W. Atlantic drift towards their tropical source.

2. A line from Ushant to Iceland, so as to cross the entrance to the English Channel and the drift of the Atlantic organisms on to the west coasts of Ireland and Scotland, and then across the entrance to the Arctic Ocean and North Sea.

3. A line from the Shetlands to the southern end of Greenland, along lat. 60° N., so as to intercept any Arctic flow which brings down the northern forms found in our fauna.

The British Islands occupy a central and commanding position, touching, as Prof. Edward Forbes pointed out long ago, upon the Arctic and Scandinavian faunas to the North, and the Germanic, South European, and Atlantic faunas to the south.

It would also be of considerable scientific interest— although having no bearing so far as can be seen at present upon practical fishery questions—to explore more systematically the seas to the north of Iceland and Norway as far north as dredging and tow-netting are practicable. Sir John Murray, it is well known, has recently revived an old idea that the faunas of the two poles are closely

similar. In opposition to this " bipolar hypothesis, " I
pointed out last session that in the Tunicata at least there
was no special resemblance between the far northern and
far southern species, and that, on the other hand, there
was a marked similarity between a series of forms from
the North Pacific and those from the North Atlantic.
Sir John Murray answers me, in a letter on the subject,
" I think your remarks quite just. I expect many cases
of north and south distribution will be knocked on the
head by further discoveries, and that many cases of
bipolar distribution not yet evident will be brought to
light."   Well, we want now to determine how far that is
the case. We both agree that what is needed is further
facts—more investigation.

There is every prospect that the Antarctic Seas will be
fully explored. Several expeditions—Borchgrevink's and
others—are there or will be soon, still others are in
prospect. The Tunicate fauna of Australian seas is fully
as well-known as that of Europe, the marine fauna round
Kerguelen Island, thanks to the " Challenger " explorations and Sir John Murray's writings, is better known
than that of Iceland, whilst of most of the inhabitants of
the seas around Spitzbergen we are practically ignorant.

Moreover, Nansen has opened up a new problem in
northern marine exploration. He has shown that a deep
sea basin occupies a part at least of the polar area. Where
does that basin begin, and how far does it extend? How
does it end towards Norway, Iceland, and Greenland?
The sea immediately to the north and west of Norway is
shallow, running out gradually to a depth of about 200
fathoms, and then, according to Mohn, descending steeply
to the bed of the deeper ocean, which reaches a depth of
2000 fathoms or more. We do not yet know the limits
nor the inhabitants of that deep basin. In advocating

Antarctic expeditions we must not forget how much still remains to be done within the Arctic circle.

All this may seem to have little connection with our L.M.B.C. work, but it is the natural extension and outcome of what we have been doing. As the problems develop we must widen our area. We commenced twelve years ago with Hilbre Island and Liverpool Bay—our work has now for several years extended all over the Irish Sea. This means practically the western fauna of Britain influenced by the Atlantic drift from the south-west and the Arctic currents from the north. Whether the relation of our north-western European fauna to that of north-western America depends upon a common circum-polar fauna as I have suggested, and whether our thorough comprehension of the Arctic fauna is bound up, as Murray thinks, with Antarctic explorations, are questions still to be answered; but this much is clear, that one enquiry in the distribution of animals leads to another, and the different faunas are like links in a chain or strands in a net—the mesh-work of life extending over the globe—just what we should expect from the consideration that all living things spread from a centre, that multiplication, struggle for existence, migration, survival of the fittest, and varying degrees of isolation have produced the difference we now find between the present inhabitants of the different regions.

I should like, then, to push our L.M.B.C. investigations further afield—out into the North Atlantic, across to the West Indies, up into the Northern Seas. And all that is wanting is a fund to meet the necessary expenses. We can find the scientific men willing to give the time and do the work. What we want is a yacht-owner willing to use his vessel, fitted up with the necessary equipment, for scientific exploration of an interesting and important nature, such as the Prince of Monaco is now doing.

There are two practices in American universities which excite the envy of professors in this country. One is the "sabbatical year"—the one year in every seven given for purposes of travel, study, and investigation. The other is the frequent endowment of an expedition—or equipment of an exploring party—by an individual man or woman who is interested in the subject, and can give a special fund for such a purpose. The Columbia University in New York, the Johns Hopkins University in Baltimore, Yale College in Newhaven, and Harvard at Cambridge, have all benefitted immensely in the past by such exploring expeditions. Nearly every year of late has seen one or more of such, due to private generosity, in the field; and the work they have done has both added to general scientific knowledge, and has also enriched with collections the laboratories and museums of the college to which the expedition belonged.

It may be that the reason why this excellent system of exploring parties from the universities has attained such slight development in our country is that our professors are not so enterprising in making known the wants of science—not so importunate in their demands upon the community in which they live—as their transatlantic brethren. It may be that some think that in this long-settled country there is nothing left to explore — no greater mistake could be made. I am sure that the geologist and the archæologist could point to innumerable problems still to be attacked on land, while our seas are far more vast, and comparatively far less known than the shores. I am sure that many marine zoologists could be usefully employed during their vacation for the next ten years in exploring such regions as I have indicated above. While, if we pass from questions of distribution to those of structure and life-histories, even the best-known regions afford abundant opportunity for research to all who enter the field.

Marine Biology has this peculiar advantage, that invoking as it does the aid of several neighbouring sciences, such as Geology, Geography, and Physiology, it is able in its turn to throw light upon these subjects, and it bristles with problems of that interesting and intricate nature that characterises the borderland between two or more sciences.

Some Recent Work.

As a contribution to the borderland between our subject and Geology, Mr. Lomas and I have, during the past year, made a careful examination of all our dredged deposits from the floor of the Irish Sea, and our results have appeared in the last volume of "Proceedings" of the Liverpool Geological Society. Amongst our results we may note that we have shown how the material now covering the floor of the Irish Sea have been the result (1) of the denudation of the coasts, (2) of the redistribution of the older deposits under the sea, (3) of vital agencies—the remains of animals and plants living in the sea. We give a revised classification of such deposits, and a detailed description of all the samples we have obtained.

As a problem on the boundary between Biology and Geography (viz., the currents of the sea), and as having, at the same time, an important practical application to the sea-fisheries, may be cited our investigation of the circulation of water in the Irish Sea by means of "drift-bottles." Our records of those bottles that have been found and returned to us show pretty conclusively that there is, in addition to both north and south in-flowing and out-going tidal currents, a slow drift to the north, which would carry, for example, floating objects, such as fish eggs and embryos, gradually further and further to the north. So that, from the spawning grounds south of the Isle of Man, they may be carried towards and along the Lancashire coast and from say the " Hole " and other parts of the central area, to Cumberland and the Solway Firth.

As another research intimately bound up with the local fisheries, I would allude to our plankton investigations. During the past year surface gatherings have been sent from various localities with more or less regularity. The result of the examination of these, on the whole, confirms very well the general conclusions we drew in the last Report as to the succession of forms throughout the year.

We find very considerable difference between gather-ings from different localities ; for example, whereas the tow-nettings from Port Erin are clear and clean and support an abundant assemblage of minute animals, those taken about the same time off Peil, in the Barrow Channel, contain much vegetable *débris*, mud, &c., due to the influence of fresh-water and the washings of the land. The true open sea at Port Erin and the absence of any body of fresh-water, and of any mud flats, ensure the presence of a much larger number of Copepoda, Dino-flagellata, and other characteristic pelagic organisms than are elsewhere. We have also noticed that the various constituents of the surface life—both larval and adult— appear earlier at Port Erin than on the Lancashire coast.

The Manx Sea-Fisheries Commission.

The report which recently appeared containing the recommendations of the Manx Industries Commission, so far as they relate to sea-fisheries, includes an important paragraph in which it is suggested that the Hatchery

which is contemplated be established in connection with
our Biological Station at Port Erin. The possibility of
some such arrangement being proposed, caused the
Committee to consider carefully the ground in the
neighbourhood of the Station with the view of determining
how far they could offer accommodation for economic
work, and what additional building and plant would be
required before they could undertake sea-fish hatching on
anything beyond an experimental scale. The chief
requirements for such a purpose are as follows :—

    1. An extension of the Aquarium house, to hold the
hatching boxes.

    2. A small boat jetty.

    3. A concrete pond on the shore.

    4. A circulation of sea-water.

    All of these are merely a matter of expense. They can
readily be provided if a grant be given for the purpose.
The hatching-house will cost about £80, the jetty £70,
and the pond £80; say, including the fittings of the
hatchery, £250 in all.

    As to a constant flow, or circulation, of sea-water, there
is no gas supply in Port Erin, and if worked by a steam
engine a very considerable expenditure and the services of
an engineer would be required. A much more economical
plan, and one that would probably be quite sufficient for
the wants of the institution, would be to pump from the
sea-well to the uppermost tank by means of a small wind-
mill, as is done in some of the Biological Stations abroad.
This would be comparatively inexpensive, and would not
require material addition to the service of the institution ;
our present Curator with the assistance of a strong lad,
and possibly a second fisher lad during the hatching
season, would be able to undertake the work.

L.M.B.C. PUBLICATIONS.

During the past year the following papers dealing with L.M.B.C. work have been published :—

1. The Eleventh Annual Report (Trans. Biol. Soc., vol. XII., p. 91).

2. The Sea-Fisheries Laboratory Report for 1897 (Trans. Biol. Soc., vol. XII., p. 176).

3. Note on a Tetramerous Specimen of *Echinus esculentus*. By H. C. Chadwick (Trans. Biol. Soc., vol. XII., p. 288).

4. Actinological Studies, I.—The Mesenteries and Œsophageal Grooves of *Actinia equina*. By J. A. Clubb, M.Sc. (Trans. Biol. Soc., vol. XII., p. 300).

5. Report on Oysters and Disease, by Prof. Boyce and Prof. Herdman (Brit. Assoc. Rep.), with an Appendix on Iron and Copper in Oysters, by Dr. Kohn.

6. The Deposits on the Floor of the Irish Sea. By Prof. Herdman and Mr. Lomas (Proc. Liverpool Geol. Soc., vol. VIII., p. 205).

It is hoped that volume V. of the "Fauna and Flora," containing reprints of these and all other papers on our local work published since the appearance of vol. IV. (in 1895) will be ready to issue during the present winter.

I am inclined to think that the time has now arrived when it may be the duty of the Committee to issue a new form of publication. Our work hitherto, during the last 12 years, has been largely faunistic and speciographic. The work of necessity must be so at first when opening up a new district. Some of our workers have published papers on morphological points, or observations on life histories and habits; but the majority of the papers in our volumes on the Fauna and Flora of Liverpool Bay have been, as was intended, occupied with the names and characteristics and distribution of the many different kinds of plants and animals in our marine district. And this Faunistic work will still go on. It is far from finished, and I hope we shall still add greatly to our records of the Fauna and Flora; but in addition it might be useful to produce a series of

L.M.B.C. MEMOIRS.

What I now propose is that each of our specialists should carefully prepare a full account, illustrated by all the necessary figures, of one or two common but important animals belonging to the group upon which he is an authority; and that these detailed and fully illustrated accounts should be issued as a series of L.M.B.C. Memoirs — obtainable at first separately, memoir by memoir as they appear, and then later bound up, say 5 or 6 together, in convenient volumes. I believe that such a series of special studies, written by those who are thoroughly familiar with the forms of which they treat, will be of great value to students in our laboratories and in Biological Stations, and will be welcomed by many working at marine biology. It is proposed that the forms selected should, as far as possible, be common L.M.B.C. (Irish Sea) animals of which no adequate account already exists in any text book. Probably all the members of our band of specialists will be willing to help in this work. The following have already promised their services, and will probably treat of the species placed opposite their names:—

Professor Harvey Gibson   *Zostera marina.*

Mr. C. E. Jones ...   *Halidrys siliquosa.*

Professor Weiss ...   Typical Diatoms (*Biddulphia*, *Chaetoceros*, *Schizonema*, and *Licmophora*).

Dr. O. V. Darbishire	A Red Sea-weed.
Mr. F. J. Cole ...	The Plaice.
Professor Herdman	*Ascidia* and *Botrylloides*.
Mr. W. E. Hoyle	A Cuttle-fish (*Eledone*).
Mr. J. Johnstone...	The Cockle.
Mr. Andrew Scott	An Ostracod and an Epizoon.
Mr. I. C. Thompson	*Calanus finmarchicus*.
Mr. H. C. Chadwick	*Echinus*, the Sea-urchin.
Mr. J. A. Clubb ...	An Actinian.
Miss L. R. Thornely	A Polyzoon Colony.
Dr. Hanitsch	A Calcareous Sponge.

Now, all these memoirs will require to be carefully illustrated with well-drawn plates, and fortunately the Committee have at their disposal a small annual sum of money (the British Association—1896—fund) which it was proposed from the first should be devoted to just such a purpose as this. That, however, leaves the printing of the memoirs still unprovided for, and with the Curator's salary and other expenses at Port Erin to meet, it will be quite impossible to spare anything for general funds. A special subscription will have to be raised to defray the expenses of the L.M.B.C. Memoirs.

I should also like to point out that papers on special points in the Zoology of Sea-Animals are being sent by our workers to London Scientific Societies, because we have not the funds to publish them satisfactorily in Liverpool. Mr. Andrew Scott has just finished a beautiful series of water-colour drawings of fish eggs, embryos and larvæ, from his investigations during the year at our Lancashire Fish Hatchery. They will be expensive to reproduce and publish, but it will be very discreditable to Committee, and City, and County, if we allow this excellent local work to be issued under the auspices of some society in London or Edinburgh. The fact is—and a very important and gratifying fact it must be to our subscribers and supporters who have helped us from year to year—that we have now in connection with our Committee an exceedingly active School of Marine Biology, every worker in which is engaged in some piece of original research. Liverpool is the right and natural place for a School of Marine Biology, and I hope that Liverpool will consider that it is creditable to the city that such local researches should be published by a Liverpool Society. An addition of about £100 a year to our funds is necessary in order to enable us to do justice to the work now being produced by our colleagues and students.

The Biological exploration of Liverpool Bay has all along been intimately connected with University College, and it is to be hoped that when the time comes—and it ought not to be long delayed—for building new biological laboratories, accommodation will be provided within the department for our School of Marine Biology. A museum devoted to exhibiting the products of the local seas, and a laboratory fitted for conducting researches upon marine life, must surely be constituent parts of the University of a great sea-port.

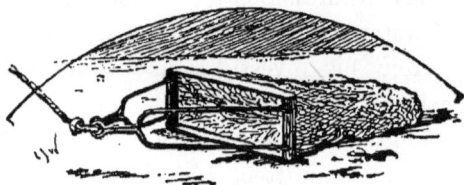

APPENDIX A.

THE LIVERPOOL MARINE BIOLOGY
COMMITTEE (1898).

R. D. DARBISHIRE, Esq., B.A., F.G.S., Manchester.

PROF. R. J. HARVEY GIBSON, M.A., F.L.S., Liverpool.

HIS EXCELLENCY LORD HENNIKER, Governor of the Isle
of Man.

PROF. W. A. HERDMAN, D.Sc., F.R.S., F.L.S., Liverpool,
Chairman of the L.M.B.C., and Hon. Director of
the Biological Station.

W. E. HOYLE, Esq., M.A., Manchester.

A. LEICESTER, Esq., formerly of Liverpool.

SIR JAMES POOLE, J.P., Liverpool.

DR. ISAAC ROBERTS, F.R.S., formerly of Liverpool.

I. C. THOMPSON, Esq., F.L.S., Liverpool, Hon. Treasurer.

A. O. WALKER, Esq., F.L.S., J.P., Colwyn Bay.

CONSTITUTION OF THE L.M.B.C.

(Established March, 1885.)

I.—The OBJECT of the L.M.B.C. is to investigate the
Marine Fauna and Flora (and any related subjects such
as submarine geology and the physical condition of the
water) of Liverpool Bay and the neighbouring parts of
the Irish Sea; and if practicable to establish and maintain
a Biological Station on some convenient part of the
coast.

II.—The COMMITTEE shall consist of not more than 12 and not less than 10 members, of whom 3 shall form a quorum; and a meeting shall be called at least once a year for the purpose of arranging the Annual Report, passing the Treasurer's accounts, and transacting any other necessary business.

III.—During the year the AFFAIRS of the Committee shall be conducted by an HON. DIRECTOR, who shall be Chairman of the Committee, and an HON. TREASURER, both of whom shall be appointed at the Annual Meeting and shall be eligible for re-election.

IV.—Any VACANCIES on the Committee, caused by death or resignation, shall be filled by the election, at the Annual Meeting, of those who, by their work on the Marine Biology of the district, or by their sympathy with science, seem best fitted to help in advancing the work of the Committee.

V.—The EXPENSES of the investigations, of the publication of results, and of the maintenance of the Biological Station shall be defrayed by the Committee, who for this purpose shall ask for subscriptions or donations from the public, and for grants from scientific funds.

VI.—The BIOLOGICAL STATION shall be used primarily for the Exploring work of the Committee, and the SPECIMENS collected shall, so far as is necessary, be placed in the first instance at the disposal of the members of the Committee and other specialists who are reporting upon groups of organisms; work places in the Biological Station may, however, be rented by the week, month, or year to students and others, and duplicate specimens which, in the opinion of the Committee, can be spared may be sold to museums and laboratories.

LIVERPOOL MARINE BIOLOGICAL STATION at PORT ERIN.

LABORATORY REGULATIONS.

I.—This Biological Station is under the control of the Liverpool Marine Biological Committee, the executive of which consists of the Hon.Director (Prof.Herdman, F.R.S.) and the Hon. Treasurer (Mr. I. C. Thompson, F.L.S.).

II.—In the absence of the Director, and of all other members of the Committee, the Station is under the temporary control of the Resident Curator or Laboratory Assistant, who will keep the keys, and will decide, in the event of any difficulty, which places are to be occupied by workers, and how the tanks, boat, collecting apparatus, &c., are to be employed.

III.—The Resident Curator will be ready at all reasonable hours and within reasonable limits to give assistance to workers at the Station, and to do his best to supply them with material for their investigations.

IV.—Visitors will be admitted, on payment of a small specified charge, to see the Aquarium and the Station, so long as it is found not to interfere with the scientific work. Occasional lectures are given by members of the Committee.

V.—Those who are entitled to work in the Station, when there is room, and after formal application to the Director, are :—(1) Annual Subscribers of one guinea or upwards to the funds (each guinea subscribed entitling to the use of a work place for four weeks), and (2) others who are not annual subscribers, but who pay the Treasurer 10s. per week for the accommodation and privileges. Institutions, such as Colleges and Museums, may become subscribers in order that a work place may be at the disposal of their staff for a certain period annually; a subscription of two guineas will secure a work place for six weeks in the year, a subscription of five guineas for four months, and a subscription of £10 for the whole year.

VI.—Each worker* is entitled to a work place opposite a window in the Laboratory, and may make use of the microscopes, reagents, and other apparatus, and of the boats, dredges, tow-nets, &c., so far as is compatible with the claims of other workers and with the routine work of the Station.

VII.—Each worker will be allowed to use one pint of methylated spirit per week, free. Any further amount required must be paid for. All dishes, jars, bottles, tubes, and other glass may be used freely, but must not be taken away from the Laboratory. Workers desirous of making, preserving, and taking away collections of marine animals and plants, can make special arrangements with the Director or Treasurer in regard to bottles and preservatives. Although workers in the Station are free to make their own collections at Port Erin, it must be clearly understood that (as in other Biological Stations) no specimens must be taken for such purposes from the Laboratory stock, nor from the Aquarium tanks, nor from the steam-boat dredging expeditions, as these specimens are the property of the Committee. The specimens in the Laboratory stock are preserved for sale, the animals in the tanks are for the instruction of visitors to the Aquarium, and as all the expenses of steam-boat dredging expeditions are defrayed by the Committee the specimens obtained on these occasions must be retained by the

---

* Workers at the Station can always find comfortable and convenient quarters at the closely adjacent Bellevue Hotel; but lodgings can readily be had by those who prefer them.

Committee (*a*) for the use of the specialists working at the Fauna of Liverpool Bay, (*b*) to replenish the tanks, and (*c*) to add to the stock of duplicate animals for sale from the Laboratory.

VIII.—Each worker at the Station is expected to lay a paper on some of his results—or at least a short report upon his work—before the Biological Society of Liverpool during the current or the following session.

IX.—All subscriptions, payments, and other communications relating to finance, should be sent to the Hon. Treasurer, Mr. I. C. Thompson, F.L.S., 53, Croxteth Road, Liverpool. Applications for permission to work at the Station, or for specimens, or any communication in regard to the scientific work should be made to Professor Herdman, F.R.S., University College, Liverpool.

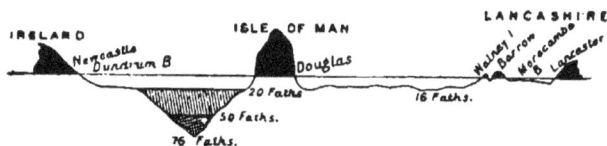

Diagrammatic Section across the Irish Sea.

APPENDIX B.

HON. TREASURER'S STATEMENT.

From the balance sheet appended it will be seen that the year ends with a small debit balance, in spite of the fact that all due economy has been observed, consistent with efficiency.

The British Association (1896) fund has, to a considerable extent, enabled the Committee to meet the necessary extra expense involved in securing the services of a competent Resident Curator at the Port Erin Station, which has already proved of great benefit to workers.

Increased workroom accommodation has also been added at the Station during the year, at a cost of about £14.

As pointed out by the Director in his Report (p. 258), an additional income of about £100 per annum could be most advantageously utilized in publishing well-illustrated papers and memoirs, embodying the results of local biological investigations such as, for want of funds, have unfortunately, in some cases, been sent, in the past, to other societies outside Liverpool.

The Hon. Treasurer, on behalf of the Committee, will be glad to receive the names of new subscribers, with the view of remedying this deficiency, and of further adding very materially to the already excellent work achieved under the auspices of the L.M.B.C. since its foundation, twelve years ago.

SUBSCRIPTIONS and DONATIONS.

	Subscriptions.			Donations.		
	£	s.	d.	£	s.	d
Ayre, John W., Ripponden, Halifax ...	1	1	0	—		
Bateson, Alfred, Harrop-road Bowdon ...	1	1	0	—		
Beaumont, W. I., Cambridge	1	1	0	—		
Bickerton, Dr., 88, Rodney-street	1	1	0	—		
Bickersteth, Dr., 2, Rodney-st.	2	2	0	—		
Brown, Prof. J. Campbell, Univ. Coll. ...	1	1	0	—		
Browne, Edward T., B.A., 141, Uxbridge-						
road, Shepherd's Bush, London ...	1	1	0	—		
Brunner, Sir J. T., Bart., M.P., Druids						
Cross	5	0	0	—		
Boyce, Prof., University College ...	1	1	0	—		
Caton, Dr., 86, Rodney-street	—			1	1	0
Clague, Dr., Castletown, Isle of Man ...	1	1	0	—		
Clague, Thomas, Bellevue Hotel, Port Erin	1	1	0	—		
Cole, F. J., (research table)	1	1	0	—		
Coombe, John N., 4, Paradise-sq., Sheffield	1	1	0	—		
Comber, Thomas, J.P., Leighton, Parkgate	1	1	0	—		
Crellin, John C., J.P., Ballachurry, An-						
dreas, Isle of Man	0	10	6	—		
Crossland, Cyril, Clare College, Cambridge	1	0	0	—		
Gair, H. W., Smithdown-rd., Wavertree	2	2	0	—		
Gamble, Col. C.B., Windlehurst, St. Helens	2	0	0	—		
Gamble, F.W., Owens College, Manchester	1	1	0	—		
Gaskell, Frank, Woolton Wood	1	1	0	—		
Gaskell, Holbrook, J.P., Woolton Wood	1	1	0	—		
Gibson, Prof. R. J. Harvey, 5, Adelaide-						
terrace, Waterloo	1	1	0	—		
Glynn, Dr., 62, Rodney-street	2	2	0	—		
Gotch, Prof., Museum, Oxford	1	1	0	—		
Forward ...£32		13	6	1	1	0

	Subscriptions. £ s. d.	Donations. £ s. d.
Forward ...	32 13 6	1 1 0
Halls, W. J., 35, Lord-street	1 1 0	—
Hanitsch, Dr., Museum, Singapore ...	1 1 0	—
Henderson, W. G., Liverpool Union Bank	1 1 0	—
Herdman, Prof., University College	2 2 0	—
Hewitt, David B., J.P., Northwich ...	1 1 0	—
Holland, Walter, Mossley Hill-road	2 2 0	—
Holt, Alfred, Crofton, Aigburth	2 2 0	—
Holt, Mrs. George, Sudley, Mossley Hill	1 0 0	—
Hoyle, W. E., Museum, Owens College, Manchester	1 1 0	—
Isle of Man Natural History and Antiquarian Society	1 1 0	—
Jarmay, Gustav, Hartford	1 1 0	—
Jones, C.W., J.P., Field House,Wavertree	1 0 0	—
Kermode, P. M. C., Hill-side, Ramsey ...	1 1 0	—
Lea, Rev. T. Simcox, 3, Wellington-fields	1 1 0	—
Leicester, Alfred, Buckhurst Farm, Edenbridge, Kent	1 1 0	—
Macfie, Robert, Airds	1 0 0	—
Meade-King, H. W., J.P., Sandfield Park	1 1 0	—
Meade-King, R. R., 4, Oldhall-street ...	0 10 0	—
Melly, W. R., 90, Chatham-street ...	1 1 0	—
Monks, F. W., Brooklands, Warrington	1 1 0	—
Mundy, Randal, Manchester	0 10 0	—
Muspratt, E. K., Seaforth Hall	5 0 0	—
Newton, John, M.R.C.S., 44, Rodney-st.	0 10 6	—
O'Kell, Robert, B.A., Sutton, Douglas ...	1 1 0	—
Paterson, Prof., University College	1 1 0	—
Poole, Sir James, Tower Buildings ...	2 2 0	—
Rathbone, Mrs. Theo., Backwood, Neston	1 1 0	—
Rathbone, Miss May, Backwood, Neston	1 1 0	—
Forward ...	£68 9 0	1 1 0

	Subscriptions. £ s. d.	Donations. £ s. d.
Forward ...	68 9 0	1 1 0
Rathbone, W., Greenbank, Allerton ...	2 2 0	—
Roberts, Isaac, F.R.S., Crowborough ...	1 1 0	—
Shackleford, Rev. L. J., Clitheroe ...	0 10 0	—
Simpson, J. Hope, Annandale, Aigburth-dr	1 1 0	—
Smith, A. T., junr., 24, King-street ...	1 1 0	—
Talbot, Rev. T. U., 4, Osborne-terrace, Douglas, Isle of Man	1 0 0	—
Thompson, Isaac C., 53, Croxteth-road	2 2 0	—
Thornely, James (the late), Baycliff, Woolton	1 1 0	—
Thornely, The Misses, Baycliff, Woolton	1 1 0	—
Toll, J. M., Kirby Park, Kirby	1 1 0	—
Torrance, Gilbert, North Quay, Douglas	1 1 0	—
Walker, A. O., Nant-y-glyn, Colwyn Bay	3 3 0	—
Walker, Horace, South Lodge, Princes-pk.	1 1 0	—
Walters, Rev. Frank, B.A., King William College, Isle of Man	1 1 0	—
Watson, A. T., Tapton-crescent, Sheffield	1 1 0	—
Weiss, Prof. F. E., Owen's College, Man'tr.	1 1 0	—
Westminster, Duke of, Eaton Hall ...	5 0 0	—
Wiglesworth, Dr., Rainhill	1 1 0	—
Yates, Harry, 75, Shude-hill, Manchester	1 0 0	—
	£95 18 0	1 1 0

SUBSCRIPTIONS FOR THE HIRE OF " WORK-TABLES," OCCUPIED
BY COLLEGES, &c.

Owens College, Manchester	£10 0 0
University College, Liverpool	10 0 0
	£20 0 0

THE LIVERPOOL MARINE BIOLOGY COMMITTEE.

In Account with ISAAC C. THOMPSON, Hon. Treasurer.

Dr.

1898.	£	s.	d.
To Balance due the Treasurer, Dec. 31st, 1897	2	15	4
,, Printing Reports and Engraving Plates	17	11	6
,, Printing and Stationery	2	7	9
,, Expenses of Dredging Expeditions	19	3	3
,, Boat Hire	4	2	1
,, Books and Apparatus at Port Erin Biological Station	16	7	7
,, Postage, Carriage of Specimens, &c.	2	6	4
,, Salaries, Curator and Laboratory Boy	68	13	4
,, Rent of Port Erin Biological Station	15	0	0
,, Cost of New Workroom	14	11	3
,, Sundries	1	7	0
	£164	5	5

Cr.

1898.	£	s.	d.
By Subscriptions and Donations actually received	87	10	0
,, Amount received from Colleges, &c., for hire of "Work Tables"	20	0	0
,, Dividend, British Workman's Public House Co., Ltd., Shares	5	18	9
,, Sale of Nat. Hist. Specimens	2	4	5
,, Sale of Reports and Volumes of Fauna	3	11	0
,, Interest on British Association (1896) Fund	32	2	2
,, Bank Interest	0	3	5
,, Admissions to Aquarium	6	7	5
,, Balance due Treasurer, Dec. 31st, 1898	6	8	3
	£164	5	5

Endowment Investment Fund :—
British Workmans Public House Co's. shares £173 1 0

ISAAC C. THOMPSON,
Hon. Treasurer.

Audited and found correct,
A. T. SMITH, Junr.

LIVERPOOL, *December* 31st 1898.

APPENDIX C.

Ïon. Director.
PROF. HERDMAN, F·R.S.,
UNIVERSITY COLLEGE,
LIVERPOOL.

L. M. B. C.

(JANUARY, 1899.)

Ïon. Treas.
I. C. THOMPSON, F.L.S.
53, CROXTETH ROAD,
LIVERPOOL

PRICE LIST

— OF —

SPECIMENS, ETC.,

SUPPLIED BY THE

MARINE BIOLOGICAL STATION

— AT —

PORT ERIN

- (ISLE OF MAN).

PORT ERIN is at the south-west corner of the Isle of Man, in the centre of the Irish Sea, and has a rich fauna, both littoral and pelagic. The Liverpool Marine Biology Committee have now appointed a resident Curator, Mr. H. C. Chadwick, who will devote a portion of his time to the collection and preparation of specimens for laboratories and museums.

The following price list may be largely added to in the future, and investigators are invited to correspond with the Curator at Port Erin, or with the Hon. Director, Professor Herdman, at Liverpool, in regard to their wants.

☞ *The prices named in this list include packages and postage.*

ALGÆ.

Cyanophyceæ.

Lyngbya majuscula ...	per tube	4d.
Calothrix confervicola ...	,,	8d.

Chlorophyceæ.

Enteromorpha clathrata and		
E. compressa	,,	4d.
Urospora bangioides	,,	4d.
Cladophora rupestris ...	,,	4d.
Bryopsis plumosa	,,	8d.
Codium tomentosum	per doz.	2/-

Phæophyceæ.

Ectocarpus confervoides ...	per tube	8d.
Elachista fucicola	,,	4d.
Sphacelaria cirrhosa	,,	8d.
Laminaria saccharina ...	per doz.	2/-
Alaria esculenta	per plant	4d.
Fucus vesiculosus ... ... -	per doz.	2/-
Himanthalia lorea ...	each	4d.
Pelvetia canaliculata	per doz.	8d.

Rhodophyceæ.

Chantransia virgatula	per tube	1/-
Gigartina mamillosa	per doz.	2/-
Plocamium coccineum	,,	2/-
Delesseria alata	,,	2/-
Odonthalia dentata	,,	2/-
Polysiphonia urceolata ...	,,	2/-
,, fastigiata	,,	1/-
,, nigrescens ...	,,	2/-
Dasya coccinea	,,	2/-
Callithamnion corymbosum ...	,,	2/-
Ptilota plumosa	,,	2/-
Ceramium rubrum	,,	1/-

Polyides rotundus	per doz.	2/-
Lithophyllum lichenoides ...	,,	2/-
Corallina officinalis	,,	1/-
,, rubens ...	,,	2/-

PROTOZOA.

Foraminifera, various	per tube	1/-
Gromia sp. ...	,,	1/-
Noctiluca miliaris ...	,,	1/-

PORIFERA.

Leucosolenia botryoides	per tube	8d.
Sycon compressum	,,	8d.
Leucandra nivea	,,	8d.
Pachymatisma johnstonii ...	each	1/6
Halichondria panicea	,,	9d.
Halisarca dujardinii	,,	9d.

HYDROIDA.

Clava
Hydractinia
Coryne
Tubularia
Obelia
Sertularia
Hydrallmania
Antennularia
Plumularia

1/- to 1/6
per tube, or
6d. to 1/-
per specimen.

HYDROMEDUSÆ.

Sarsia
Amphicodon
Thaumantias
Obelia

1/- to 1/6
per tube.

SCYPHOMEDUSÆ.

Cyanea capillata	each 2/- to 2/6
Aurelia aurita	,, 1/- to 1/6
Chrysaora isosceles	,, 2/- to 2/6
Depastrum cyathiforme ...	9d., three 2/-

CTENOPHORA.

Pleurobrachia pileus	per tube 8d.

ALCYONARIA.

Sarcodictyon catenata	per colony 1/6
Alcyonium digitatum ...per colony 9d., three 2/-
Do., polyps expanded per colony 1/- to 1/6

ACTINIARIA.

Sagartia sp., various ...	each 4d., three 10d.
Actinia mesembryanthemum	,, 4d., ,, 10d.
Bunodes gemmacea ...	,, 9d., ,, 1/6
Adamsia palliata	,, 9d., ,, 2/-

ECHINODERMATA.

Antedon bifida	each 1/-, three 2/6
Asterias glacialis ...	... ,, 1/6
,, rubens ...	... ,, 4d., ,, 10d.
Henricia sanguinolenta ...	... ,, 6d., ,, 1/3
Solaster papposus	... ,, 9d., ,, 2/-
Asterina gibbosa ...	... ,, 4d., ,, 10d.
Palmipes placenta	... ,, 1/6, ,, 4/-
Porania pulvillus ...	... ,, 6d., ,, 1/3
Stichaster roseus ...	... ,, 2/6
Echinus esculentus	... ,, 9d., ,, 2/-
Spatangus purpureus ...	... ,, 1/6, ,, 4/-
Synapta inhærens	... ,, 1/-, ,, 2/6
Ophiura ciliaris	... ,, 6d., ,, 1/3

Amphiura elegans .... each 4d., three 10d.
Ophiocoma nigra ... ...   ,, 9d.,   ,, 2/-
Ophiothrix fragilis   ...   ,, 6d.,   ,, 1/3

VERMES.

Leptoplana tremellaris ...   each 6d., three 1/3
Lineus marinus ... ...   ,, 9d.,   ,, 2/-
Tetrastemma sp., various   ,, 9d.,   ,, 2/-
Sagitta bipunctata   ...   ,, 8d. per tube
Pontobdella muricata   ...   ,, 6d., three 1/3
Dinophilus tæniatus   ...   ... 9d. per tube
Myzostomum sp.   ...   ... 1/- ,,
Aphrodite aculeata   each 1/-, three 2/6
Polynoë sp., various   ...   ,, 6d.,   ,, 1/-
Nephthys cœca ... ... 1/- per tube
Nephthys hombergi   ...   each 6d., three 1/3
Nerine vulgaris ... ... per tube 1/-
Nereis sp., various   each 4d., three 10d.
Arenicola marina   ...   ,, 4d.,   ,, 10d.
                                   (50 for 5/-)
Terebella nebulosa   ...   ,, 4d.,   ,, 10d.
Spirorbis borealis   ...   ... 8d. per tube

POLYZOA.

Flustra
Bugula
Crisia
Vesicularia   6d. to 2/-
Scrupocellaria   per colony,
Amathia   according to
Pedicellina   size and condition
Alcyonidium   of polypides.
Cellaria
Membranipora
Bowerbankia

CRUSTACEA.

Carcinus mœnas ...	...	three 1/-
Portunus puber	each 8d.,	,, 1/9
Stenorhynchus, Inachus, &c., ...	,, 9d.,	,, 2/-
Eupagurus bernhardus ...	6d. to 9d.,	,, 1/3
Galathea sp., various	each 6d.,	,, 1/3
Nephrops norvegicus ...	... ,, 4d.,	six 1/9
Crangon, Pandalus, Virbius, &c.		six 8d.
Mysis sp., various	8d. per tube	
Cumacea sp., various	8d. ,,	
Ligia oceanica	8d. ,,	
Idotea marina	8d. ,,	

Talitrus locusta
Orchestia littorea } 9d. per dozen.
Gammarus locusta

Protella phasma } 9d. per tube
Caprella linearis

Ostracoda, various	1/- per tube
Copepoda ,, ...	8d. ,,

PYCNOGONIDA.

Nymphon gracile } 4d. to 6d. each
Pycnogonum littorale

MOLLUSCA.

Anomia ephippium	three 9d.
Pecten maximus	each 9d.
,, opercularis	three 9d
Pectunculus glycimeris	,, 9d.
Cardium	,, 9d.
Venus	... ...	three 9d.
Solen } sp. div.	each 6d., three 1/3
Chiton	...	three 9d.

MOLLUSCA.—*Continued.*

Patella vulgata	four 6d.
Helcion pellucidum ...	four 6d.
Trochus zizyphinus	three 6d.
Littorina	per dozen 8d.
Natica	three 1/-
Aporrhaïs pes pelicani	,, 1/-
Purpura lapillus	per dozen 8d.
Buccinum undatum	} each 4d., or
Fusus gracilis	} three 10d.
Aplysia punctata... ...	each 6d., three 1/3

NUDIBRANCHIATA.

Archidoris tuberculata 6d. to 9d., three 1/3 to 2/-
Dendronotus arborescens	each 1/-, three 2/6
Doto fragilis ... ... ,, 9d., ,, 2/-
Eolis papillosa ... ,, 9d., ,, 2/-
Eolis coronata ... ,, 9d., ,, 2/-

CEPHALOPODA.

Eledone cirrosa	each 2/-
Sepiola atlantica	,, 1/-

TUNICATA.

Oikopleura flabellum (Appendicularia) per tube 8d.

Botryllus	} per colony
Amaroucium	} 1/- to 2/-,
Leptoclinum	} according to
Diplosoma	} size and species.

TUNICATA.—*Continued.*

Clavelina lepadiformis	... per cluster 1/6
Ascidia mentula	each 6d., three 1/3
Polycarpa glomerata	... per cluster 1/3
Molgula occulta	each 9d., three 1/9

Larger quantities of most of the above species can be supplied at proportionately lower rates. Many larval forms, notably those of the Crustacea and Polychæta, can be obtained during the spring and summer months. Reasonable notice of wants should be given.

Microscopic preparations, illustrating the Fauna and Flora of the district, can be supplied at 6/- per dozen. Tow-nettings, preserved in accordance with the purchaser's requirements, can be supplied by special arrangement with the Hon. Director, or through the Curator. Naturalists who desire to receive supplies of interesting living or preserved material for microscopic study can order weekly or monthly tubes, at 1/- per tube, post free, containing what the Curator considers to be in best condition at the time. For terms as to the hire of work-tables in the Laboratory at Port Erin, or for copies of the L.M.B.C. publications (Annual Reports, and "Fauna," vols. I. to IV.), application must be made to Professor Herdman, in Liverpool. Subscriptions and all other payments should be made to the Hon. Treasurer, Mr. I. C. Thompson, 53, Croxteth Road, Liverpool.

L.M.B.C. NOTICES.

The public are admitted by ticket to inspect the Aquarium at suitable hours daily, when the Curator will be, as far as possible, in attendance to give information. Tickets of admission, price threepence each, to be obtained at the Biological Station or at the Bellevue Hotel. The various tanks are intended to be illustrative of the marine life of the Isle of Man. It is intended also that short lectures on the subject should be given from time to time by Prof. Herdman, or by others of the Committee.

Applications to be allowed to work at the Biological Station, or for specimens (living or preserved) for Museums, Laboratory work, and Aquaria, should be addressed to Professor Herdman, F.R.S., University College, Liverpool.

Subscriptions and donations should be sent to Mr. I. C. Thompson, F.L.S., 53, Croxteth Road, Liverpool.

The surplus copies of the five Annual Reports upon the Marine Biological Station formerly on Puffin Island (1888 to 1892, the complete set) have been collated and bound up to form an 8vo. volume of about 180 pages, illustrated with cuts and plates, and containing the original lithographed covers. There are 20 copies of this vol., which are now offered by the Committee at 3/- each nett.

Copies of the Annual Reports from 1893 onwards, can also be had, price sixpence each (all post free).

The L.M.B. Committee are publishing their Reports upon the Fauna and Flora of Liverpool Bay in a series of 8vo. volumes at intervals of about three years. Of these there have appeared :—

Vol. I.   (372 pp., 12 plates), 1886, price 8/6.
Vol. II. (240 pp., 12 plates), 1889, price 7/6.
Vol. III. (400 pp., 24 plates), 1892, price 10/6.
Vol. IV. (475 pp., 53 plates), 1895, price 10/6.

Copies of any of the above publications may be ordered from the Liverpool Marine Biology Committee, University College, Liverpool, or from the Hon. Treas., 4, Lord Street, Liverpool.

Isaac C. Thompson,

Hon. Treas.

[WORK FROM THE PORT ERIN BIOLOGICAL STATION.]

LIST of the ARANEIDA of PORT ERIN and DISTRICT.

By A. R. JACKSON, B.Sc. (VICT.),
Victoria University Scholar in Zoology.

WHILE occupying the University College work-place in the Port Erin Biological Station during the session 1897-98, in addition to my study of marine life, I paid some attention to collecting the spiders of the neighbourhood.

Prof. Herdman suggested that I should follow this up, on a further visit to the station, with the object of completing a list of the Araneida found at Port Erin ; and I am now, as the result of my observations, able to record the following 57 species. They were all found within a radius of 1 mile of the Biological Station, except *Amaurobius ferox* and *Epeïra quadrata*.

ARANEIDA.

I. DYSDERIDÆ.

Dysdera crocota, rare.
Harpactis hombergii, fairly common.
Segestria senoculata, very common.

II. DRASSIDÆ.

Micaria pulicaria, fairly common.
Prosthesima latreillei, fairly common.
P. pusilla, rare, only two specimens were taken.
Drassus lapidosus, very common.
D. cupreus, very common.
D. troglodytes, fairly common.
Clubiona terrestris, fairly common.
C. reclusa, fairly common.

III. AGELENIDÆ.

Tegenaria derhami, very common.
Textrix denticulata, fairly common.

IV. DICTYNIDÆ.

Amaurobius fenestralis, very common.
A. similis, very common.
A. ferox, very local, found in Ballasalla quarries.

V. THERIDIIDÆ.

Theridion sisyphum, common, and only taken in
  Colby glen.
Phyllonethis lineata, very common.
Asagena phalerata, one specimen, on the moors.
Ero furcata, only cocoons found.
Pholcomma gibbum, one specimen.
Linyphia triangularis, very common.
L. pusilla, one specimen, on the cliffs.
L. lineata, fairly common.
L. clathrata, fairly common.
Leptyphantes leprosus, very common.
L. minutus, very common.
L. blackwallii, fairly common.
L. tenuis, fairly common.
Bathyphantes concolor, very common.
B. variegatus, fairly common.
Neriene rubens, fairly common.
N. bipunctata, fairly common.
Erigone atra, very common.
E. dentipalpis, very common.
Pedanostethus lividus, fairly common, in gorse
  bushes.
Tiso vagans, rare.

Ceratinella breve, fairly rare.
Peponocranium ludicrum, very common.
Pachygnatha degeeri, common.

VI. EPEIRIDÆ.

Meta menardii, in Bradda mines, very local.
M. merianæ, very common.
M. segmentata, very common.
Zilla x-notatum, very common.
Epeïra umbratica, very common and local, on
    cliffs.
E. diademata, very common, local.
E. quadrata, very common, and found at Fox-
    dale.

VII. THOMISIDÆ.

Xysticus cristatus, very common everywhere.

VIII. LYCOSIDÆ.

Lycosa ruricola, very common.
L. terricola, very common.
L. pulverulenta, very common.
Pardosa nigriceps, fairly common.
P. pullata, very common.
P. amentata, very common.
P. palustris, very common.

IX. ATTIDÆ.

Epiblemium scenicum, rare, two specimens found
    on laboratory wall.
Heliophanus cupreus, fairly common.

THIRTEENTH ANNUAL REPORT of the LIVERPOOL MARINE BIOLOGY COMMITTEE and their BIOLOGICAL STATION at PORT ERIN.

By Professor W. A. HERDMAN, D.Sc., F.R.S.

THE past year has been notable in several respects. In the amount and quality of the students' work at the Biological Station it is, I consider, the best year we have ever had ; several new lines of investigation have been started; a new set of publications—the L.M.B.C. Memoirs —has been successfully launched, and we have renewed our tenancy of the buildings at Port Erin under new landlords. All these and other matters will be dealt with in order in the following pages. We may begin, as usual, with the formal statistics as to the occupation of the work tables at the Laboratory, and then pass to the Curator's report upon the events of the year at Port Erin. Work at Liverpool, the publications of the Committee, and wider considerations, such as the result of the Stockholm Conference on Marine Exploration, will close the Report.

The Station Record.

During the past year the following naturalists have worked at the Biological Station, in addition to the Curator (Mr. H. C. Chadwick) who has been in constant attendance with the exception of a fortnight's holiday in May.

DATE.	NAME.	WORK.
January.	Prof. W. A. Herdman	General.
—	Mr. I. C. Thompson	Copepoda.
March.	Miss Sollas, Newnham Coll., Cambridge	General.
—	Mr. H. S. Harrison, Univ. Coll., Cardiff	Hydrozoa.

DATE.	NAME.	WORK.
March.	Mr. E. T. Townsend, Owens Coll.	General.
—	Prof. W. A. Herdman	General.
—	Mr. I. C. Thompson	Copepoda.
—	Mr. R. L. Ascroft	Fishes.
—	Mr. E. J. W. Harvey, Liverpool... ...	General.
April.	Mr. A. R. Jackson, Univ. Coll., Liverpool	Arachnida.
—	Mr. J. C. Mann, Univ. Coll.	General.
—	Mr. R. Kelly, Univ. Coll....	General.
—	Mr. F. J. Cole, Univ. Coll.	Comp. Ascidians.
May.	Mr. C. E. Jones, Univ. Coll.	Marine Algæ.
—	Mr. H. Yates, Manchester ...	Polychæta, &c.
June.	Mr. E. Schuster, New Coll., Oxford	Polychæta.
July.	Prof. W. A. Herdman	General.
—	Mr. I. C. Thompson	Copepoda.
—	Mr. E. Schuster	Polychæta.
—	Mr. J. T. Jenkins, Univ. Coll., Abery-	Mollusca, Fishes, and
	stwyth and Liverpool	Plankton.
August.	Mr. C. Crossland, Clare Coll., Cambridge	General.
—	Mr. J. T. Jenkins, Univ. Coll., Liverpool	Mollusca, Fishes, &c.
—	Mr. R. Simon, Manchester	General.
—	Mr. J. Corker, Manchester	General.
—	Mr. Buckley, Manchester	General.
—	Mr. D. R. Mackay, Manchester ...	Hydroida.
September.	Mr. J. T. Jenkins, Univ. Coll.	Mollusca, Fishes, &c.
—	Mr. C. Crossland, Clare Coll., Cambridge	General.
—	Mr. H. Yates, Manchester	Polychæta, &c.
October.	Mr. I. C. Thompson ...	Copepoda.
—	Mr. P. M. C. Kermode	General.
—	Prof. W. A. Herdman	General.
December.	Mr. I. C. Thompson	Official.
—	Prof. W. A. Herdman ...	Official.

This record shows an increase not merely in the number of individuals, but what is more important, in the length of time that each worker stayed, and, therefore, in the seriousness and extent of his work. The list, it will be noticed, includes students from Oxford, Cambridge, and the Welsh Colleges, in addition to those from the Colleges at Manchester and Liverpool which have secured work-tables at the Station.

THE CURATOR'S REPORT.

Mr. CHADWICK reports as follows :—

"The dissecting and compound microscopes provided early in the year have proved most useful. All the instruments and apparatus required by workers have been carefully used, and, with the exception of certain of the dredges, are in good condition. The small iron trawl, provided last year, has been frequently used, and its loss, together with the rope, towards the close of the season, was most unfortunate. The nets attached to the 'Agassiz' and shrimp trawls were repaired after the last steamboat dredging excursion, but renewal of both will probably be necessary ere long. All the smaller dredges are in service-able condition. Mr. H. Yates, of Manchester, has kindly lent, for an indefinite period, a sliding microtome by Reichert, specially adapted for cutting sections of material imbedded in celloidin.

"The Shellbend boats, though in need of slight repair, are little the worse for the frequent use to which they have been put.

"The Library has been well used, and the additions to the stock of books, including a nearly complete set of the Journal of the Marine Biological Association, have consider-ably enhanced its value. Several zoologists have kindly continued to contribute copies of their papers, for the reception of which, and other additions, a third bookcase has been provided. Further donations from authors, and others, will be very welcome.

"A small beginning has been made with the formation of a collection of microscope slides, illustrating the local fauna, which it is hoped will be of constantly increasing value to future workers at the Station.

"The sale of living and preserved specimens has not assumed noteworthy proportions, but orders from various

sources keep coming in, and a number of carefully preserved specimens have been supplied to colleges and
private workers.

"The record of meteorological observations begun last
year has been continued, and in view of the exceptionally
stormy character of the winter months, and the equally
exceptional amount of sunshine experienced during the
summer, a record of the average temperature of the air
and sea may not be without interest.

MONTH.	9-30 a.m. AIR.	3-30 p.m.	9-30 a.m. SEA.	3-30 p.m.
October, 1898	... 54° F.	55°	55°	56°
November	... 48°	49°	51°	52°
December	... 48°	48°	50°	50°
January, 1899	... 44°	45°	46·5°	47·5°
February	... 43°	46°	45°	47°
March	... 42·5°	45·5°	39°	47°
April	... 46·5°	49°	48·5°	50°
May	... 50·5°	54°	52°	54°
June	... 59°	65°	58°	61°
July	... 60°	61°	59·5°	60·5°
August	... 65°	65·5°	61·5°	60°
September	... 57°	59°	58·5°	60°

"The aquarium has not proved quite so attractive to
visitors this year as last, only 310 persons having paid for
admission. This is probably in large measure due to the
exceptionally fine weather which prevailed throughout the
summer, enabling visitors to Port Erin to spend most of
their time in boating and other out-door amusements.
Still, the interest shown by the visitors has been well
maintained, and the preserved specimens on the shelves
have been scarcely less attractive to many than the living
occupants of the tanks. Cracks extending from top to
bottom of the plate glass fronts of both wall tanks in the.

basement, appeared, one at the beginning of the winter and the other early in the present year; the result, most probably, of slight settlement of the building. Fortunately, they have not interfered with the usefulness of the tanks. Each still holds a depth of 18 inches of water without appreciable leakage.

"During the early part of March all the tanks were emptied, thoroughly cleansed and, where necessary, re-stocked. A considerable number of the animals introduced at Easter, 1898, survived the winter, but during the excessive heat experienced at the end of August, the mortality amongst them was great, and few now remain. A specimen of *Sabella pavonia*, placed in one of the shallow table tanks in July, 1898, is still (December, 1899), alive. Several Shannies (*Blennius pholis*) and a Sting-fish (*Cottus scorpius*) spawned in one of the tanks during the first week in March, but the eggs were not fertilised. The last named fish has well maintained its reputation for pugnacity and voracity by attacking and killing every fish that has been placed in the tank, with the exception of a young Dragonet (*Callionymus lyra*), which escaped the fate of three others placed in the tank at the same time, and still survives. I have several times seen the Sting-fish swallow so many young plaice that it could only with some effort raise itself from the bottom of the tank.

"A small lobster, about 3 inches long, was captured and placed in one of the wall tanks, about the middle of February, and it now appears to be thoroughly accli-matised. One morning in March I found the dead body of a Sting-fish, which had been in a sickly condition for several days, buried in the gravel at the bottom of the tank. On removing it I found that it had been partly eaten. A fortnight later a hermit crab was killed, and its body similarly buried. As the lobster was the only animal in the tank capable of doing this, I did not disturb it, and found, on the following morning, that it had been disturbed and re-buried. The next morning nothing remained but the chelæ and walking legs. The lobster has now established itself in a particular crevice in the rock-work, to which all food given to it is carried.

"In February I was fortunate enough to see a large number of ova extruded by a specimen of *Nereis pelagica*. I kept the worm in a small dish, and one day noticed that the parapodia on the left side of the ninth and tenth segments had become much swollen. On the next day the swelling had increased, and while watching the movements of the worm I saw the ova extruded from the swollen region, probably from the nephridiopores of the affected segments. I then fixed and preserved the worm, and it has since been photographed by Mr. E. Schuster, of New College, Oxford.

"In regard to Faunistic work, I have availed myself of every opportunity afforded by low tides to collect on the shores of Port Erin and the neighbouring bays, and thirteen dredging excursions have been undertaken in small boats. The courses dredged over, and the animals collected, were carefully recorded. The harbour buoy was brought ashore for its annual cleaning on August 11th this year, and again yielded over thirty species. Amongst them was an abnormal specimen of the Nudibranch, *Eolis glottensis* (?), in which the foot is divided into two quite distinct portions of exactly similar shape, one in front of the other. Amongst the additions to our lists are the Hydroid *Campanulina repens*, and the Nudibranch *Hermæa dendritica*, both discovered in 'Pat's Dub' by Miss Sollas.

"I have devoted some time during the past year to the study of the Polychæta of the bay, and have gained a good deal of knowledge of the habitats and times of occurrence of the different species. During the ensuing year it is my intention to devote special attention to some of the other less known groups, so that at its close, the detailed report on the shore and bottom fauna of the bay (illustrated by charts), which has been alluded to in former annual reports, may be completed and issued.

" Experiments with various reagents have been made from time to time, with a view of discovering new and improving old methods of narcotising and fixing marine animals. Menthol, recommended by Dr. H. C. Sorby, has proved very useful for Hydroida, Siphonophora, and Polyzoa. By dropping a few crystals on the surface of the water in a trough in which I examined *Agalmopsis elegans*, I was able to keep the specimens quite motionless and in a fully extended condition for some time. The usually refractory Sertularians give very good results. A 1 % solution of Cocaine has been used with complete success in narcotising the Lamellibranch, *Tapes pullastra*, with its siphons fully extended."

EASTER COLLECTING PARTY.

The usual dredging expedition of the Liverpool Marine Biology Committee in the Easter vacation opened very successfully, but was brought to a sad and untimely end on the third day by an unfortunate boat accident in Port Erin Bay. On March 31st dredging and trawling to the east of the Calf Island were carried on from the fisheries steamer " John Fell," and on the following forenoon the working of the " Tanner" closing net, and the method of pumping plankton from the bottom by means of a hose-pipe, were demonstrated to the Students on the steamer. On the afternoon of Saturday, April 1st, two of the workers at the Biological Station went out to collect surface plankton in the smaller station boat. While hauling in the tow-net, when returning, the boat capsized, and both men were thrown into the water. One of them (Mr. E. J. W. Harvey, of Liverpool) was picked up by the other boat from the Biological Station, but his companion (Mr. Eric T. Townsend, of Prestwich) was unfortunately drowned before assistance could reach him. The body was eventually recovered. Mr. Townsend was a student of the Owens' College, and was the occupant of the College work-table at the Biological Station for the Easter vacation. He was a promising Student, keenly interested in his biological work, and much esteemed by his fellow-workers. The party broke up after this sad occurrence. It is the first serious accident of any kind that has taken place on the L.M.B.C. expeditions or at their Biological Station during the fifteen years of work.

Notes on Work done at the Station.

A good deal of the Laboratory work this year has been not so much that of Specialists as of Advanced Students, which is not intended for publication, and in regard to which little that is of interest can be said. Two of the other workers, Mr. I. C. Thompson, F.L.S., and Mr. J. T. Jenkins, B.Sc., have given me special notes about Copepoda and Fishes, which will be inserted at the end of this section. Of those that remain :—Mr. A. R. Jackson, B.Sc., was engaged in adding to his list of Arachnids of the district, and met with some success ; Mr. F. J. Cole continued his studies on the budding of Compound Ascidians, and collected and preserved much material for future work in Liverpool ; and Mr. H. S. Harrison, B.Sc., has written me a letter telling of the Hydroid and Medusoid material that he collected, and ending with the

following words :—" I hope some time to make a longer
stay at Port Erin at a more favourable time of year, for
at that time I felt that I had not a fair opportunity of
doing justice to the splendid facilities for study and research offered by the fauna of the district. However, in
spite of the shortness of my stay and the unfavourable
weather, I found everything very instructive."

NEW COPEPODA.

Mr. THOMPSON has kindly drawn up the following note
upon his work :—

" During the year I have regularly received for examination from our Curator, Mr. H. C. Chadwick, bottles of
plankton material collected by him in and about Port
Erin Bay. In addition to these, I have myself, on several
visits to Port Erin, used the tow-net for collecting about
the district. As Mr. Chadwick will discuss the other
organisms noted in the collections, I will refer here only to
the COPEPODA, which, however, usually form the chief
proportion of the organisms in the tow-net.

" The most notable feature of the year has been the
appearance of two species of Copepoda new to the district,
viz., *Candacia pectinata*, and *Corycæus anglicus*, each of
which has been taken on several separate occasions. The
former occurred during January, June, and July, and was
taken both at or near the surface, and also at a depth of
33 fathoms. The latter species was taken during the early
winter (Nov. 26th, 1898), and a shoal of it was also
captured on May 29th.

" *Candacia pectinata* appears to be generally, but very
sparingly, distributed about the British Isles. It was
first found by Drs. Brady and Robertson, at a depth of 40
fathoms, off the Scilly Islands. I have on several occasions
found it on the west coast of Scotland, and Mr. Thomas

Scott, F.L.S., reports it from the Frith of Forth. About the Channel Islands I have taken it sparingly, and more plentifully about Valencia in Ireland. Dr. Brady reports the "Challenger" captures of this species about Australia, the Phillipine Islands, and between Ascension and the Azores. I found it common about the Canary Islands, and have lately detected it in plankton taken near Madagascar, by Capt. Fred. Wyse, a record of whose collecting results I hope shortly to publish.

"This and the other species of the genus *Candacia* are easily distinguishable by their dark-coloured antennæ, spines and plumes, and the terminal spines of the swimming feet.

"*Corycæus anglicus* is fairly common about the south and south-west coasts of England and Ireland, but with the exception of its occasional appearance in the Forth,

Plan of the L.M.B.C. District.

as reported by Mr. Thomas Scott, it has, I believe, been hitherto unknown on our northerly coasts.

"During an inspection cruise of the spawning grounds off the west of the Isle of Man, on the fishery steamer "John Fell," in January last, when, by the kindness of Mr. Dawson, the Superintendent of the Lancashire Fishery District, some of the L.M.B.C. were invited to be present, some interesting observations were made as to the quantity and comparative character of the plankton collected under similar conditions at the surface and at the sea bottom. I have already recorded the results in a paper 'Notes on Mid-Winter Surface and Deep Tow-Nettings in the Irish Sea,' see L.B.S. Trans., vol. XIII., pp. 156—62. It is hoped that we may be able at other seasons of the year to continue observations on a similar plan.

"In addition to the forms noted above, Mr. Andrew Scott has found several other species of Copepoda new to our district and one, at least, new to science, during the past year in his work at the Piel Hatchery, on the Lancashire coast. These will be described in a joint paper by Mr. Scott and myself, which will be laid before the Biological Society, and published in the 'Transactions.'"

THE FISH OF PORT ERIN.

Mr. J. T. JENKINS, as the result of one part of his work at the Biological Station, has drawn up the following report upon :—"The Distribution of Fish in and around Port Erin Bay during August and September, 1899."

"The total number of species captured during these two months was 32. Taking the families in order, we get :—

SPARIDÆ. — *Pagellus centrodontus* (Sea Bream) was abundant outside the Bay, more especially to the north of Fleshwick. It was taken with mackerel bait a short distance off the ground.

COTTIDÆ.—Four species were captured. *Cottus scorpius* and *C. bubalis* (Bullheads) were taken in rock pools at low tide all round the Bay, from the breakwater to the north side. They may be taken either with a hand net or with a hook, using Littorina as bait. Young ones were also captured in the dredge; they are voracious feeders.

Trigla gurnardus (Grey Gurnard) are very common. They can be taken within the Bay and outside from the Calf to two or three miles north of Fleshwick. They are voracious feeders, will bite readily at Arenicola, herring or mackerel bait. *Trigla cuculus* (Red Gurnard) are not so common as *T. gurnardus*, but fairly abundant; distribution similar; some very large specimens were captured.

SCOMBRIDÆ.—*Scomber scomber* (Mackerel) is occasionally taken in the Bay. It is plentiful outside from Bradda Head to the Calf. It is taken with artificial bait, india rubber tubing with a piece of skin of a fish. Five dozen were taken in one afternoon (Sept. 12th) by "whiffing" with two lines from a yacht.

GOBIIDÆ.—*Gobius niger* (the Goby). Distribution as for the *Cottus* species. *Callionymus lyra* (Dragonet). Frequently taken in the dredge from Bradda Head to Bay Fine.

DISCOBOLI.—*Cyclopterus lumpus* (Lump Sucker). A single specimen was found adhering to the buoy.

Two species of *Lepadogaster* (Suckers) were obtained: viz., *L. gouanii* and *L. decandolii*. These were obtained in rock pools at low tide and also in the dredge within the Bay.

BLENNIIDÆ.—*Blennius pholis* (Shanny) is a littoral form; in rock pools between the Biological Station and Spaldrick Bay at low tide. Also, along with it, *Centronotus gunnellus* (Butter fish), taken in rock pools throughout the Bay.

GASTEROSTEIDÆ. — *Gasterosteus spinachia* (15 spined Stickleback) is a littoral form ; distribution as for the last species.

LABRIDÆ.—*Labrus mixtus* (Cook Wrasse) is abundant round the breakwater, and was occasionally taken off Bradda Head and in Bay Fine.

GADIDÆ.—A large number of young forms of various species of *Gadus* are found around the breakwater.

Gadus virens (Saithe): young taken in the dredge within the Bay. *Gadus pollachius* (Pollack) is abundant : taken within the Bay near the buoy; also taken outside frequently from north of Fleshwick to Halfway rock. *Gadus morrhua* (Cod) and *Gadus æglefinus* (Haddock) : generally taken a few miles outside the Bay.

Molva vulgaris (Ling) is taken to south of Port Erin (Bay Fine and off the Calf). *Merluccius vulgaris* (Hake) : distribution as for *G. morrhua*. *Motella mustela* and *M. tricirrata* (Rocklings) are taken in the vicinity of the breakwater.

OPHIDIIDÆ.—One species only, *Ammodytes tobianus* (Lesser Sand Eel). This is taken at low tide in pools on the south side of the Bay near the landing stage.

PLEURONECTIDÆ.—I only succeeded in capturing two species of flat fish—*Pleuronectes platessa* (Plaice) and *P. limanda* (Dab). I should be inclined to think that other Pleuronectids are common. Both species mentioned were taken frequently in the Bay and outside the Bay from the breakwater to Halfway rock. Young forms were frequently taken in the dredge, both inside and outside the Bay. On Aug. 21st, a shoal of young Pleuronectids were observed at low water opposite the Laboratory.

SCOMBRESOCIDÆ.—*Belone vulgaris* (Garfish) is taken when "whiffing" for mackerel. Probably accompanies mackerel shoals.

CLUPEIDÆ.—*Clupea harengus* (Herring) is taken abundantly a few miles south-west of the Calf.

MURÆNIDÆ.—*Conger vulgaris* (Conger Eel) is taken outside the Bay from north of Fleshwick to the Calf. Evidently prefers mackerel to any other form of bait, and undoubtedly selects it in preference to herring.

SYNGNATHIDÆ.—*Syngnathus acus* (Pipe fish) is captured at low tide and in rock pools. Seems to prefer neighbourhood of *Chorda filum*, to detached or floating portions of which it presents some resemblance.

ELASMOBRANCHII.—*Raja maculata* (Skate) and *Scyllium canicula* (Dog fish) were taken outside the Bay when ground fishing with mackerel bait for other forms."

WORK ON OYSTERS AND DISEASE.

As the progress of the investigation upon the conditions under which oysters flourish, and the connection between unhealthy oysters and disease in man, carried on for some years now by Prof. Boyce and myself, has been noted from time to time in these Annual Reports, it may interest some readers to know that the work is now concluded, and that the final results have been published (October, 1899) under the auspices of the Lancashire Sea-Fisheries Committee, as the first "Lancashire Sea-Fisheries Memoir." This is a thin quarto volume* of about 70 pages, illustrated by eight partly coloured plates, and giving the details of the histological, chemical, and bacteriological evidence upon which the conclusions as to oysters printed in last year's report were based. Since that report was published last year it is satisfactory to know that (1) the "British Oyster Industries Association," on very much the lines that we suggested, has been

* "Oysters and Disease," published by George Philip and Son, London and Liverpool, at 7s. 6d. nett.

formed, and (2) an Oyster Bill has been laid before Parliament and referred to a Select Committee of the House of Lords. The Bill, although capable of improvement in some details, does much to meet the present difficulties; if, however, the Oyster Association takes a sufficiently high view of its responsibilities and duties, the provisions of the Bill may become unnecessary.

I may, finally, give here a few sentences of the practical conclusions at which Prof. Boyce and I have arrived as the result of the bacteriological work :—

" It is evident from the result of these experiments, and a consideration of all the facts brought to light in recent years in regard to the bacteriology of shell-fish and its influence on public health, that we must regard oysters, mussels, cockles, and the like as nutritious food matters which, from their nature and the circumstances of their cultivation and sale, are liable to become contaminated with organisms—pathogenic or otherwise—and their deleterious products.

" Once this is recognised, the practical applications are largely a matter of common sense. Shell-fish must not be taken as food from grounds where there is any possibility of sewage contamination ; after removal from the sea, while in transit, in store, or in market, they should be carefully protected from any possibility of insanitary environment ; they should not be kept longer than is absolutely necessary in shops, cellars, &c., in towns, where, even if not running the risk of fresh contamination, they are under conditions favourable to the reduction of their vitality, and the growth of their bacterial contents—the fresher they are from the sea the more healthy they are likely to be.   Finally, only absolutely fresh shell-fish should be eaten uncooked, and

those that are cooked must be *sufficiently cooked*, raised to
boiling point, and kept there at least ten minutes."

INTERNATIONAL EXPLORATION OF THE SEA, AND THE
STOCKHOLM CONFERENCE.

In our last Report the further exploration of the North
Atlantic and the Arctic Seas, in the interests both of
fisheries work and of scientific investigation, was discussed
at some length, and suggestions were made as to the lines
along which series of observations were highly desirable.
This matter has been before the minds of scientific men
both in this country and abroad for some years now, and
many biologists would, no doubt, gladly do the work if
the necessary observing vessels were provided by the
European governments concerned in the coast fisheries.
When it was announced, early in 1899, that our government had accepted the invitation from the Swedish
government to take part in an International Conference
on the whole subject, to be held at Stockholm during the
summer, hope ran high that at last the opportunity had
come when strong representations would be made to
the governments of north-western Europe, such as
would lead to the provision of the required boats and men
for a number of years sufficient to carry out the biological
investigation of the seas around the British Isles.

The Conference was held at Stockholm in the latter part
of June, and a great deal of the credit of having brought
the matter to that point must be given to the Secretary
of the Swedish Committee, Dr. Otto Pettersson, the distinguished Hydrographic expert. The Hydrographic
element was probably very strong amongst the delegates
present, and certainly seems to be the dominant note in
the published account of the deliberations. Probably,

that is the reason why the official Report*, now published, is a somewhat disappointing document in the eyes of biologists. It consists of a number of resolutions in regard to what is desirable, or what is required in the investigation of the sea, and of one recommendation as to the formation of a central bureau to control the work and conduct a laboratory.

One section is called a " Programme" of work, but there is no practical programme of biological work—laying down how, when and where the investigations are to be carried out—such as was expected, and is required. For, surely, what we need most at the present time, in the interests of more exact fisheries† knowledge, is the nearest possible approximation to a " census " of our seas, beginning with the territorial waters. Most fisheries disputes and differences of opinion are due to the absence of such exact knowledge. If such an approximate census, or record of really reliable statistics, had been taken fifty years ago it would now be invaluable to fisheries inspectors, superintendents and local authorities, as well as to biologists. Our descendants will justly reproach us if, with our increased knowledge and opportunity, we let the twentieth century commence without inaugurating some such system of statistics.

The Report of the Conference says nothing of all this. In place of asking for boats and men, it expresses many admirable sentiments and pious wishes as to what is desirable and what should be done—sentiments and wishes that are quite unexceptionable, but which have been before the public now for some years, and which are in the main

---

* " Conférence Internationale pour l'Exploration de la Mer, réunie à Stockholm, 1899."

† The British Government, we are told, only joined in the Conference in the interests of the fisheries applications of marine exploration.

agreed to on all hands. We looked for something more from this Conference. Their one definite recommendation (see "Résolutions Textuelles," p. 12) to the governments concerned, is in regard to the establishment of the central bureau, in which the work will apparently, in large part, be that of a physico-chemical laboratory. I do not think that, after what I have written in previous L.M.B.C. Reports, I can be accused of under-valuing the importance of hydrographic work, in its connection with the fisheries, as carried out of late years chiefly by the Scandinavians ; but it seems curious, to say the least of it, that the obvious biological investigations of primary importance have been passed so lightly over, while the secondary hydrographic investigations are strongly urged. The impression given by the Report is, certainly, that it has been drawn up by hydrographers, and not by biologists.

There are points of detail in the report that might be criticised if it were worth while—such as that the sea-area, proposed by the Conference, to be covered by the hydrographic investigations should certainly be extended so as to include the English Channel and the Irish Sea— and probably, also, the west coast of Ireland. At any rate, the omission of the whole of the Irish Sea from a scheme undertaken (so far as our government delegates were concerned) in the interests of the British fisheries requires explanation.

It has been pointed out, by another critic, that the best course for the British government to pursue, in order to give effect to the report of the Stockholm Conference, is to develop, and, as far as possible, co-ordinate the work of the various institutions already in existence, such as the Marine Biological Association and the Scottish Fishery Board, at the same time encouraging the formation of local laboratories at various points around the coast, such as those of the Liverpool Committee (at Port Erin), of the Lancashire Sea-Fisheries (at Piel), and of the North-umberland Committee (at Cullercoats). In place of any such scheme of biological collaboration between existing marine stations, the report of the Conference urges the institution of the new central bureau and laboratory—an elaborate and expensive organisation, which will probably appear to most biologists a matter of quite secondary importance, against which objections may be made. The following one has already been raised*:—"With an elaborate organisation, such as that suggested by the Conference, there is a danger that the work of the biological stations would degenerate into the mere taking and recording of routine observations, whilst original work and the develop-ment of new methods of research, which are in reality of far greater importance, would receive a check. Good men would certainly not be attracted to work which con-sisted merely in recording observations taken according to a stereotyped plan dictated by a central bureau. A large amount of individual freedom to the workers is absolutely essential in order to secure the best results from scientific research."

In my opinion, what we want at the present time is not conferences or committees, or a central bureau, so much as boats, men, and *work at sea*.

The Future of the Biological Station.

Since last Report the Bellevue Hotel (and with it the Biological Station, built on the grounds of the hotel), like so many other hotels in the Isle of Man, has passed into the hands of a syndicate. After some negotiations, the

* "Nature," Nov. 16th, 1899.

Port Erin Biological Station and its surroundings, at low tide, from the sea.   [From a photo by Rev. T. S. Lea.]

Committee have arranged to continue their tenancy of the buildings : so our landlord is no longer Mr. Thomas Clague, but a Company.

We have no desire to make a change at present, but we feel that the question of expansion will have to be faced soon. We have the doubtful advantage of being, perhaps, the smallest Biological Station (so far as regards buildings) in the world, our space is much cramped, each year we are, at holiday times, over-crowded with workers, and there are many additions and improvements in laboratory and aquarium that are badly wanted. Our last Annual Report was most favourably commented upon in an editorial review in the well-known journal " Natural Science," and the article ended with the following sentences* :—

" We cannot, indeed, but regard it as remarkable
" that a wealthy city like Liverpool, with all its
" traditions as a sea-port, should be unable to offer
" its Marine Station more than the very small sum
" at present at its disposal. If successful and per-
" severing work with small means deserves encour-
" agement, it certainly should not be lacking to the
" Port Erin Station. We wish Professor Herdman
" and his colleagues all success in the carrying out of
" their enlarged conception of local research."

We thank our unknown friend, and hope that his view of the matter may be adopted by the citizens of Liverpool.

But we do not ask merely for money to enable us to carry on more work. We want more than material support. We wish to have the moral support afforded by knowing that we are of use to the community. At present we educate (in the laboratory) a small number of Science and Medical Students : we should like to educate all kinds

* " Natural Science " for March, 1899, p. 186.

of students. We do, in a sense, through the aquarium convey information to all visitors who enter the building, and through our Reports to all who read them. But much more use might be made of our institution, even on its present small scale, and there is scarcely a limit to the uses to which it might be put if re-built on a larger scale and adequately equipped. When one thinks of the *hundreds* of school teachers and students of arts and theology who, in America, flock to the Marine Biological Stations during the summer vacation, in order that they may have the opportunity of becoming acquainted with biological thoughts and methods, and of studying under expert guidance the facts and ways of living Nature, one cannot but be struck by the contrast here; and one is led to wonder how long it will be before the elements of nature-knowledge are recognised as an essential part of a liberal education.

It is evident to some of us, from experience of and conversations with school teachers, that the demand for vacation work at Biological Stations is with us—it is the means of satisfying the demand that is absent. The immense success of the movement in America (*e.g.*, at the Woods Holl Biological Station, Massachusetts) not only justifies but requires us to urge its adoption here. If any Technical Instruction Committee, or other educational body (or individual), will build me a new biological station from my plans, within reasonable distance of Liverpool, and containing a laboratory in which, say 30, students can carry on work, I will undertake to admit school teachers free, and teach them in a free vacation class, and I feel sure, from some experience I have had, that the room would always be full, and that the students would find they were spending not merely an instructive, but also a most enjoyable vacation—learning new things every day, but in a manner so different from their ordinary school life that it would be a rest and a mental refreshment.

The importance of studying living Nature, and the uses that ought to be made of Biological Stations in education, are rapidly coming to be recognised in many other parts of the world. Here these institutions are not yet recognised as university laboratories, and time spent in them does not count in the curriculum for degrees. But in far Japan, Professor Mitsukuri tells me all the Biological Students of Tokio University are required to spend *at least* one season at the Marine Station of Misaki, which is a recognised institution of the University, while those who propose to graduate in Biology must spend a considerably longer period in such work. Many of the American Universities have Biological Stations as a necessary part of their equipment; and, in some cases, professors, staff, and students regularly migrate for the summer term to the sea-side laboratory.

Although England led the way in the past in Marine Biology, we are now behind other countries in the facilities and arrangements which cannot be provided by private enterprise. But all these things will come in time. We shall have University Biological Stations and Municipal Biological Stations some day. The pity is that we cannot have them *now* instead of 10, 20, or 30 years hence. Sir Michael Foster, in a recent address, has said :—" It is a matter of regret that the enthusiasm of the young learner should be spent wholly on the museum and the laboratory, that he should be pushed by compulsion and drawn by rewards into morphological and physiological studies of the more formal and mechanical kind, while no encouragement is given to him to look Nature face to face in the field, and to catch direct from her lips the Catholic teaching which she alone can give."

PLANKTON.

It was shown in last year's Report that one of the most important things the Biologist can do to add to our knowledge of life in the sea, with a practical view to the explanation of the movements and distribution of fish, is to make a systematic survey of the microscopic floating and drifting life of the sea, and its relation on the one hand to the physical conditions at the time (especially the temperature and salinity of the water), and on the other to the food materials found in the stomachs of the fish.

We have continued our efforts during the past year to contribute our share towards a knowledge of this "plankton" life of the sea round the British coasts. At some considerable expense we purchased a small hand pump, fitted on a stand for convenience of work on deck, and 20 fathoms of india-rubber hose-pipe for lowering to the bottom. This apparatus was used on several occasions early in the year, with rather disappointing results. It may be the more exact method in so far that it only gives organisms from the definite depth to which the end of the hose-pipe is lowered, while an ordinary open tow-net, lowered to the depth in question and raised again, may show some small admixture of organisms from the water above; but, on the other hand, the open tow-net catches far more material, and therefore gives a more complete knowledge of the fauna.

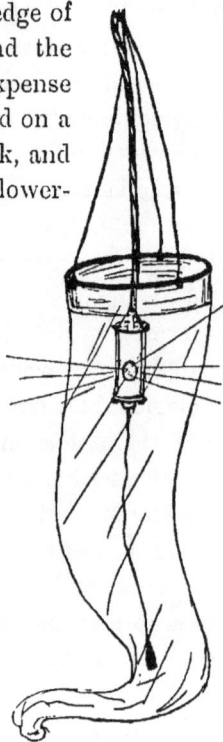

A Tow-Net.

There is a tendency, in trying to attain to more exact methods in biology, towards arguing and collecting, as if organisms were evenly distributed through the particular layer of water or region they inhabit—as if, in fact, they were like the salts dissolved in the water, so that one sample from a locality ought to be exactly like any other sample. This is far from being the case.   One gallon or cubic foot of water, though, in the main, like its neighbours, may contain one of the rarer organisms not present in the next gallon examined.   Consequently, in a net hauled up vertically, or in a comparatively small volume of water drawn up from a special layer through the hose-pipe, there is considerable risk of missing some of the less frequent organisms; while the ordinary tow-net, through which a very much greater mass of water passes, will probably contain a more completely representative gathering.   This is, in the main, our objection to the pump-plankton method which we tried at Port Erin at Easter on board the fisheries steamer " John Fell."   Equal times of working with the ordinary open tow-net and with the pump and hose gave results, both at bottom and at surface, which were absurdly disproportionate.   The open tow-net collected in the same period many times as much material, and far more species than the pump showed.

Mr. Thompson, in his note on a previous page as to Copepoda, has referred to the interesting differences we found between surface and bottom (30 fathoms) plankton in mid-winter.   We hope to continue such observations in the deeper water lying between the Isle of Man and Ireland.   That, however, can only be done on our comparatively rare steamboat expeditions.   In the meantime our Curator at the Station has endeavoured, during the whole of the past year, to take at least one ordinary surface tow-net gathering across Port Erin Bay in the week. Of course, such a scheme is dependent to some extent upon weather, and cannot always be carried out with absolute regularity. A few weeks have been missed now and then when it was practically impossible to work from a small boat in the Bay; but in all about thirty gatherings are fairly equally spread over the year, and every month is represented.

Apparatus brought back from a dredging expedition.

Mr. Chadwick has examined all these gatherings in the fresh condition, and they have then, after preservation, been carefully investigated by Mr. Thompson for Copepoda, while Mr. E. T. Browne, a former worker at Port Erin, has kindly identified the Medusæ sent to him. Mr. Chadwick has compiled MS. lists of all the gatherings, and has compared and summarised these with the object of ascertaining how far they bear out, correct, or supplement the conclusions we gave in a previous Report (No. XI., p. 17). The following summary of the results obtained from the examination of the past year's plankton, at Port Erin, has been drawn up by Mr. Chadwick for this Report. Mr. Chadwick has also provided us with the two carefully drawn plates which illustrate this section of the Report, and represent some of the more interesting larval forms captured during the year.

SUMMARY OF PORT ERIN PLANKTON IN 1899.

The following summary is the result of careful examination of the series of tow-nettings taken in Port Erin Bay during the year 1899. It has not been possible to identify all the organisms taken, nor yet, in some few cases, to refer them to their respective classes, and this is especially the case with some of the larval forms. All the organisms taken during an hour's tow-netting were examined alive, under a low power, immediately after capture, and the figures on Plates VI. and VII., with which these notes are illustrated, were drawn from living specimens under appropriate magnification.

The occurrence of *Pelagia perla*, *Gattiola finmarchica*, *Autolytus incertus*, and the larvæ of *Phoronis* and *Balanoglossus* in Port Erin Bay is here recorded for the first time. With reference to the Medusæ, Mr. E. T. Browne writes :—"The occurrence at Port Erin of *Pelagia*, *Mitrocomella*, *Melicertidium*, and *Laodice* tends to show that you have had a northerly current this summer, bringing down animals through the north channel from the Atlantic shores of Ireland and Scotland." We hope to publish a more complete account of the times of occurrence and distribution of the Medusæ at a future time.

As an example of what may be looked for in a summer tow-netting across Port Erin Bay, the following list of organisms, taken on June 22nd, will be interesting. The

time was 9-45 to 11-30 a.m.; temperature of sea, 59°;
wind, E.S.E.; light, sky overcast.

Diatoms, many.
Ceratium tripos, few.
Phialidium cymbaloideum, many.
P. temporarium, many.
Sarsia tubulosa, several.
Margelis bella, few.
Mitrocomella polydiadema, several.
Aurelia aurita, many.
Hormiphora, many.
Pleurobrachia, many.
Asterid larvæ, few.
Echinoid plutei, several.
Ophiuroid plutei, many.
Pilidium larva, several.
Mitraria larva, several.
Tornaria larva, several (figs. 7, 9, 10, Pl. VII.).
Trochophores, several.
Post-larval stage of *Polygordius* (?), few.
Post-larval stage of *Pectinaria*, few.
Post-larval stage of *Amphitrite* (?), several.
Post-larval stage of Polynoe, few.
Cyphonautes larva, many.
Actinotrocha larva of *Phoronis*, several.
Nauplii, very few.
Evadne nordmanni, numerous.
Podon intermedium, many.
Calanus finmarchicus, numerous.
Pseudocalanus elongatus, numerous.
Temora longicornis, many.
Centropages hamatus, few.
Acartia clausii, numerous.
Oithona spinifrons, few.

Candacia pectinata, few.

Zoeæ, many.

Oikopleura, few.

A few notes on the prevalent forms in the different months will now be given.

In the only tow-netting taken in January, the most notable forms were a Radiolarian (*Acanthometra*), of which there were many; the Hydromedusa *Hybocodon prolifer*, which, curiously enough, has not again been noted during the year; a Holothurian larva (fig. 1, Pl. VII.), and the three stages of the development of an Asterid larva, figured on the same plate (figs. 4 to 6).

There were also several trochophore and post-larval stages of Polychæta, and single specimens of *Tomopteris onisciformis* and *Autolytus cornutus.* *Sagitta,* recorded in the Eleventh Annual Report as abundant in January, was not seen. There were a few Molluscan veligers and several specimens of *Oikopleura.* Lastly, a single specimen of the curious larva represented on Pl. VI. (fig. 5), was seen, and kept under observation for some time.

In the February tow-nettings the Acanthometrids were well represented, and the Hydromedusæ *Obelia* and the first stage, with four tentacles, of *Phialidium temporarium*, appeared. Bipinnaria and Auricularia larvæ were in force, and a few Echinoid plutei, of two species, were noted. *Sagitta,* the larva of a Bopyrid, Nauplii, and Zoeæ appeared in increasing numbers as the end of the month drew near. Fish eggs appeared in small numbers on February 3rd, and fish larvæ, of two species, were taken on the 23rd.

A tow-netting, taken on March 13th, showed the Radiolarians to be present in diminishing numbers. Several very young specimens of *Hormiphora* appeared, and many Echinoid plutei, of three species, along with small numbers of Auricularia and Bipinnaria larvæ, and segmenting eggs of *Asterina gibbosa*, were seen. One specimen of *Autolytus incertus* was noted. On March 24th Cirripede nauplii were present in such large numbers that the water in the shore pools appeared to be muddy, and the quantity of Diatoms and other Algæ reached its maximum about the same date.

On April 14th two specimens of a minute Planarian worm (Plate VI., fig. 1), and a single specimen of the larva of *Loxosoma*, were noted. Post-larval stages of Polychæta (Pl. VI., figs. 7 and 8), and nauplii of Cirripedia and Copepoda were very numerous. A general increase in numbers was noted on the 27th, and the Hydromedusæ *Saphenia mirabilis* and *Phialidium cymbaloideum* appeared.

On May 23rd a Diatom *(Rhizosolenia)* was present in enormous numbers. On the 29th the Medusæ *Phialidium temporarium*, *P. cymbaloideum*, *Laodice calcarata*, *Sarsia tubulosa*, and *Pilema octopus*, and the Siphonophore *Agalmopsis elegans* were all represented, as also was the interesting larva of *Cerianthus* known as *Arachnactis bournei*. On the same date *Autolytus cornutus* was represented by half-a-dozen specimens, and the post-larval stage of *Pectinaria* was noted for the first time this year. *Sagitta*, represented by a few specimens on the 23rd, was fairly common on the 29th. On the latter date Nauplii of several species were still abundant, and several Megalopas were noted. *Oikopleura* was common.

On June 1st the Hydromedusæ, especially *Phialidium temporarium*, and the Siphonophore *Agalmopsis elegans* were more numerous, and *Pleurobrachia* appeared in great force. *Sagitta* and *Tomopteris* were common. Post-larval stages of Polychæta and Nauplii had almost disappeared, while Zoœa of several species were tolerably common. On this date the Actinotrocha larva of *Phoronis*, one of the most interesting additions to our lists, was seen for the first time. Fish eggs and larvæ were still present in small numbers. Echinoid plutei of two, possibly three, species were present in large numbers in a tow-netting taken shortly before sunset on June 5th, and during the following week Ophiuroid plutei became almost equally numerous. Auricularia and Bipinnaria larvæ reappeared in small numbers. In addition to the Medusæ previously recorded, *Margelis bella*, *Euchilota pilosella*, and *Aurelia aurita* were present in fair numbers, and the medusa of *Corymorpha nutans* and the Actinula larva of *Tubularia* were represented by several specimens of each. *Sagitta* was common, as also was the post-larval stage of *Pectinaria*. *Evadne nordmanni* and *Podon intermedium* were found in increasing numbers in every tow-netting taken during this month. The number of larval forms reached its maximum during the last week in June.

The tow-nettings taken in the early part of July were similar to those of the latter end of June, the only addition to the organisms then recorded being the post-larval stage of *Chætopterus variopedatus*. On July 16th and during the following week *Aurelia aurita* and *Beroe ovata* were present in the Bay in enormous numbers, and *Tiara pileata* was common at the surface during the calm evenings of the following week. A species of *Cyanea* also occurred sparingly at this time.

During the early part of August several of the Hydromedusæ disappeared from the tow-nettings, while *Melicertidium octocostatum* was noted for the first time on the 11th. A species of *Obelia* was then common, and *Agalmopsis elegans* was still represented. Amongst a few Echinoid plutei that of *Echinocyamus pusillus* was noted. On the 19th Ophiuroid plutei were again numerous, and on the 22nd the Zoea and later larval stages of Macrura and Brachyura reached their maximum in numbers.

The early part of September brought unsettled weather, and with it a general diminution in the number of organisms present in the tow-nettings. With the exception of Nauplii, very few larval forms were seen. On the 15th a single specimen of the free-swimming Polychæte, *Gattiola finmarchica*, was taken. *Oikopleura* attained its maximum numbers at the end of this month.

Early in October large numbers of *Pelagia perla* appeared in the Bay, and with them *Pleurobrachia* and *Hormiphora* reappeared, as also did the Radiolarians noted in January and February. *Sagitta* and *Oikopleura* were again common on the 18th. Two specimens of the trochophore represented on Plate VI. (fig. 2) were seen on the 28th, when post-larval stages of two Polychætes were common. Molluscan veligers and Lamellibranch fry were present in small numbers on the same date.

The November tow-nettings revealed a marked increase in the numbers of Diatoms and Molluscan and Tunicate larvæ. The gastrulating blastospheres represented on Plate VII. (figs. 2 and 3) appeared towards the end of the month, as did two stages of the Holothurian larva (fig. 1).

In December the gastrulating blastospheres just mentioned increased in numbers, but no further stages of their development were seen ; and the Holothurian larva again occurred. *Sagitta* was represented on the 17th and 30th by two or three specimens. Several specimens of the trochophore represented on Plate VI. (fig. 3) were noted on the latter date. Post-larval stages of four species of Polychæta had each a few representatives.

Comparison of the foregoing summary with that contained in the Eleventh Annual Report, already cited, shows that, in the main, the various organisms appear and disappear about the same times year after year.

Finally, we print a complete list of the organisms found

in the tow-nettings in 1899, with the times at which they occurred.

Diatoms, throughout the year.

Ceratium tripos, C. fusus, and *C. furca* throughout the year.

Tintinnus, September to December.

Codonella, November.

Dictyocysta, December.

Radiolarians, October to March.

Actinula larva of *Tubularia,* June.

Phialidium temporarium, January and May to August.

P. cymbaloideum, May, June, and July.

Hybocodon prolifer, January.

Sarsia tubulosa, May to August.

Obelia, February to November.

Corymorpha nutans, June and July.

Mitrocomella polydiadema, June and July.

Margelis bella, May to July.

Tiara pileata, July, August, and December.

Saphenia mirabilis, April.

Euchilota pilosella, June to August.

Laodice calcarata, May.

Melicertidium octocostatum, August.

Aurelia aurita, June to August.

Pelagia perla, October.

Pilema octopus, March, May, and October.

Cyanea, July and August.

Agalmopsis elegans, May to August.

Hormiphora, March to November.

Pleurobrachia, June to October.

Beroe ovata, July and August.

Arachnactis bournei, May and June.

Gastrulating blastospheres, probably Asterid, November to March and June.

Bipinnaria and Auricularia larvæ, Feb., Mar., & June.

Echinoid plutei, January, March, June to December.

Pluteus of *Echinocyamus pusillus*, August.

Ophiuroid plutei, February, March, June to Sept.

Holothurian larvæ, January, August, Nov., and Dec.

Pilidium larva, June.

Mitraria larva, June and November.

Tornaria larva, June and July.

Planarian worm, March and April.

Sagitta bipunctata, throughout the year.

Trochophore larvæ, February, June, Nov. to Dec.

Post-larval stages of Polychæta, throughout the year.

Post-larval stage of Polynoe, June to September.

Post-larval stage of *Chætopterus variopedatus*, July.

Post-larval stage of *Pectinaria*, May to July.

Post-larval stage of (?) *Amphitrite*, June and July.

Autolytus cornutus, January, May, July, and Sept.

Autolytus incertus, March.

Gattiola finmarchica, September.

Tomopteris onisciformis, June to January.

Larva of *Loxosoma*, April.

Cyphonautes larva, Jan., Feb., June to Dec.

Actinotrocha larva of *Phoronis*, June to August.

Nauplii of Copepoda, throughout the year.

Nauplii of Cirripedia, March and April.

Pupa stage of Cirripedia, March to May.

Evadne nordmanni, June to September.

Podon intermedium, June to September.

Calanus finmarchicus, throughout the year.

Pseudocalanus elongatus, throughout the year.

Acartia clausii, throughout the year.

Oithona spinifrons, throughout the year.

Euterpe gracilis, Jan., Feb, April, and Oct.

Temora longicornis, March to October.

Centropages hamatus, March to September.
Isias clavipes, May, June, and September.
Corycæus anglicus, June and November.
Candacia pectinata, June to August.
Metridia longa, August.
Anomalocera patersoni, October.
Lernæa branchialis, November.
Monstrilla sp., November.
Larva of Bopyrid, December to February.
Zoeas, January to June, August and November.
Megalopas, May to August.
Molluscan veligers, January to April, Sept. to Dec.
Lamellibranch fry, September to February.
Gastropod fry, January to March, July to October.
Tunicata larvæ, Jan., Feb., July, August, and Nov.
Oikopleura, throughout the year.
Fritillaria, October and November.
Fish eggs, February, March, June.
Fish larvæ, February to July.

It will probably be convenient to readers if the EXPLAN-
ATION OF THE PLATES is given here rather than at the end
of the Report.

PLATE VI.

Fig. 1. Planarian worm, Oct. 28th, 1898.
Fig. 2. Trochophore larva, Oct. 28th, 1898.
Fig. 3. Trochophore larva, showing early stage of meta-
         meric segmentation, Dec. 17th and 30th, 1898.
Fig. 4. Trochophore-like larva, with three bands of cilia.
         Pale-green, highly refractive bodies in interior.
         Dec. 30th, 1898, and Jan. 24th, 1899.
Fig. 5. Unknown larva. Organ *o* is paired, and the
         granules in the interior vesicle were in active
         rotation. Jan. 24th, 1899.

Fig. 6. Trochophore larva, Dec. 30th, 1898, and Feb. 3rd, 1899.

Figs. 7 and 8. Post-larval stages of Polychæta, of which many were taken during April, 1899.

PLATE VII.

Fig. 1. Holothurian larva, Jan. 24th, 1899.

Fig. 2. Echinoderm Blastosphere, at close of segmentation, Dec. 17th, 1898.

Fig. 3. Gastrula stage of fig. 2, in optical section. *a.c.*, Amœboid cells budded from the invagination.

Fig. 4. Asterid Gastrula, Jan. 24th, 1899.

Fig. 5. Later stage of fig. 4, showing the archenteron, *ar.*

Fig. 6. Still later stage, from ventral surface. *ar.*, archenteron, from which the enterocœlic vesicles, *v.p.d.*, are derived. The water vascular rosette is derived, according to Metschnikoff, from the left vesicle, that on right of figure.

Fig. 7. Tornaria larva of *Balanoglossus*, June 5th, 1899.

Fig. 8. Bipinnaria larva, Feb. 23rd, 1899, from dorsal surface. *œ.*, Oesophagus; *a.*, Anus; *hy.*, Hydropore; *v.p.d.*, anterior enterocœlic vesicles.

Figs. 9 and 10. Tornaria larvæ of *Balanoglossus*, June 23rd, 1899. *p.p.*, Proboscis pore; *p.c.*, Proboscis cavity. Figure 10 is most probably identical with the Tornaria figured by Mr. G. C. Bourne (Journ. Mar. Biol. Assoc., vol. I., p. 63, Pl. VIII., fig. 13).

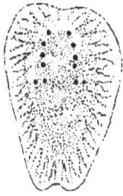

Fig. 1.

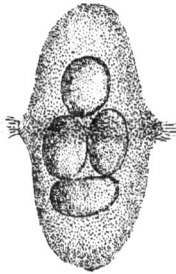

Fig. 2.

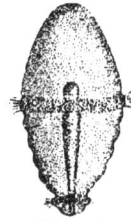

Fig. 3.

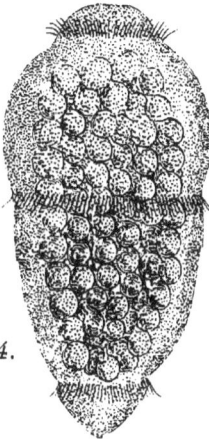

Fig. 4.

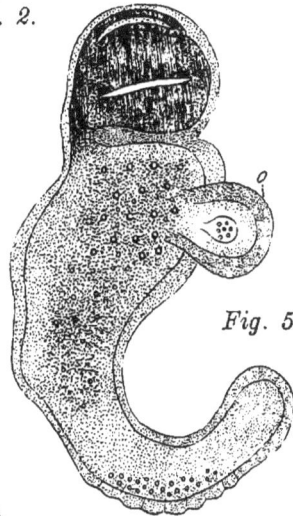

Fig. 5.

Fig. 6.

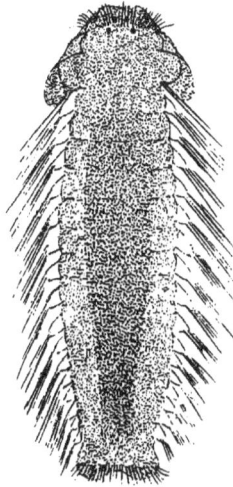

Fig. 7.

100ᵗʰˢ of an inch.

Scale for figs. 1 to 3 & 6 to 8.

100ᵗʰ of an inch.   Fig. 8.

Scale for figs. 4 & 5.

H.C.C. del.

Fig. 1.

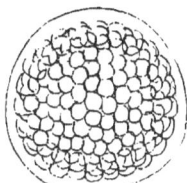

Fig. 2.

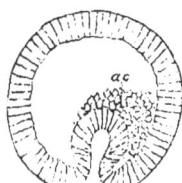

Fig. 3.

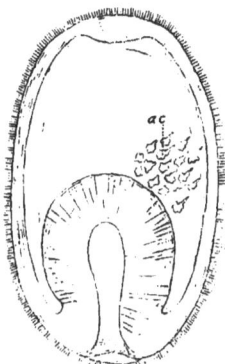

Fig. 4.

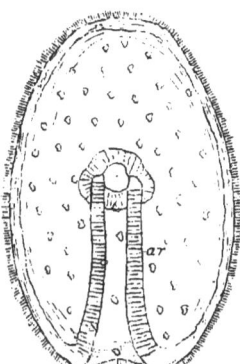

Fig. 5.

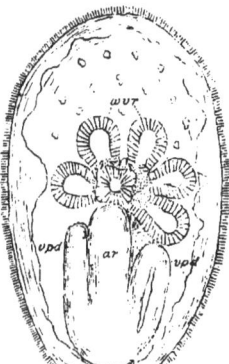

Fig. 6.

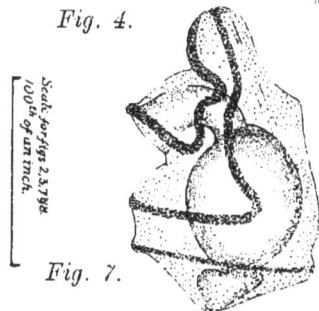

Fig. 7.

Scale for figs 2,3,7,9.
100th of an inch.

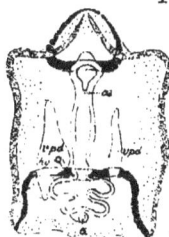

Fig. 8.

Scale for figs 4,5,10.
100th of an inch.

Scale for figs 9,10.
100th of an inch.

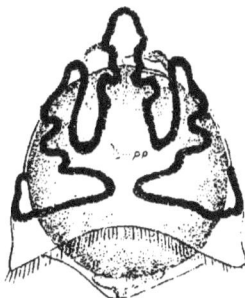

J.C.C. del

Fig. 9.

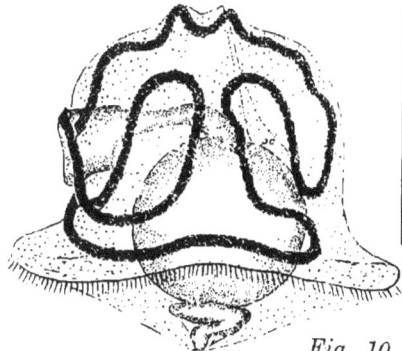

Fig. 10.

PUBLICATIONS.

During the past year the following papers dealing with biological work in the L.M.B.C. District have been published :—

1. The Twelfth Annual Report (Trans. Biol. Soc., vol. XIII., pp. 21—65), with illustrations, 1898.

2. The Sea-Fisheries Laboratory Report for 1898 (Trans. Biol. Soc., vol. XIII., pp. 69—155, and Plates I. and II). By W. A. Herdman, Andrew Scott and James Johnstone, with additional Articles by R. L. Ascroft, C. A. Kohn, F. W. Gamble and F. W. Keeble, 1899.

3. List of the Araneida of Port Erin and District. By A. R. Jackson (Trans. Biol. Soc., vol. XIII., pp. 66—68), 1899.

4. Note on Mid-Winter Surface and Deep Tow-Nettings in the Irish Sea. By I. C. Thompson (Trans. Biol. Soc., vol. XIII., pp. 156—162), 1899.

5. Lancashire Sea-Fisheries Memoir, No. I. Oysters and Disease. By W. A. Herdman and R. Boyce; pp. 60, and eight Plates; quarto. G. Philip and Sons, London and Liverpool, 1899.

6. L.M.B.C. Memoirs, No. I. Ascidia. By W. A. Herdman; pp. viii. and 60, and five Plates; octavo. Dobb and Co., Liverpool, 1899.

This last item brings us to the consideration of our last new series of publications :—

THE L.M.B.C. MEMOIRS.

It will be remembered that in last year's Report I proposed a scheme for the issue of a series of special Memoirs upon typical British marine plants and animals.

In our twelve years experience of a Biological Station (five years at Puffin Island and seven at Port Erin), where college students and young amateurs formed a large proportion of the workers, the want has been constantly felt of a series of detailed descriptions of the structure of certain common typical animals and plants, chosen as representatives of their groups, and dealt with by specialists. The same want has probably been felt in other similar institutions and in many College laboratories.

The suggestion has been received so cordially that we have been encouraged to carry out the scheme without delay, and the first papers of the series are now being issued. They will be called the "L.M.B.C. Memoirs," each will treat of one type, and they will be issued separately as they are ready, and will be obtainable Memoir by Memoir as they appear, or later bound up in convenient volumes. It is hoped that such a series of special studies, written by those who are thoroughly familiar with the forms of which they treat, will be found of value by students of Biology in our laboratories and in Marine Stations, and will be welcomed by many others working privately at Marine Natural History.

It is proposed that the forms selected, should, as far as possible, be common L.M.B.C. (Irish Sea) animals and plants of which no adequate account already exists in any text-book. Probably most of the specialists who have taken part in the L.M.B.C. work in the past, will prepare accounts of one or more representatives of their groups. The following have already promised their services, and in some cases the Memoir is already far advanced. The first Memoir (No. I., Ascidia) was published in October, price 1s. 6d.; the second (No. II., Cardium) is now in type, and will be issued in a few weeks; the third (No. III., Echinus) will appear before the end of 1899. A couple of Botanical Memoirs will, it is hoped, be ready early in 1900, and others will follow in rapid succession.

The complete list, so far as arranged, is as follows :—

Memoir I. ASCIDIA, W. A. Herdman, 60 pp., 5 Pl., 1/6.
    II. CARDIUM, J. Johnstone, 92 pp., 7 Pl., 2/-.
    III. ECHINUS, H. C. Chadwick, ... pp., 3 Pl., 1/-.
    IV. CODIUM, R. J. H. Gibson and Helen Auld.
        HIMANTHALIA, C. E. Jones.
        PERIDINIAN, George Murray & F. G. Whitting.
        ALCYONIUM, S. J. Hickson.
        DENDRONOTUS, J. A. Clubb.
        ZOSTERA, R. J. Harvey Gibson.
        DIATOMS, F. E. Weiss.
        GIGARTINA, O. V. Darbishire.
        PLAICE, F. J. Cole and J. Johnstone.
        CUTTLE-FISH, W. E. Hoyle.
        OSTRACOD, Andrew Scott.
        BOTRYLLOIDES, W. A. Herdman.
        PATELLA, J. R. Ainsworth Davis.
        CALANUS, I. C. Thompson.
        ACTINIA, J. A. Clubb.
        BUGULA, Laura R. Thornely.
        CALCAREOUS SPONGE, R. Hanitsch.
        PORPOISE, A. M. Paterson.
        ARENICOLA, J. H. Ashworth.
        OYSTER, W. A. Herdman and J. T. Jenkins.

I desire now to acknowledge that a generous donation from Mr. F. H. Gossage, of Woolton, has met the
expense of preparing the plates in illustration of the first few
Memoirs, and so has enabled the Committee to commence
the publication of the series sooner than would otherwise
have been possible. At the end of last year's Report, after
stating my suggestion as to the L.M.B.C. Memoirs, I
made an appeal for funds to meet the expense. in the
course of which I happened to say :—" Liverpool is the

right and natural place for a School of Marine Biology, and I hope that Liverpool will consider that it is creditable to the city that such local researches should be published by a Liverpool Society. An addition of about £100 a year to our funds is necessary in order to enable us to do justice to the work now being produced by our colleagues and students."

Shortly afterwards I received from Mr. F. H. Gossage a most welcome and gratifying letter, enclosing a cheque for £100 to be used at my discretion for the benefit of the Marine Biological work. A part of this kind gift has been devoted to the purchase of much-needed books and instruments, and the remainder is now meeting the expense of the plates in these L.M.B.C. Memoirs.

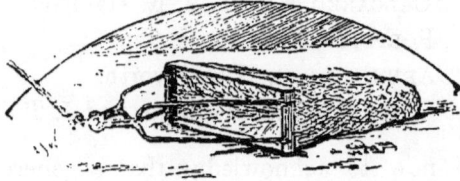

APPENDIX A.

THE LIVERPOOL MARINE BIOLOGY
COMMITTEE (1899).

R. D. DARBISHIRE, Esq., B.A., F.G.S., Manchester.

PROF. R. J. HARVEY GIBSON, M.A., F.L.S., Liverpool.

HIS EXCELLENCY LORD HENNIKER, Governor of the Isle
of Man.

PROF. W. A. HERDMAN, D.Sc., F.R.S., F.L.S., Liverpool,
Chairman of the L.M.B.C., and Hon. Director of
the Biological Station.

W. E. HOYLE, Esq., M.A., Owens College, Manchester.

P. M. C. KERMODE, Esq., Secy. Nat. Hist. Soc., Ramsey,
Isle of Man.

A. LEICESTER, Esq., formerly of Liverpool.

SIR JAMES POOLE, J.P., Liverpool.

DR. ISAAC ROBERTS, F.R.S., formerly of Liverpool.

I. C. THOMPSON, Esq., F.L.S., Liverpool, Hon. Treasurer.

A. O. WALKER, Esq., F.L.S., J.P., Maidstone.

CONSTITUTION OF THE L.M.B.C.

(Established March, 1885.)

I.—The OBJECT of the L.M.B.C. is to investigate the
Marine Fauna and Flora (and any related subjects such
as submarine geology and the physical condition of the
water) of Liverpool Bay and the neighbouring parts of
the Irish Sea; and, if practicable, to establish and maintain
a Biological Station on some convenient part of the
coast.

II.—The COMMITTEE shall consist of not more than 12 and not less than 10 members, of whom 3 shall form a quorum ; and a meeting shall be called at least once a year for the purpose of arranging the Annual Report, passing the Treasurer's accounts, and transacting any other necessary business.

III.—During the year the AFFAIRS of the Committee shall be conducted by an HON. DIRECTOR, who shall be Chairman of the Committee, and an HON. TREASURER, both of whom shall be appointed at the Annual Meeting and shall be eligible for re-election.

IV.—Any VACANCIES on the Committee, caused by death or resignation, shall be filled by the election at the Annual Meeting, of those who, by their work on the Marine Biology of the district, or by their sympathy with science, seem best fitted to help in advancing the work of the Committee.

V.—The EXPENSES of the investigations, of the publication of results, and of the maintenance of the Biological Station shall be defrayed by the Committee, who, for this purpose, shall ask for subscriptions or donations from the public, and for grants from scientific funds.

VI.—The BIOLOGICAL STATION shall be used primarily for the Exploring work of the Committee, and the SPECIMENS collected shall, so far as is necessary, be placed in the first instance at the disposal of the members of the Committee and other specialists who are reporting upon groups of organisms ; work places in the Biological Station may, however, be rented by the week, month, or year to students and others, and duplicate specimens which, in the opinion of the Committee, can be spared may be sold to museums and laboratories.

LIVERPOOL MARINE BIOLOGICAL STATION
at PORT ERIN.

LABORATORY REGULATIONS.

I.—This Biological Station is under the control of the Liverpool Marine Biological Committee, the executive of which consists of the Hon. Director (Prof. Herdman, F.R.S.) and the Hon. Treasurer (Mr. I. C. Thompson, F.L.S.).

II.—In the absence of the Director, and of all other members of the Committee, the Station is under the temporary control of the Resident Curator or Laboratory Assistant, who will keep the keys, and will decide, in the event of any difficulty, which places are to be occupied by workers, and how the tanks, boats, collecting apparatus, &c., are to be employed.

III.—The Resident Curator will be ready at all reasonable hours and within reasonable limits to give assistance to workers at the Station, and to do his best to supply them with material for their investigations.

IV.—Visitors will be admitted, on payment of a small specified charge, to see the Aquarium and the Station, so long as it is found not to interfere with the scientific work. Occasional lectures are given by members of the Committee.

V.—Those who are entitled to work in the Station, when there is room, and after formal application to the Director, are :—(1) Annual Subscribers of one guinea or upwards to the funds (each guinea subscribed entitling to the use of a work place for three weeks), and (2) others who are not annual subscribers, but who pay the Treasurer 10s. per week for the accommodation and privileges. Institutions, such as Colleges and Museums, may become subscribers in order that a work place may be at the disposal of their staff for a certain period annually ; a subscription of two guineas will secure a work place for six weeks in the year, a subscription of five guineas for four months, and a subscription of £10 for the whole year.

VI.—Each worker* is entitled to a work place opposite a window in the Laboratory, and may make use of the microscopes, reagents, and other apparatus, and of the boats, dredges, tow-nets, &c., so far as is compatible with the claims of other workers, and with the routine work of the Station.

VII.—Each worker will be allowed to use one pint of methylated spirit per week free. Any further amount required must be paid for. All dishes, jars, bottles, tubes, and other glass may be used freely, but must not be taken away from the Laboratory. Workers desirous of making, preserving, or taking away collections of marine animals and plants, can make special arrangements with the Director or Treasurer in regard to bottles and preservatives. Although workers in the Station are free to make their own collections at Port Erin, it must be clearly understood that (as in other Biological Stations) no specimens must be taken for such purposes from the Laboratory stock, nor from the Aquarium tanks, nor from the steam-boat dredging expeditions, as these specimens are the property of the Committee. The specimens in the Laboratory stock are preserved for sale, the animals in the tanks are for the instruction of visitors to the Aquarium, and as all the expenses of steam-boat dredging expeditions are defrayed by the Committee, the specimens obtained on these occasions must be retained by the

---

* Workers at the Station can always find comfortable and convenient quarters at the closely adjacent Bellevue Hotel ; but lodgings can readily be had by those who prefer them.

Committee (*a*) for the use of the specialists working at the Fauna of Liverpool Bay, (*b*) to replenish the tanks, and (*c*) to add to the stock of duplicate animals for sale from the Laboratory.

VIII.—Each worker at the Station is expected to lay a paper on some of his results—or at least a short report upon his work—before the Biological Society of Liverpool during the current or the following session.

IX.—All subscriptions, payments, and other communications relating to finance, should be sent to the Hon. Treasurer, Mr. I. C. Thompson, F.L.S., 53, Croxteth Road, Liverpool. Applications for permission to work at the Station, or for specimens, or any communication in regard to the scientific work should be made to Professor Herdman, F.R.S., University College, Liverpool.

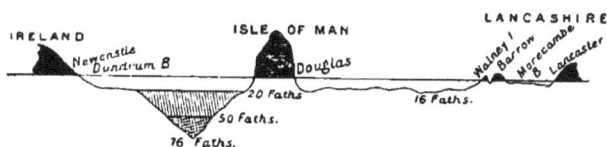

Diagrammatic Section Across the Irish Sea.

APPENDIX B.

HON. TREASURER'S STATEMENT.

From the balance sheet appended it will be seen that the year again ends with a small debit balance, although all due economy has been observed, consistent with efficiency.

The British Association (1896) fund has, to a considerable extent, enabled the Committee to continue the extra expense involved in securing the services of a competent Resident Curator at the Port Erin Station, which has already proved of marked benefit to workers and students, the number of whom have thereby considerably increased.

The appeal made by the Director in his last year's Report for an additional income of about £100 per annum, to be utilised in publishing well-illustrated papers and memoirs, embodying the results of local biological investigations, has been responded to by Mr. Gossage by a donation for the present year. This fund is being separately applied for the publication of a valuable series of well-illustrated Biological Memoirs now appearing, the accounts of which will be published later.

The Hon. Treasurer will be glad to receive the names of new subscribers, with the view of continuing these publications, and further adding very materially to the already excellent work achieved under the auspieces of the L.M.B.C. since its foundation, fifteen years ago.

SUBSCRIPTIONS and DONATIONS.

	Subscriptions.			Donations.		
	£	s.	d.	£	s.	d.
Ayre, John W., Ripponden, Halifax ...	1	1	0	—		
Bateson, Alfred, Harrop-road Bowdon ...	1	1	0	—		
Beaumont, W. I., Cambridge	1	1	0	—		
Bickerton, Dr., 88, Rodney-street ...	1	1	0	—		
Bickersteth, Dr., 2, Rodney-st.	2	2	0	—		
Brown, Prof. J. Campbell, Univ. Coll. ...	1	1	0	—		
Browne, Edward T., B.A., 141, Uxbridge-						
road, Shepherd's Bush, London ...	1	1	0	—		
Brunner, Sir J. T., Bart., M.P., L'pool.	5	0	0	—		
Boyce, Prof., University College ...	1	1	0	—		
Buckley, W., Hulme, Manchester	0	10	0	—		
Caton, Dr., 86, Rodney-street	—			1	1	0
Clague, Dr., Castletown, Isle of Man ...	1	1	0	—		
Clubb, J. A., Public Museums, Liverpool	0	10	6	—		
Cole, F. J., (research table)	1	1	0	—		
Coombe, John N., 4, Paradise-sq., Sheffield	1	1	0	—		
Comber, Thomas, J.P., Leighton, Parkgate	1	1	0	—		
Corker, J., Moss Side, Manchester ...	0	10	0	—		
Crellin, John C., J.P., Andreas, I. of Man	0	10	6	—		
Crossland, Cyril, Clare College, Cambridge	2	10	0	—		
Dickinson, Dr., 8, Croxteth-rd., Liverpool	—			2	2	0
Gair, H. W., Smithdown-rd., Wavertree	2	2	0	—		
Gamble, Col. C.B., Windlehurst, St. Helens	2	0	0	—		
Gamble, F.W., Owens College, Manchester	1	1	0	—		
Gaskell, Frank, Woolton Wood	1	1	0	—		
Gaskell, Holbrook, J.P., Woolton Wood	1	1	0	—··		
Gibson, Prof. R. J. Harvey, Waterloo ...	1	1	0	—		
Gotch, Prof., Museum, Oxford	1	1	0	—		

Forward ...£32 11 0   3   3   0

	Subscriptions.			Donations.		
	£	s.	d.	£	s.	d.
Forward ...	32	11	0	3	3	0
Halls, W. J., 35, Lord-street	1	1	0	—		
Hanitsch, Dr., Museum, Singapore ...	1	1	0	—		
Harmer, F. W., Cringleford, Norwich...	1	1	0	—		
Harrison, H. S., Univ. Coll., Cardiff ...	0	10	0	—		
Henderson, W.G., Liverpool Union Bank	1	1	0	—		
Herdman, Prof., University College ...	2	2	0	—		
Hewitt, David B., J.P., Northwich ...	1	1	0	—		
Holland, Walter, Mossley Hill-road	2	2	0	—		
Holt, Alfred, Crofton, Aigburth	2	2	0	—		
Holt, Mrs. George, Sudley, Mossley Hill	1	0	0	—		
Holt, R. D., 54, Ullet-road, Liverpool ...	2	0	0	—		
Hoyle, W. E., Museum, Owens College	1	1	0	—		
Isle of Man Natural History Society ...	1	1	0	—		
Jarmay, Gustav, Hartford	1	1	0	—		
Jenkins, J. T., Univ. Coll., Liverpool ...	3	10	0	—		
Jones, C.W.,J.P., Field House,Wavertree	1	0	0	—		
Kermode, P. M. C., Hill-side, Ramsey...	1	1	0	—		
Lea, Rev. T. Simcox, 3, Wellington-fields	1	1	0	—		
Leicester, Alfred, Aston Clinton, Bucks...	1	1	0	—		
Macfie, Robert, Airds	1	0	0	—		
Mackay, J. R., Longsight, Manchester ...	1	1	0	—		
Meade-King, H. W.,J.P., Sandfield Park	1	1	0	—		
Meade-King, R. R., 4, Oldhall-street ...	0	10	0	—		
Melly, W. R., 90, Chatham-street ...	1	1	0	—		
Monks, F. W., Brooklands, Warrington	1	1	0	—		
Muspratt, E. K., Seaforth Hall	5	0	0	—		
Newton, John, M.R.C.S., 44, Rodney-st.	0	10	6	—		
Okell, Robert, B.A., Sutton, Douglas ...	1	1	0	—		
Paterson, Prof., University College ...	1	1	0	—		
Poole, Sir James, Liverpool	2	2	0	—		
Rathbone, Mrs. Theo., Backwood, Neston	1	1	0	—		
Forward ...	£74	17	6	3	3	0

	Subscriptions.			Donations.		
	£	s.	d.	£	s.	d.
Forward ...	74	17	6	3	3	0
Rathbone, Miss May, Backwood, Neston	1	1	0	2	2	0
Rathbone, W., Greenbank	2	2	0		—	
Roberts, Isaac, F.R.S., Crowborough ...	1	1	0		—	
Schuster, E. H., New College, Oxford ...	2	10	0		—	
Simon, R., Broughton, Manchester ...	0	10	0		—	
Simpson,J.Hope, Annandale, Aigburth-dr	1	1	0		—	
Smith, A. T., junr., 24, King-street ...	1	1	0		—	
Sollas, Miss, Oxford	1	1	0		—	
Talbot, Rev. T. U., Douglas, I. of Man...	1	1	0		—	
Thompson, Isaac C., 53, Croxteth-road	2	2	0		—	
Thornely, The Misses, Aigburth-Hall-Rd.	1	1	0		—	
Timmis, T. Sutton, Cleveley, Allerton ...	2	2	0		—	
Toll, J. M., Kirby Park, Kirby	1	1	0		—	
Torrance, Gilbert, North Quay, Douglas	1	1	0		—	
Walker, A. O., Maidstone	3	3	0		—	
Walker, Horace, South Lodge, Princes-pk.	1	1	0		—	
Walters, Rev. Frank, B.A., the late ...	1	1	0		—	
Watson, A. T., Tapton-crescent, Sheffield	1	1	0		—	
Weiss, Prof. F. E.,Owen's College, Man'tr.	1	1	0		—	
Westminster, Duke of, Eaton Hall ...	5	0	0		—	
Wiglesworth, Dr., Rainhill	1	1	0		—	
Yates, Harry, 75, Shude-hill, Manchester	1	11	0		—	
	£108	11	6	5	5	0

SUBSCRIPTIONS FOR THE HIRE OF COLLEGE " WORK-TABLES."

Owens College, Manchester ...			£10	0	0
University College, Liverpool			10	0	0
			£20	0	0

THE LIVERPOOL MARINE BIOLOGY COMMITTEE.

In Account with ISAAC C. THOMPSON, Hon. Treasurer.

Dr.

1899.	£	s.	d.
To Balance due Treasurer, Dec. 31st, 1898	6	8	3
,, Printing Reports and Engraving Plates	17	7	9
,, Printing and Stationery	0	12	6
,, Expenses of Dredging Expeditions	26	13	8
,, Boat Hire	1	16	9
,, Books and Apparatus at Port Erin Biological Station	21	3	1
,, Postage, Carriage of Specimens, &c.	4	6	3
,, Salary, Curator	90	0	0
,, Rent of Port Erin Biological Station	15	0	0
,, New Shellbend Boat	9	0	0
,, Sundries	2	5	5
	£194	13	8

Cr.

1899.	£	s.	d.
By Subscriptions and Donations received	114	17	6
,, Amount received from Colleges, &c., for hire of "Work Tables"	20	0	0
,, Dividend, British Workman's Public House Co., Ltd., Shares	7	8	6
,, Sale of Nat. Hist. Specimens	2	7	8
,, Sale of Reports and Volumes of Fauna	5	3	1
,, Interest on British Association (1896) Fund	38	13	4
,, Bank Interest	0	3	3
,, Admissions to Aquarium	3	16	6
,, Balance due Treasurer, Dec. 31st, 1899	2	3	10
	£194	13	8

Endowment Investment Fund :—
British Workmans' Public House Co's. shares £173 1 0

Audited and found correct,

A. T. SMITH, JUNR.

ISAAC C. THOMPSON,
Hon. Treasurer.

LIVERPOOL, *December 31st*, 1899.

SOME RECENT ADDITIONS to the COPEPODA of
LIVERPOOL BAY.

By I. C. THOMPSON, F.L.S., and ANDREW SCOTT.

With Plate VIII.

[Read November 10th, 1899.]

THE collection and examination of tow-net and dredged
material, which is carried on more or less continuously at
Port Erin, Isle of Man, and in the vicinity of Piel, Lancashire, still proves a means of adding fresh names to the
extensive list of species from the district, which was
published in connection with the meeting of the British
Association in Liverpool in 1896. That list represents all
the species of marine animals and plants recorded by the
workers of the L.M.B.C. during the first ten years of its
history, and practically brings us down to the end of
1896. We now wish to place on record the various additions
to the list of Copepoda that have come under our notice
during the three years that have passed since the complete
list was published.

The present report represents an addition of fifteen
species new to the district, including one new to science,
Leptopsyllus herdmani (Pl. VIII.), and five species of
Copepod fish parasites. Of these, two are open sea free-
swimmers, the other (non-parasitic) species have either been
obtained from dredged material or by digging holes in the
shore between tide marks, the water being allowed to collect,
and then bailed out, pouring it out through a fine sieve.

1. *Candacia pectinata*, Brady, Copepoda of the British
   Islands, vol. I., p. 49, Pl. VIII. and X., 1878.

   This species was taken in the open sea near Port Erin
in January last, and on several occasions during June and

July, 1899, both at or near the surface, as well as at a depth of 33 fathoms. It appears to be generally, but very sparingly, distributed about the British Isles, and throughout the Atlantic, Pacific, and Indian Oceans.

It and the other species of the genus *Candacia* are easily distinguishable by their dark-coloured antennules, spines and plumes, and the terminal spines of the swimming feet.

2. *Corycœus anglicus*, Lubbock, On Eight New Species of Entomostraca found at Weymouth, Ann. and Mag. of Nat. Hist., 2nd Ser., vol. XX., Pl. XI., 1857.

A single specimen of this species was taken by tow-net off Port Erin on November 26th, 1898, and a shoal of it was captured there on May 29th, 1899. It is a fairly common species about the south and south-west coasts of England and Ireland, and Mr. Thomas Scott has reported it from the Forth, and more recently, we understand, he took it in the Clyde, but it is entirely new to the L.M.B.C. district.

3. *Ameira exilis*, T. and A. Scott, Ann. and Mag. Nat. Hist., Ser. 6, vol. XIII., Feb., 1894. Length 1·5 mm.

This slender and characteristic species was taken amongst material collected from holes dug in the soft mud near the remains of the old steamboat pier, Piel; not uncommon. March, 1899.

4. *Stenhelia intermedia*, T. Scott, 15th Ann. Rept. Scot. Fish. Board, part 3, p. 169. Length ·7 mm.

In the same locality as the last. Aug., 1898; rare.

5. *Delavalia mimica*, T. Scott, 15th Ann. Rept. S.F.B.. part 3, p. 150. Length ·65 mm.

This peculiar little species occurred sparingly in material dredged from a depth of 33 faths. off the Isle of Man. Jan. 29, 1899.

6. *Laophonte denticornis*, T. Scott, 12th Ann. Rept.
S.F.B., part 3, p. 246. Length ·85 mm.

A slender species, not unlike *L. serrata* at first sight,
but on closer examination, is seen to be quite distinct.
From the same gathering as the last, which contained 31
species of Copepoda.

7. *Leptopsyllus intermedius*, T. and A. Scott, Ann. and
Mag. Nat. Hist., Ser. 6, vol. XV., 1895. Length ·57 mm.

This little species belongs to a curious genus, the members of which, from the structure of their appendages,
appear to live entirely amongst the mud and sand, and
have not apparently so far been obtained at any depth.
In material collected from holes dug in the sand in front
of the Laboratory at Port Erin ; common. Jan., 1897.

Leptopsyllus herdmani, n. sp., Pl. VIII., figs. 1—9.

Description of the female :—Length, exclusive of caudal
setæ, ·65 mm. ($\frac{1}{42}$ of an inch). Body, seen from above,
elongate and moderately robust. Antennules short and
stout, composed of eight joints, of which the sixth and
seventh are the shortest, as shown by the following
formula—

Proportional length of

the joints	34	. 22	. 18	. 14	. 10	. 8	. 7	. 15
Number of joints	1	2	3	4	5	6	7	8

Antennæ and mouth organs (figs. 1—4) nearly similar to
those of *L. robertsoni*, T. and A. Scott. Both branches of
the first pair of swimming feet (fig. 5) composed of two
joints, the basal joint of the inner branch reaches considerably beyond the end of the outer branch ; second
joint very short, being only about one-sixth the length of
the basal joint, and having at its apex one short and one
moderately long seta. The second joint of the outer
branch is furnished with four moderately long setæ, all

placed on the apex. Second (fig. 6) and third pairs of swimming feet nearly alike, the outer branches being composed of three joints, and the inner of one only, which has a slight constriction on its outer margin near the apex. In the fourth pair (fig. 7) the outer branch is also composed of three joints, but the inner branch is distinctly two-jointed, there being a small apical joint carrying one seta, and a moderately long basal joint. The fifth pair of feet (fig. 8) consists of one broad lamelliform shield, due to the complete coalescence of the basal joint of each pair. The end of the coalesced basal joints is rounded, with a slight concavity in the middle, on either side of which are two short setæ. The length of the plate is equal to nearly two-thirds of the breadth at the widest part; outer branches very small, longer than broad, and bearing one marginal and two terminal setæ. Abdomen composed of four segments, the first or genital segment being the largest, the second or third segments are nearly of equal length, the fourth segment is very small, being about half the length of the second or third. Caudal stylets (fig. 9) short and stout, the breadth being equal to about two-thirds of the length. There are three setæ on the extremity of the inner angle, and one on the outer angle, with an intervening space. There is also one small seta on the outer margin near the base. The male is yet unknown. In the same gathering as the last; not common.

This new member of the genus *Leptopsyllus* is easily distinguished from previously described species by the elongate joint of the inner branch of the first pair of swimming feet, and by the structure of the fifth pair.*

We have much pleasure in naming this species after Professor Herdman.

---

* This species has recently been taken in the Clyde, off Millport, by T. Scott.

9. *Lichomolgus hirsutipes*, T. Scott, 11th Ann. Rept. S.F.B., part 3, p. 206.   Length 1·4 mm.

This well-marked species was obtained from collections made in the Zostera beds near Piel.   July, 1899.

10. *Hersiliodes littoralis* (T. Scott), 10th Ann. Rept. S.F.B., part 3, p. 260.   Length 1·1 mm.

This species, which is readily recognised by the structure of the fifth pair, occurred sparingly in gatherings made on the mud flats near Piel.   June and August, 1899.

11. *Caligus diaphanus*, Nordmann, Microgr. Beitr., 11, 26.

A parasitic Copepod on the Cod.   April 19th, 1897.

12. *Caligus gurnardi*, Kr., Bidrag til Kundskab om Snyltek- rebsene, p. 150, Pl. II., fig. 3, *a—g*.

One specimen was taken in the dredge, at 26 faths., off Port Erin Bay.   Nov. 4th, 1898.

13. *Trebius caudatus*, Kr., Tidsskrift., II. ; 30, t. I., fig. 4.

A parasitic Copepod on the Hake.   April 19th, 1897. This species and *Caligus diaphanus* have already been referred to in the 11th Ann. Rept. of the Marine Biologi- cal Station, Port Erin, p. 11.

14. *Chondracanthus radiatus*, (see Bidrag Kundskab om Snyltekrebsene of Kroyer).

Found on the fins of a Codfish.   Dec. 6th, 1897.

Nicothæ astaci, Aud. & M. Edw., Ann. Sc. Nat., Ser. I., vol. IX., Pl. XLIX.

This peculiar parasitic Copepod, which has all its appendages fully developed, is found occasionally in con- siderable numbers on the branchiæ of the common lobster, caught on our coasts.   We have noted its occurrence on lobsters from Holyhead, Port Erin, and Piel.   The wing- like projections of the fourth thoracic segment give it an unusual appearance.

Although this species does not appear to have been taken apart from its host, yet it looks quite capable of leading a free existence.

EXPLANATION OF THE PLATE.

Leptopsyllus herdmani, n. sp.

Fig. 1.	Female seen from above.	× 106.
Fig. 2.	Antennule of female.	× 500.
Fig. 3.	Antenna.	× 500.
Fig. 4.	Posterior foot-jaw.	× 500.
Fig. 5.	First pair of swimming feet.	× 500.
Fig. 6.	Second ,, ,,	× 500.
Fig. 7.	Fourth ,, ,,	× 500.
Fig. 8.	Fifth ,, ,,	× 500.
Fig. 9.	Last two segments of abdomen and caudal stylets.	× 260.

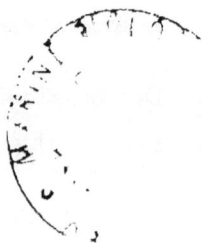

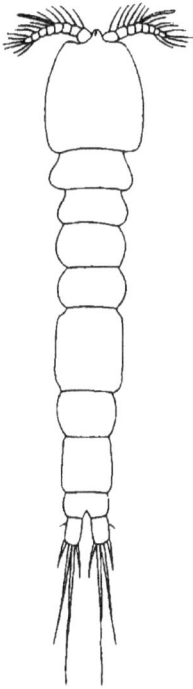

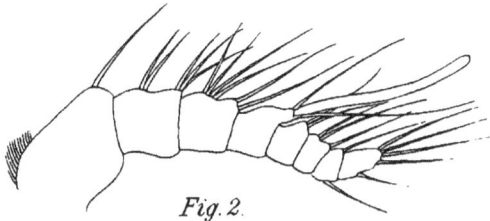

Fig. 2.

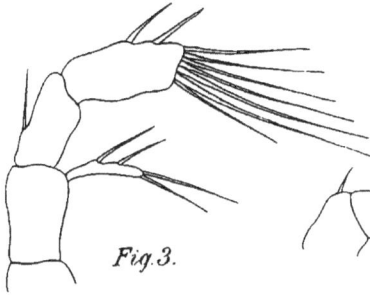

Fig. 3.

Fig. 4.

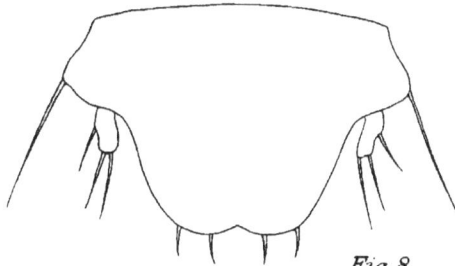

Fig. 1.

Fig. 8.

Fig. 6.

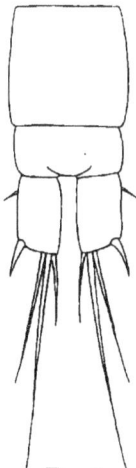

Fig. 5.

Fig. 9.

Fig. 7.

S. J. H. del.

S. B. Lith.

LEPTOPSYLLUS HERDMANI, n. sp.

LIST OF THE SPECIES RECORDED FROM THE IRISH SEA AREA.

The species in this list are given in zoological order, commencing with the Algæ and the Protozoa, and each name is followed by a brief reference to the volume and page of the.L.M.B.C. publications in which the species was recorded or described. The following contractions have been made use of :—The four published volumes of the 'Fauna of Liverpool Bay' are indicated as i., ii., iii., iv. The L.M.B.C. 'Annual Reports' are indicated as 1st to 10th A.R. The 'Transactions' of the Liverpool Biological Society are referred to as T.L.B.S., I., &c. Species which have been found recently, but of which the record has not yet been published, are followed by 10th A.R. to indicate the Annual Report which will appear in December 1896.

The Committee are indebted to some of the Liverpool Marine Biology Committee and other naturalists, who have worked at Port Erin, and have written reports upon the marine fauna, for compiling or supervising the compilation of the following lists :—

LIST OF THE DIATOMACEA.

[See Report by HENRY STOLTERFOTH, M.D., &c., in ' Fauna,' vol. ii.]

Achnanthes brevipes, Ag.
  A. longipes, Ag.
  A. subsessilis, Ehr.
Actinocyclus crassus, W. Sm.
  A. Ralfsii, W. Sm.
Actinoptychus splendens (Shad), Ralfs.
  A. undulatus, Ehr.
Amphiprora alata, Kütz.
  A. paludosa, Greg.
  A. plicata, Greg.
  A. pusilla, Greg.
  A. vitrea, Greg.
Amphora affinis, Kütz.
  A. binodis, Greg.
  A. commutata, Grun.
  A. complexa, Greg.
  A. hyalina, Kütz.
  A. lævis, Greg.
  A. litoralis, Dn.
  A. membranacea, W. Sm.
  A. minutissima, Gray
  A. salina, W. Sm.
  A. spectabilis, Greg.
  A. ventricosa, Greg.
Asterionella Bleakleyii, W. Sm.
  A. Ralfsii, W. Sm.
Atheya decora, West.
Bactereastrum varians, Lauder.
Berkleya obtusa, Grev.
Biddulphia aurita, Breb.
  B. Baileyii, W. Sm.
  B. obtusa, Kütz.
  B. granulata, Roper.
  B. radiatus, Greg.
  B. rhombus, W. Sm.
  B. suborbicularis, Grun.
  B. turgida, W. Sm.

Campylodiscus bicostatus, W. Sm.
  C. cribrosus, W. Sm.
Cestodiscus johnsonianum, Greg.
Chœtoceros armatum, West.
  C. boreale, Bail.
  C. paradoxum, Cleve.
  C. Wighamii, Brightw.
Cocconeis scutellum.
  C. britannica, Nægeli.
  C. eccentrica, Dn.
Coscinodiscus asteromphalus, Grun.
  C. concinnus, W. Sm.
  C. eccentricus, W. Sm.
  C. fimbriatus, Ehr.
  C. obscurus, Schmidt.
  C. radiatus, Ehr.
Cymbella scotica, W. Sm.
Dickeia alvoides, Berk.
Dimeregramma nanum, Greg.
Epithemia constricta, Greg.
  E. gibba, Kütz.
  E. turgida, W. Sm.
Eucampia zodiacus, Ehr.
  E. striata, Stolt.
Eupodiscus argus, Ehr.
Gomphonema marinum, W. Sm.
Grammalophora marina, Kütz.
  G. serpentaria, Kütz.
Hantzschia virgata, Grun.
Hyalodiscus stelliger, Bail.
  H. scoticus, Grun.
Lauderia delicatula, Peragello.
Licmophora gracilis, Grun.
  L. anglica, Grun.
  L. dalmatica, Kütz.
Mastogloia lanceolata, Th.
  M. Smithii, Th.

Melosira borreri, Grev.
  M. nummuloides, Bory.
  M. sulcata, Ehr.
  M. Westii, W. Sm.
Navicula abrupta, Greg.
  N. æstiva, Dn.
  N. affinis, Ehr.
  N. aspera, Ehr.
  N. Boeckii, Herberg.
  N. bombus, Ehr.
  N. carassius, Ehr.
  N. clepsydra, Ehr.
  N. crabro, Ehr.
  N. cyprinus, Ehr.
  N. didyma, Ehr.
  N. distans, W. Sm.
  N. fortis, Greg.
  N. fusca, Greg.
  N. fusiformis, Grun.
  N. granulata, Breb.
  N. interrupta, Kütz.
  N. Johnsonii, Greg.
  N. litoralis, Dn.
  N. lyra, Ehr.
  N. marina, Greg.
  N. northumbrica, Dn.
  N. numerosa, Dn.
  N. palpebralis, Breb.
  N. peregrina, Dn.
  N. pusilla, W. Sm.
  N. pygmæa, Kütz.
  N. rectangulata, Greg.
  N. rostrata, Ehr.
  N. semiplena, Greg.
  N. suborbicularis, Greg.
  N. subsalina, Dn.
  N. venata, Kütz.
  N. Westii, Greg.
Nitzschia bilobata, W. Sm.
  N. birostrata, W. Sm.
  N. closterium, W. Sm.
  N. distans, Greg.
  N. granulata, W. Sm.
  N. lanceolata, W. Sm.
  N. notabilis, Grun.
  N. obtusa, W. Sm.
  N. panduriformis, Greg.
  N. (Bacillaria) paradoxa, Gm.
  N. plana, W. Sm.
  N. punctata, Grun.
  N. reversa, W. Sm.
  N. sigma, W. Sm.
  N. scalaris, W. Sm.
  N. tænia, W. Sm.
  N. tryblionella, Hantz.
Plagiogramma gregorianum, Grev.

Plagiogramma van-Heurckii, Grun.
Pleurosigma æstuarii, W. Sm.
  P. angulatum, W. Sm.
  P. balticum, W. Sm.
  P. delicatulum, W. Sm.
  P. distortum, W. Sm.
  P. elongatum, W. Sm.
  P. fasciola, W. Sm.
  P. formosum, W. Sm.
  P. hippocampus, W. Sm.
  P. litorale, W. Sm.
  P. marinum, W. Sm.
  P. obscurum, W. Sm.
  P. prolongatum, W. Sm.
  P. scalprum, W. Sm.
  P. strigilis, W. Sm.
  P. strigosum, W. Sm.
  P. tenuissimum, Greg.
  P. transversale, Roper.
Rhabdonema arcuatum, Kütz.
  R. minutum, Kütz.
Rhaphoneis amphiceros, Ehr.
  Do. many varieties of this species.
Rhizosolenia imbricata, Brightw.
  R. setigera, Brightw.
  R. styliformis, Brightw.
  R. Wighamia, Brightw.
Schizonema crucigera, W. Sm.
  S. eximium, Th.
  S. helmintosum, Greg.
  S. vulgare, Th.
Scoliopleura latistriata, Breb.
  S. tumida, Breb.
Skeltonema costatum, Grun.
Stauroneis acuta, W. Sm.
  S. salina, W. Sm.
  S. linearis, W. Sm.
Stephanopyxis turris, Grev.
Striatella unipunctata, Ag.
Surirella constricta, W. Sm.
  S. crumena, Breb.
  S. gemma, Ehr.
  S. fastuosa, Ehr.
  S. salina, W. Sm.
  S. splendida, Kütz.
  S. striatula, Turp.
Synedra affinis, Kütz., var. *arcus*, Kütz.
  S. fulgens, Kütz.
  S. Gallionii, Ehr.
  S. obtusa, W. Sm.
  S. pulchella, var. *acicularis*, Kütz.
Toxonidia gregoriana, Dn.
  T. insignis, Dn.
Triceratium Brightwellii, West.
  T. farus, Ehr.
  T. striolatus, Ehr.

LIST OF THE MARINE ALGÆ.

[See Reports by Professor R. J. HARVEY GIBSON, M.A., F.L.S., in 'Fauna,' vol. ii.
p. 1. and vol. iii. p. 65.]

CYANOPHYCEÆ.

Ord. CHROOCOCCACEÆ.
Glæocapsa crepidinum, Thur. ii. 27,
iii. 90.
Ord. CHAMÆSIPHONACEÆ.
Dermocarpa prasina, Born. iii. 7, A. R.
iv., iii. 86, 91.
D. schousbæi, Born. iii. 86, 91.
Ord. OSCILLARIACEÆ.
Spirulina tenuissima, Kütz. iii. 86, 91.
S. pseudotenuissima, Crn. iii. 86, 91.
Oscillaria nigroviridis, Thw. ii. 27,
iii. 91.
O. corallinæ, Gom. ii. 27, iii. 91.
Phormidium papyraceum, Gom. ii. 27
(as *Osc. spiralis*), iii. 91.
Lyngbya semiplena, J. Ag. ii. 27, iii. 91.

Lyngbya æstuarii, Liebm. ii. 27, iii. 91.
L. majuscula, Harv. ii. 27, iii. 91.
L. spectabilis, Thur. in herb. iii. 91.
Symploca hydnoides, Kütz. iii. 91.
Microcoleus chthonoplastes, Thur. ii. 27,
iii. 92.
Rivularia biasolettiana, Menegh. ii. 26,
iii. 92.
R. atra, Roth. R. ii. 26, iii. 92.
Calothrix confervicola, C. Ag. ii. 26,
iii. 92.
C. pulvinata, C. Ag. iii. 92.
C. scopulorum, C. Ag. ii. 26, iii. 92.
Ord. NOSTOCACEÆ.
Anabæna torulosa, Lagerh. ii. 26, iii. 92.
Nodularia harveyana, Thur. iii. 92.

CHLOROPHYCEÆ.

Ord. BLASTOSPORACEÆ.
Prasiola stipitata, Suhr. A. R. iv. 8,
iii. 92.
Ord. ULVACEÆ.
Monostroma grevillei, J. Ag. ii. 22,
iii. 92.
Diplonema confervoides, Holm. and
Batt. iii. 92.
Enteromorpha clathrata, J. Ag. ii. 23,
iii. 93.
E. ralfsii, Harv. ii. 23, iii. 93.
E. erecta, J. Ag. ii. 23, iii. 93.
E. ramulosa, Harv. i. 24, ii. 23, iii. 93.
E. percursa, C. Ag. var. *ramosa.* J. Ag.
ii. 23 (as *E. percursa*), iii. 93.
E. compressa, Grev. i. 24, ii. 23, iii. 93.
E. linza, J. Ag. i. 25, ii. 23, iii. 93.
E. intestinalis, Link. i. 24, ii. 23, iii. 93.
E. canaliculata, Batt. iii. 93.
Ulva latissima. i. 314, ii. 23 (as *U.
lactuca,* var. *genuina*), iii. 93.
Ord. ULOTHRICHACEÆ
Ulothrix implexa, Kütz. ii. 24 (as
Rhizoclonium), iii. 93.
U. isogona, Thur. ii. 24.
Ord. CHÆTOPHORACEÆ.
Entoderma wittrockii, Wille. A. R. iv. 7,
iii. 93.
E. flustræ, Rke. A. R. iv. 7, iii. 93.
Ord. CLADOPHORACEÆ.
Urospora penicilliformis, Aresch. ii. 26
(as *Conferva youngana*), iii. 94.
U. flacca, H. and B. ii. 24 (as *U.
flacca*), iii. 94.
U. bangioides, H. and B. iii. 94.

Urospora collabens, H. and B. iii. 94.
Chætomorpha tortuosa, Kütz. ii. 24,
iii. 96.
Ch. linum, Kütz. ii. 24 (as *Conf.
crassa*), 26 (as *Conferva sutoria,*
Phyc. Brit.), iii. 96.
Ch. melagonium, Kütz. ii. 23, iii. 96.
Ch. ærea, Kütz. ii. 24, iii. 97.
Ch. litorea, H. and B. ii. 26 (as *Con-
ferva litorea.*)
Rhizoclonium riparium, Harv. i. 24,
ii. 24, iii. 97.
Rh. tortuosum, Kütz. ii. 24.
Rh. arenosa, Kütz. ii. 25 (as *Conferva
arenosa*).
Rh. casparyi, Harv. iii. 118.
Cladophora pellucida, Kütz. A. R. iv. 8,
iii. 97.
C. hutchinsiæ, Kütz. ii. 24, 25 (as *C.
diffusa*), iii. 97.
C. utriculosa, Kütz. var. *lætevirens,*
Hauck. i. 25 (as spec.), ii. 25, iii. 97.
C. rupestris, Kütz. i. 25, ii. 24, iii. 12, 97.
C. glaucescens, Griff. iii. 97.
C. fracta, Kütz. ii. 25, iii. 97.
C. flexuosa, Griff. i. 24, ii. 25, iii. 97.
C. albida, Kütz. ii. 25, iii. 97. var.
refracta, H. and B. ii. 25 (as spec.),
iii. 97.
C. arcta, Kütz. ii. 24, iii. 97.
C. lanosa, Kütz. ii. 25, iii. 97. var.
uncialis, Thur. ii. 25 (as spec.),
iii. 97.
C. rudolphiana, Kütz. ii. 25, iii. 118.
C. gracilis, Kütz. ii. 25.

Ord. BRYOPSIDACEÆ.
Bryopsis hypnoides, Lamx. ii.25, iii. 98.
B. plumosa, C. Ag. i. 25, ii. 25, iii. 98.
Ord. VAUCHERIACEÆ.
Vaucheria dichotoma, Lyngb. var.
marina, C. Ag. ii. 22, iii. 98.

Vaucheria Thuretii, Wor. ii. 22, iii. 98.
Ord. CODIACEÆ.
Codium tomentosum, Stackh. ii. 22, iii. 98.

PHÆOPHYCEÆ.

Ord. DESMARESTIACEÆ.
Desmarestia viridis, Lamx. i.313, ii. 21, iii. 98.
D. aculeata, Lamx. i. 25, 313, ii. 21, iii. 98.
D. ligulata, Lamx. iii. 98.
Ord. DICTYOSIPHONACEÆ.
Dictyosiphon fœniculaceus, Grev. ii. 20, iii. 98.
Ord. PUNCTARIACEÆ.
Litosiphon pusillus, Harv. ii. 21, A. R. iv. 8, iii. 98.
L. laminariæ, Harv. ii. 21, iii. 118.
Stictyosiphon subarticulatus, Hauck. iii. 99.
Punctaria plantaginea, Grev. ii. 20, iii. 99.
P. latifolia, Grev. i. 313, ii. 20, iii. 99. var. *zosteræ*, Le Jol. iii. 99, A. R. iv. 8 (as *P. tenuissima*).
Striaria attenuata, Grev. iii. 118.
Ord. ASPEROCOCCACEÆ.
Myriotrichia claræformis, Harv. ii.19, iii. 99. var. *filiformis*, Farl. ii. 19 (as spec.), iii. 99.
Asperococcus echinatus, Grev. i. 25, ii. 21, iii. 99. var. *vermicularis*, Griff. iii. 99.
A. bulbosus, Lamx. ii. 21 (as *A. turneri*), iii. 118.
Streblonema velutinum, Thur. ii. 18 (as *Ectocarpus*), iii. 99.
Ectocarpus terminalis, Kütz. A. R. iv. 8, iii. 99.
E. confervoides, Le Jol. var. *siliculosus*, Kjell. ii. 18, iii. 99.
E. fasciculatus, Harv. ii. 18, iii. 99.
E. tomentosus, Lyngb. ii. 18, iii. 99.
E. granulosus, C. Ag. ii. 19, iii. 99.
E. crinitus, Carm. ii. 18.
E. hincksiæ, Harv. ii. 19.
Isthmoplea sphærophora, Kjell. A. R. iv. 8, iii. 100.
Pylaiella litoralis, Kjell. ii. 19 (as *Ectocarpus*), iii. 100.
Ord. ARTHROCLADIACEÆ.
Arthrocladia villosa, Duby. iii. 100.
Ord. ELACHISTACEÆ.
Elachista scutulata, Duby. ii. 20, iii. 100.
E. fucicola, Fries. ii. 20, iii. 100.
E. flaccida, Aresch. ii. 20, iii. 100.
Ord. SPHACELARIACEÆ.
Sphacelaria radicans, Harv. ii. 19, iii. 100, A. R. iv. 8.

Sphacelaria cirrhosa, C. Ag. i. 25, ii. 19, iii. 100. var. *fusca*, H. and B. i. 25, ii. 19 (as spec.), iii. 100.
S. plumigera, Holm. iii. 100.
Chætopteris plumosa, Kütz. i. 25, 313 (as *Sphacelaria*), ii. 19, iii. 100.
Halopteris filicina, Kütz. iii. 101.
Stypocaulon scoparium, Kütz. i. 25, ii. 7 (as *Sphacelaria*), 19, iii. 101.
Cladostephus spongiosus, C. Ag. i. 24, 313, ii. 6, 7, 19, iii. 101.
C. verticillatus, C. Ag. i. 24, ii. 19, iii. 101.
Ord. MYRIONEMACEÆ.
Myrionema strangulans, Grev. ii. 18 (as *M. vulgare*), iii. 101. var. *punctiforme*, Thur. ii. 18 (as spec.), iii. 101.
Ascocyclus leclancherii, Magn. ii. 18 (as *Myrionema*), iii. 101.
A. reptans, Rke. A. R. iv. 8, iii. 101.
Ralfsia verrucosa, Aresch. ii. 22, A. R. iv. 7, iii. 101.
Ord. CHORDARIACEÆ.
Chordaria flagelliformis, C. Ag. i. 25, ii. 20, iii. 101.
Mesogloea vermiculata, Le Jol. ii. 20, iii. 101.
M. verticillata, Ag. ii. 20.
Castagnea virescens, Thur. ii.20, iii.101.
Leathesia difformis, Aresch. ii. 20 (as *L. umbellata*), iii. 101.
Ord. SCYTOSIPHONACEÆ.
Phyllitis zosterifolia, Rke. iii. 101.
Ph. fascia, Kütz. ii. 21.
Scytosiphon lomentarius, J. Ag. ii. 21, iii. 102.
Ord. CHORDACEÆ.
Chorda filum, Stackh. ii. 21, iii. 102.
Ord. LAMINARIACEÆ.
Laminaria saccharina, Lamx. i. 313, ii. 22, iii. 102.
L. hieroglyphica, J. Ag. var. *phyllitis*, Le Jol. ii. 22, iii. 102.
L. digitata, Edm. i. 313, ii. 6, 21, iii. 102.
L. hyperborea, Fos. iii. 8 (A. R. iv. 8), 102.
Saccorhiza bulbosa, De la Pyl. iii. 102.
Alaria esculenta, Grev. ii. 22, iii. 102.
Sargassum linifolium, C. Ag. ii. 17, iii. 119.

Ord. SPOROCHNACEÆ.
Sporochnus pedunculatus, C. Ag. iii.
102.
Ord. CUTLERIACEÆ.
Cutleria multifida, Grev. ii. 22, iii.
103.
Aglaozonia parvula, Zan. ii. 22, iii.
119.
Ord. FUCACEÆ.
Fucus ceranoides, Linn. iii. 103.
F. vesiculosus, Linn. i. 312, ii. 17, iii.
10 20, 103.
F. serratus, Linn. i. 312, ii. 17, iii.
10, 20, 103.
F. platycarpus, Thur. ii. 17, iii. 103.
Ascophyllum nodosum, Le Jol. i. 312.
(as *Fucus*), ii. 15, 17, iii. 10, 11, 20

(as *Fucus*), 103. var. *scorpioides*,
Hauck. ii. 17, iii. 119.
Himanthalia lorea, Lyngb. ii. 17, 18,
20, iii. 11, 103.
Halidrys siliquosa, Lyngb. i. 24, 112,
312 ; ii. 11 (as *Fucus*), 17, iii. 103.
Pelvetia canaliculata, Decne et Thur.
ii. 17 (as *Fucus*), iii. 103.
Cystoseira, sp. ii. 17, 20, iii. 119.
Ord. DICTYOTACEÆ.
Dictyota dichotoma, Lamx. i. 313, ii.
18, iii. 104. var. *implexa*, J. Ag.
iii. 104. var. *intricata*. iii. 8 (A. R.
iv.).
Taonia atomaria, J. Ag. iii. 104.
Dictyopteris polypodioides, Lamx. iii.
119.

RHODOPHYCEÆ.

Ord. PORPHYRACEÆ.
Porphyra laciniata, C. Ag. i. 24, ii.
5, 8 (as *P. vulgaris*), iii. 104.
Bangia fuscopurpurea, Lyngb. ii. 3,
iii. 104.
Ord. HELMINTHOCLADIACEÆ.
Chantransia virgatula, Thur. ii. 5, iii.
104.
Ch. secundata, Thur. iii. 8 (A. R. iv.),
104.
Ch. daviesii, Thur. ii. 7 (as *Calli-
thamnion*), iii. 104.
Helminthocladia purpurea, J. Ag. iii.
104.
Helminthora divaricata, J. Ag. iii. 105.
Nemalion multifidum, J. Ag. ii. 6, iii
119.
Ord. GELIDIACEÆ.
Naccaria wiggii, End. ii. 6, iii. 105.
Gelidium corneum, Lamx. i. 24, ii.
12, iii. 105.
G. crinale, J. Ag. ii. 13, iii. 105.
Ord. GIGARTINACEÆ.
Chondrus crispus, Stackh. i. 25, 313,
ii. 9, iii. 12, 105.
Gigartina mamillosa, J. Ag. ii. 10,
iii. 105.
Phyllophora rubens, Grev. i. 24, ii. 10,
15, iii. 105.
P. membranifolia, J. Ag. i. 24, ii. 10,
iii. 105.
P. traillii. H. and B. iii. 8 (A. R.
iv.), 105.
P. palmettoides, J. Ag. iii. 105.
Gymnogongrus griffithsiæ, Mart. ii.
10, iii. 105.
G. norvegicus, J. Ag. ii. 10 (as *Chon-
drus*), iii. 105.
Ahnfeldtia plicata, Fries. iii. 7 (A. R.
iv.), 105.
Callophyllis laciniata, Kütz. ii. 11
(as *Rhodymenia*), iii. 106.

Ord. RHODOPHYLLIDACEÆ.
Cystoclonium purpurascens, Kütz. i.
24 (as *Hypnæa*), ii. 11, iii. 106.
Catenella opuntia, Grev. i. 313, ii. 12,
iii. 8 (A. R iv.), 106.
Rhodophyllis bifida, Kütz. ii. 10 (as
Rhodymenia), iii. 107.
Ord. SPHÆROCOCCACEÆ.
Sphærococcus coronopifolius, Grev. ii.
12, iii. 107.
Gracilaria confervoides, Grev. i. 25,
ii. 12, iii. 107.
Calliblepharis ciliata, Kütz. i. 24, ii.
10 (as *Rhodymenia*), iii. 8 (A. R. iv.),
107.
C. jubata, Kütz. iii. 7 (A. R. iv.),
107.
Ord. RHODYMENIACEÆ.
Rhodymenia palmata, Grev. ii. 11, 18,
iii. 108.
Rh. palmetta, Grev. i. 313, ii. 11,
iii. 108.
Lomentaria articulata, Lyngb. ii. 8
(as *Chylocladia*), 11, iii. 12, 108.
L. clavellosa, Gaill. iii. 108.
Champia parvula, Harv. iii. 108.
Chylocladia kaliformis, Grev. ii. 13
(as *Lomentaria*), iii. 108.
Ch. ovalis, Hook. iii. 108.
Plocamium coccineum, Lyngb. i. 313,
ii. 10, iii. 108.
Microcladia glandulosa, Grev. iii. 120.
Euthora cristata, J. Ag. iii. 120.
Ord. DELESSERIACEÆ.
Nitophyllum punctatum, Grev. ii. 12,
iii. 108.
N. laceratum, Grev. ii. 12, iii. 108.
Delesseria alata, Lamx. i. 24, ii. 12,
iii. 108.
D. sinuosa, Lamx. i. 24, ii. 12, iii. 108.
D. hypoglossum, Lamx. ii. 12, iii. 109.
D. ruscifolia, Lamx. ii. 12, iii. 109.

Ord. NEMASTOMACEÆ.
Furcellaria fastigiata, Lamx. ii. 10,
24 (as *Fastigiaria*), iii. 115.
Ord. RHIZOPHYLLIDACEÆ.
Polyides rotundus, Grev. ii. 13 (as
P. lumbricalis), iii. 7 (A. R iv.),
115.
Ord. SQUAMARIACEÆ.
Petrocelis cruenta, J. Ag. ii. 5 (as
P. pellita), iii. 115.
Peyssonnelia dubyi, Crn. i. 313, ii. 5,
iii. 115.
Hildenbrandtia prototypus, Nard. var.
rosea, Kütz. iii. 7 (A. R. iv.). 116.
Ord. CORALLINACEÆ.
Schmitziella endophlœa, Born. et Batt.
iii. 8 (A. R. iv.), 116.
Melobesia confervoides, Kütz. ii. 15,
iii. 116.

Melobesia pustulata, Lamx. iii. 116.
M. farinosa, Lamx. ii. 15, iii. 117.
M. membranacea, Lamx. ii. 15, iii.
117.
M. verrucata, Lamx. ii. 16, iii. 117.
Lithophyllum lichenoides, Phil. ii. 16,
iii. 117.
L. lenormandi, Rosan. iii. 8 (A. R.
iv.), 117.
Lithothamnion polymorphum. Aresch.
ii. 16 (as *Melobesia*), iii. 117.
L. calcareum, Aresch. ii. 16, iii. 117.
L. fasciculatum, Aresch. ii. 120.
Corallina officinalis, Linn. i. 24, 97,
313, 321, ii. 16, iii. 11, 20, 117.
C. rubens, Ellis et Sol. i. 24, ii. 16,
ii. 117.

LIST OF THE FORAMINIFERA.

[See Mr. J. D. SIDDALL'S Report in ' Fauna,' vol. i., and papers since by Mr.
PEARCEY, vol iii., p. 41, Mr. BURGESS, vol. iii., p. 59, Mr. CHAFFER, 7th A. R.,
40, and Dr. CHASTER, Southport Soc. N. Sci., 1892, and 10th A. R.]

Lieberkühnia Wageneri, Clap. i. 42.
Shepheardella tæniformis, Sid.
Gromia dujardinii, Sch.
G. oviformis, Duj.
Squamulina lævis, Schul.
Nubecularia lucifuga, Defr.
Biloculina ringens, Lamk.
B. depressa, D'Orb.
B. elongata, D'Orb.
Spiroloculina limbata, D'Orb.
S. planulata, D'Orb.
S. excavata, D'Orb.
S. acutimargo, Brady.
S. depressa, D'Orb.
Miliolina trigonula, Lamk.
M. tricarinata, D'Orb.
M. oblonga, Montagu.
M. boueana, D'Orb.
M. seminulum, Linn.
M. venusta, Karrer.
M. subrotunda, Mont.
M. secans, D'Orb.
M. bicornis, W. & J.
M. ferussacii, D'Orb.
M. fusca, Brady.
M. agglutinans, D'Orb.
M. spiculifera, Sid.
M. contorta, D'Orb.
M. auberiana, D'Orb.
M. pulchella, D'Orb.
M. sclerotica, Karr.
Ophthalmidium inconstans, Brady.
Sigmoilina tenuis, Cz.
S. celata, Costa.
Cornuspira involvens, Reuss.
Astrorhiza limicola, Sand.
Dendrophrya radiata, S. Wright.

Dendrophrya erecta, S. Wright.
Technitella legumen, Norman.
Psammosphœra fusca, Schul.
Hyperammina elongata, Brady.
H. arborescens, Norm. 10th A. R.
Haliphysema Tumanowiczii, Bow.
Reophax fusiformis, Will.
R. scorpiurus, Montf.
R. Scottii, Chaster.
R. findens, G. M. Dawson.
R. moniliforme, Sid.
R. nodulosa, Brady.
Haplophragmium globigeriniforme, P. & J.
H. canariense, D'Orb.
H. agglutinans, D'Orb.
H. anceps, Brady.
H. glomeratum, Brady.
Placopsilina bulla, Brady.
P. Kingsleyi, Sid.
P. rarians, Carter.
Ammodiscus incertus, D'Orb.
A. gordialis, P. & J.
A. charoides, P. & J.
A. shoneanus, Sid.
A. spectabilis, Brady.
Trochammina nitida, Brady.
T. squamata, P. & J.
T. ochracea, Will.
T. plicata, Terq.
T. inflata, Mont.
T. macrescens, Brady.
Textularia sagittula, Defr.
T. agglutinans, D'Orb.
T. porrecta, Brady.
T. variabilis, Will.
T. trochus, D'Orb.
T. gramen, D'Orb.

Textularia fusiformis, Chaster.
Spiroplecta sagittula, Defrance.
S. biformis, P. & J.
Gaudryina filiformis. Berth.
Verneuilina polystropha, Reuss.
V. spinulosa, Reuss.
Clavalina obscura, Chaster.
Bigenerina digitata, D'Orb.
Bulimina pupoides, D'Orb.
B. elongata, D'Orb.
B. marginata, D'Orb.
B. aculeata, D'Orb.
B. ovata, D'Orb.
B. elegans, D'Orb.
B. elegantissima, D'Orb.
B. squamigera, D'Orb.
B. fusiformis, Will.
Virgulina schreibersiana, Czjzek.
V. bolivina, D'Orb.
Bolivina punctata, D'Orb.
B. plicata, D'Orb.
B. pygmæa, D'Orb.
B. difformis, Will
B. cenariensis, Costa.
B. dilatata, Reuss.
B. lævigata, Will.
B. variabilis, Will.
Cassidulina lævigata, D'Orb.
C. crassa, D'Orb.
Lagena sulcata, W. & J.
L. interrupta, Will.
L. costata, Will.
L. Williamsoni, Alcock.
L. caudata, P. & J.
L. Lyelli, Seguenza.
L. Feildeniana, Brady.
L. striato-punctata, P. & J.
L. lævis, Mont.
L. gracillima, Seg.
L. apiculata, Reuss.
L. globosa, Mont.
L. striata, D'Orb.
L. clavata, D'Orb.
L. lævigata, Reuss.
L. protea, Chaster.
L. hertwigiana, Brady.
L. crinata, P. & J.
L. lineata, Will.
L. botelliformis, Br.
L. semilineata, Wr.
L. spiralis, Br.
L. quadrata, Will.
L. millettii, Chaster.
L. falcata, Chaster.
L. inæquilateralis, Wr.
L. bicarinata, Terq.
L. semi-alata, B. & M.
L. castrensis, Sch
L. lagenoides, Will.
L. tenuistriata, Br.
L. depressa, Chaster.
L. gracilis, Will.
L. semistriata, Will.
L. distoma, P. & J.

Lagena aspera, Reuss.
L. marginata, W. & B.
L. Orbignyana, Seg.
L. trigona-marginata, P. & J.
L. lucida, Will.
L. trigono-oblonga, Seg. & Sid.
L. ornata, Will.
L. trigono-ornata. Brady.
L. pulchella, Brady.
L. melo, D'Orb.
L. squamosa, Mont.
L. hexagona, Will.
L. hispida, Reuss.
Nodosaria scalaris, Lamk.
N. radicula, Linn.
N. Calomorpha, Reuss.
N. hispida, D'Orb.
N. pyrula, D'Orb.
N. communis, D'Orb.
N. obliqua, D'Orb.
Lingulina carinata, D'Orb.
L. herdmani, Chaster.
Vaginulina legumen, Linn.
V. linearis, Mont.
Marginulina costata, Batsch.
M. glabra, D'Orb.
Cristellaria rotulata, Lamk.
C. crepidula, F. & M.
C. italica, Defr.
C. variabilis, Reuss.
C. elongata, Will.
C. cultrata, Montfort.
C. gibba, D'Orb.
C. vortex, F. & M.
Polymorphina lactea, W. & J.
Do., var. *oblonga*, Will.
P. oblonga, D'Orb.
P. gibba, D'Orb.
P. subæqualis, D'Orb.
P. communis, D'Orb.
P. thouini, D'Orb.
P. compressa, D'Orb.
P. lanceolata, Reuss.
P. concava, Will.
P. spinosa, D'Orb.
P. orbignyii, Zborzewskii.
P. sororia, Reuss.
P. rotundata, Born.
P. concava, Will.
Uvigerina pygmæa, D'Orb.
U. angulosa, Will.
U. canariensis, D'Orb.
Globigerina bulloides, D'Orb.
Do., var. *triloba*, Reuss.
G. inflata, D'Orb.
G. æquilateralis, Br.
G. rubra, D'Orb.
Orbulina universa, D'Orb.
Pullenia sphæroides, D'Orb.
Sphæridina dehiscens, P. & J.
Spirillina vivipara, Ehrenb.
S. margaritifera, Will.
S. tuberculata, Brady.
S. limbata, Brady.

Patellina corrugata, Will.
Discorbina rosacea, D'Orb.
D. ochracea, Will.
D. globularis, D'Orb.
D. orbicularis, Terquem.
D. biconcava, P. & J.
D. turbo, D'Orb.
D. parisiensis, D'Orb.
D. nitida, Wright.
D. Wrightii, Br.
D. Bertheloti, D'Orb.
D. minutissima, Chaster.
D. tuberculata, Balkwill & Wright.
Planorbulina mediterranensis, D'Orb.
Truncatulina Haidengerii, D'Orb.
T. ungeriana, D'Orb.
T. lobatula, Walker & Jacob.
T. refulgens, Montf.
T. reticulata, Czjzek.
Pulvinulina repanda, F. & M.

Pulvinulina repanda, var. *concamerata*, Mont.
P. auricula, F. & M.
P. canariensis, D'Orb.
P. karsteni, Reuss.
P. nitidula, Chaster.
Rotalia Beccarii, Linn.
R. nitida, Will.
Gypsina inhærens, Schul.
G. vesicularis, P. & J.   10th A. R.
Nonionina asterizans, F. & M.
N. pauperata, B. & W.
N. turgida, Will.
N. scapha, F. & M.
N. umbilicatula, Mont.
N. depressula, W. & J.
N. stelligera, D'Orb.
N. boueana, D'Orb.
Polystomella crispa, Linn.
P. striato-punctata, F. & M.

LIST OF THE PORIFERA.

[See Reports by Mr. T. HIGGIN and Dr. R. HANITSCH in ' Fauna,' vol. i. p. 72, vol. ii. p. 28, vol. iii. p. 192, and Annual Reports.]

CALCAREA.

HOMOCŒLA.
Leucosolenia botryoides, Ellis and Solander. iii. p. 233.
L. contorta, Bowerbank. iii. p. 233.
L. coriacea, Fleming. iii. p. 232.
L. lacunosa, Johnston. iii. p. 233.
HETEROCŒLA.
Sycon asperum, Gibson. i. p. 365.

Sycon compressum, auct. ii. p. 45.
S. coronatum, E. & Sol. iii. p. 237.
Aphroceras ramosa, Carter. i. p. 92.
Leucandra fistulosa, J. i. p. 92.
L. Gossei, Bow. iii. p. 236.
L. impressa, Hanitsch. iii. p. 234.
L. Johnstoni, Carter. iii. p. 236.
L. nivea, Grant. iii. p. 236.

SILICEA.

HEXACERATINA.
Halisarca Dujardini, J. ii. 32.
Aplysilla rubra, Hanitsch. iii. p. 196 ; 'Irish Sponges,' T.L.B.S., V., p. 219.
TETRACTINELLIDA.
Dercitus Bucklandi, Bow. iii. p. 221.
Stryphnus ponderosus, Bow. i. p. 88.
Do., var. *rudis*. iii. p. 221.
Stellatta Grubei, O. Sch. iii. p. 227.
Pachymatisma Johnstonia, Bow. iii. p. 229.
MONAXONIDA.
Chalina oculata, Pall. i. p. 76.
Acervochalina gracilenta, Bow. iii. p. 199.
A. limbata, Mont. ii. p. 34.
Chalinula Montagui, Flem. iii. p. 201.
Halichondria panicea, Pall. ii. p. 32.
H. albescens, J. i. p. 79.
H. coccinea, Bow. i. p. 79.
Reniera clara, Bow. i. p. 84.
R. densa, Bow. i. p. 83.
R. elegans, Bow. i. p. 82.
R. fistulosa, Bow. i. p. 83.
R. ingalli, Bow. iii. 199.

Reniera pallida, Bow. i. p. 83.
R. simulans, J. i. p. 83.
R. varians, Bow. iii. p. 198.
Esperiopsis fucorum, J. i. p. 84.
Esperella ægagropila, J. iii. p. 202.
Desmacidon fruticosum, Mont. 7th A.R. p. 22.
Dendoryx incrustans, Esper. iii. p. 204.
Jophon expansum, Bow. 6th A. R. p. 44.
Myxilla irregularis, Bow. 8th A. R. p. 18.
Pocillon Hyndmani, Bow. Irish sponges, T.L.B.S., V., p. 217.
Plumohalichondria plumosa, Mont. i. p. 78.
Microciona atrasanguinea, Bow. iii. p. 207.
Raspailia ventilabrum, Bow. iii. p. 212.
Vibulinus rigidus, Mont. iii. p. 213.
Echinoclathria seriata, Grant. iii. p. 205.
Hymeniacidon caruncula, Bow. i. p. 79.
H. sanguineum, Bow. i. p. 87.

Axinella mammillata, Hanitsch. iii.
   p. 209.
Subcrites carnosus, J. i. p. 86.
S. domuncula, Olivi. iii. p. 214.
N. ficus, J. iii. p. 216.
Cliona celata, Grant, iii. p. 216.
Polymastia mammillaris, Bow. iii. p.
   220.

Polymastia robusta, Bow. iii. p. 220.
Tethya lyncurium, Linn. iii. p. 220.
MONOCERATINA.
Leiosella pulchella, Sow. 8th A. R.
   p. 18.
Spongelia fragilis, Mont. iv. p. 198.

LIST OF THE CŒLENTERATA.

A. HYDROZOA.   I. HYDROIDA.

[See Report by Prof. HERDMAN and others in ' Fauna,' vol. i., and Report by
Miss L. R. THORNELY in ' Fauna,' vol. iv.]

ATHECATA.
Clava multicornis, Forskal. i. 97, iv.
   225.
C. leptostyla, Agassiz. i. 97, iv. 225.
Tubiclava cornucopiæ, Norm. 9th A. R.,
   p. 10.
Hydractinia echinata, Fleming. i. 97,
   iv. 225.
Coryne van-Benedeni. iv. 222, 225.
C. vaginata, Hincks. 10th A. R.
C. pusilla, Gaertner. i. 98, iv. 225.
Syncoryne eximia, All. 8th A. R., p. 19.
Eudendrium ramcum, Pallas. i. 98,
   iv. 225.
E. ramosum, Linn. i. 98, iv. 225.
E. capillare, Alder. i. 98, iv. 225.
Hydranthea margarica, Hincks. iv.
   222, 223, 225.
Garveia nutans, T. S. W. i. 99, iv. 226.
Bimeria vestita, T. S. W. i. 100, iv. 226.
Perigonimus repens, T. S. W. 9th A. R.,
   p. 11.
Dicoryne conferta, Ald. 8th A. R., p. 19.
Bougainvillia muscus, Allman. i. 100,
   iv. 226.
B. ramosa, V. Ben. iv. 222, 226.
Tubularia indivisa, Linn. i. 100, iv.
   226.
   T. coronata, Abildgaard. i. 100, iii.
      49, iv. 226.
   T. simplex, Ald. i. 100, iv. 226.
   T. larynx, Ellis and Solander. i. 101,
      iv. 222, 226.
   T. britannica, Pennington. i. 101, iv.
      226.
   T. attenuato, Allm. iii. 49, iv. 222, 226.
Ectopleura Dumortierii, van Beneden.
   i. 101, iv. 226.
Corymorpha nutans, Sars. i. 101, iv.
   226.
THECAPHORA.
Clytia Johnstoni, Ald. i. 101, iv. 226.
Obelia geniculata, Linn. i. 102, iv.
   222, 226.
   O. gelatinosa, Pall. i. 102, iv. 226.
   O. longissima, Pall. i. 102, iv. 226.

Obelia flabellata, Hincks. i. 102, iv.
   226.
   O. dichotoma, Linn. i. 103, iv. 226.
   O. plicata, Hincks. iv. 222, 226.
Campanularia volubilis, Linn. i. 103,
   iv. 226.
   C. Hincksii, Ald. i. 104, iv. 226.
   C. fragilis, Hincks. iv. 222, 226.
   C. caliculata, Hincks. i. 104, iv. 226.
   C. verticillata, Linn. i. 104, iv. 226.
   C. flexuosa, Hincks. i. 104, iv. 226.
   C. angulata, Hincks. i. 105, iv. 226.
   C. neglecta, Ald. i. 105, iv. 226.
   C. raridentata, Ald. iii. 49, iv. 222,
      226.
Gonothyræa Lovéni, Allm. i. 105, iv.
   226.
   G. gracilis, Sars. iv. 222, 226.
   G. hyalina, Hincks. iv. 222, 223, 226.
Opercularella lacerata, Johnston. i.
   105, iv. 226.
Lafoëa dumosa, Fleming. i. 106, iv.
   226.
   Do., var. *robusta,* Sars. iv. 222, 226.
   L. fruticosa, Sars. iv. 222, 226.
Calycella syringa, Linn. i. 106, iv. 227.
   C. fastigiata, Ald. iv. 222, 224, 227.
   C. pigmæa, Ald. iv. 222, 224, 227.
   C. grandis, Hincks. iv. 222, 227.
   C. costata, Hincks. iv. 222, 227.
   C. humilis, Hincks. iv. 222, 227.
Filellum serpens, Hassall. i. 106, iv.
   222, 224, 227.
Coppinia arcta, Dalyell. i. 106, iv. 227.
Halecium halecinum, Linn. i. 107, iv.
   227.
   H. Beanii, Johnst. i. 107, iv. 227.
   H. tenellum, Hincks. iii. 49, iv. 225,
      227.
   H. muricatum, Ellis & Sol. iii. 49.
Sertularella polyzonias, Linn. i. 108,
   iv. 227.
   S. rugosa, Linn. i. 108, iv. 227.
   S. Gayi, Lamx. iv. 222, 227.
   S. tenella, Ald. iii. 49, iv. 222, 227.
   S. fusiformis, Hincks. iv. 222, 227.

Diphasia rosacea, Linn. i. 108, iv. 227.
D. attenuata, Hincks. i. 109, iv. 227.
D. pinaster, Ellis & Sol. i. 109, iv. 227.
D. tamarisca, Linn. i. 109, iv. 227.
D. fallax, Johnst. i. 109, iv. 227.
Sertularia pumila, Linn. i. 109, iv. 227.
S. gracilis, Hassall. i. 110, iv. 227.
S. operculata, Linn. i. 110, iv. 227.
S. filicula, Ellis & Sol. i. 110, iv. 227.
S abietina, Linn. i. 110, 227.
S. argentea, E. & Sol. i. 110, iv. 227.
S. cupressina, Linn. i. 111, iv. 227.
Hydrallmania falcata, Linn. i. 111, iv. 227.
Thuiaria articulata, Pall. i. 111, iv. 222, 227.

Thuiaria thuja, Linn. iv. 222, 227.
Antennularia antennina, Linn. i. 112, iv. 227.
A. ramosa, Lam. i. 112, iv. 227.
Aglaophenia pluma, Linn. i. 112, iv. 228.
A. myriophyllum, Linn. i. 112, iv. 222, 228.
A. tubulifera, Hincks. iv. 222, 228.
A. pennatula, E. & Sol. iv. 222, 228.
Plumularia pinnata, Linn. i. 113, iv. 228.
P. frutescens, E. & Sol. iv. 222, 228.
P. setacea Ellis. i. 113, iv. 228.
P. Catharina, J. i. 113, iv. 228.
P echinulata, Lam. iv. 222, 225, 228.
P. similis, Hincks. i. 113. iv. 228.

II. MEDUSÆ.

[See ' List of Medusæ and Ctenophora of the L.M.B.C. District,' by J. A. CLUBB, ' Fauna,' vol. i. p. 114, and ' Report on the Medusæ of the L.M.B.C. District,' by E. T. BROWNE, ' Fauna,' vol. iv. p. 371.]

HYDROMEDUSÆ.

ANTHOMEDUSÆ.
Codonium pulchellum, Forb. iv. 374.
Corymorpha nutans, Sars. 10th A. R.
Sarsia tubulosa, Sars. i. 115, iv. 375.
Dipurena halterata, Forb. iv. 375.
Steenstrupia rubra, Forb. iv. 375.
Euphysa aurata, Forb. iv. 376.
Hybocodon prolifer, Agassiz. 10th A. R.
Amphicodon fritillaria, Steenstr. iv. 379
Tiara pileata, Forskal. iv. 386.
Turris neglecta, Lesson. i. 115, iv. 388.
Dysmorphosa carnea, M. Sars. iv. 388.
D. minima, Hæckel. iv. 388.
? Cytæandra areolata, Ald. iv. 390.
Lizzia blondina, Forb. iv. 393.
Margelis principis, Steenstrup. iv. 394.
M ramosa, Van Beneden.
M. britannica, Forb. i. 115, iv. 395.
Margellium octopunctatum, Sars. i. 117, iv. 398.
Podocoryne carnea, Sars. 10th A. R.
Thaumantias hemisphærica, Müll. i. 116, 117, iv. 403.

Laodice cruciata, L. Agassiz. i. 115, iv. 401.
L. calcarata, L. Agassiz. iv. 404.
LEPTOMEDUSÆ.
Melicertidium octocostatum, Sars. iv. 405.
Clytia Johnstoni, Alder. iv. 406.
Eucope octona, Forb. i. 115, iv. 406.
Obelia lucifera, Forb. iv. 406.
Tiaropsis multicirrata, Sars. iv. 406.
Epenthesis cymbaloidea, Slabber. i. 116, iv. 407.
Mitrocomella polydiadema, Romanes. iv. 407.
Phialidium variabile, Hæckel. iv. 408.
Ph. temporarium, Browne, 10th A. R.
Ph. cymbaloidium, Van Beneden.
Entima insignis, Keferstein. iv. 410.
Saphenia mirabilis, Wright iv. 410.
Tiaropsis multicirrata, Sars. 10th A. R.
Thaumantias convexa, Forb. i. 116.
T. lucida, Forb. i. 116.

SCYPHOMEDUSÆ.

STAUROMEDUSÆ.
Depastrum cyathiforme, Sars. iv. 152, 411.
Haliclystus auricula, Rathke. iv. 157, 411.

DISCOMEDUSÆ.
Chrysaora isosceles, Linn. iv. 412.
Cyanea capillata, Linn. i. 117. iv. 412.
Aurelia aurita, Lam. i. 117, iv. 412.
Pilema octopus, Linn. i. 118, iv. 413.

ˊIII. SIPHONOPHORA.

Agalmopsis elegans, Sars. 10th A. R. | *Physalia pelagica*, Esch. i. 118.

IV. CTENOPHORA.

SACCATA.
Pleurobrachia pileus, Flem. i. 118.
P. pomiformis, Pat. i. 119.

EURYSTOMATA.
Beroë ovata, Lam. i. 119.
LOBATA.
Bolina hibernica, Pat. i. 119.

B. ACTINOZOA.   I. ALCYONARIA.

[See Report on the Alcyonaria of the L.M.B.C. District, by Professor HERDMAN
' Fauna,' vol. i. p. 120, and also note upon yellow variety of Sarcodictyon catenata
in ' Fauna,' vol. iv. p. 322.]

ALCYONIDA.
Sarcodictyon catenata, Forb. i. 120,
   iv. 322.

Alcyonium digitatum, Linn. i. 122.
PENNATULIDA.
Virgularia mirabilis, Lamk., A. R.

II. ACTINIARIA.

[See Report on the Actiniaria of the L.M.B.C. District, by Dr. J. W. ELLIS,
' Fauna,' vol. i. p. 123 (nomenclature revised since by Prof. HADDON).]

PROTANTHIDÆ.
Corynactis viridis, All. i. 129.
Capnea sanguinea, Forb. i. 129.
HEXACTINIDÆ.
Halcampa chrysanthellum, Peach. i.
   123.
Metridium dianthus, Ellis. i. 123.
Cereus pedunculatus, Penn. i. 124.
Sagartia miniata, Gosse. i. 125.
   S. rosea, Gosse.
   S. venusta, Gosse. i. 125.
   S. nivea, Gosse.
   S. lacerata, Dall. 7th A. R., 22.
   S. sphyrodeta, Gosse. i. 127.

Cylista viduata, Müll. i. 125.
C. undata, Müll. i. 125.
Do., var. candida, Müll. i. 126.
Adamsia palliata, Bohadsch. i. 127.
Actinia equina, Linn. i. 127.
Anemonia sulcata, Penn. i. 128.
Urticina crassicornis, Müll. i. 128.
Bunodes verrucosa, Penn. i. 129.
Paraphellia expansa, Hadd. 9th
   A. R., 9.
ZOANTHIDÆ.
Epizoanthus arenacea, Delle C. i. 130.
CERIANTHIDÆ.
Cerianthus Lloydii, Gosse. i. 130.

LIST OF THE ECHINODERMATA.

[See Professor HERDMAN's Report upon the Crinoidea, Asteroidea, Echinoidea, and
Holothuroidea, and Mr. H. C. CHADWICK's Report upon the Ophiuroidea in the
' Fauna,' vol. i., and Mr. H. C. CHADWICK's Second Report on the Echinodermata
in the ' Fauna,' vol. ii., and papers in vol. iv.]

CRINOIDEA.

Antedon bifida, Penn. (rosaceus, Auct.). i. 131, ii. 48.

ASTEROIDEA.

Asterias rubens, Linn. i. 132, ii. 49.
A. glacialis, Linn. i. 133, ii. 50.
A. hispida, Penn. i. 133.
Stichaster roseus, Müll. ii. 49.
Henricia sanguinolenta, O.F.M. i. 133,
   ii. 50.
Solaster endeca, Linn. i. 134, ii. 50.
   S. papposus, Fabr. i. 134, ii. 50.
Asterina gibbosa, Penn. i. 134.

Palmipes placenta, Penn. i. 135, iv.
   23.
Porania pulvillus, O.F.M. i. 13., ii.
   51.
Astropecten irregularis, Penn. i. 135,
   ii. 51.
Luidia ciliaris, Phil. i. 136, ii. 52,
   iv. 271.

ECHINOIDEA.

DESMOSTICHA.
Echinus esculentus, Linn. i. 136.
E. miliaris, Linn i. 136.
CLYPEASTRIDA.
Echinocyamus pusillus, O. F. M. i. 137.

PETALOSTICHA.
Spatangus purpureus, Müll. i. 137.
Echinocardium cordatum, Penn. i. 138.
E. flavescens, O. F. Müll. i. 138.
Brissopsis lyrifera, Forb. iv. 23, 175.

HOLOTHURIOIDEA.

APODA.
Synapta inhærens, O. F. M. iv. 363.
PEDATA.
Phyllophorus Drummondi, Thomps. 8th A. R., 12. i. 138.
Ocnus brunneus, Forb. i. 139.

Cucumaria pentactes, Müll. i. 139.
C. Hyndmani, Thomp. i. 139.
C. Planci, Marenz. ii. 53.
Thyone fusus, O. F. M. i. 138, iv. 178.
T. raphanus, D. & K. iv. 175, 178.

OPHIUROIDEA.

Ophiura ciliaris, Linn. i. 140.
O. albida, Forb. i. 141.
Ophiopholis aculeata, Linn. i. 141.
Amphiura elegans, Leach. i. 142.

Amphiura Chiajii, Forb. 9th A. R., p. 17.
Ophiocoma nigra, Abild. i. 142.
Ophiothrix fragilis, Abild. i. 143.

LIST OF THE VERMES.

TURBELLARIA.

[See Report by F. W. GAMBLE, M.Sc., in 'Fauna,' vol. iv.]

I. POLYCLADIDA.

Leptoplana tremellaris, O. F. Müll. iv. 72.
Cycloporus papillosus, Lang. iv. 74.

Oligocladus sanguinolentus, Quatref. iv. 76.
Stylostomum variabile, Lang. iv. 77.

II. RHABDOCŒLIDA.

Aphanostoma diversicolor, Oe. iv. 59.
Convoluta paradoxa, Oe. iv. 59.
C. flavibacillum, Jens. iv. 61.
Promesostoma marmoratum, Schultze. iv. 61.
P. ovoideum, Schm. iv. 62.
P. lenticulatum, Schm. iv. 62.
Byrsophlebs intermedia, v. Graff. iv. 63.
Proxenetes flabellifer, Jens. iv. 63.
Pseudorhynchus bifidus, McInt. iv. 64.
Acrorhynchus caledonicus, Clap. iv. 65.
Macrorhynchus Naegelii, Köll. iv. 66.
M. helgolandicus, Metsch. iv. 66.

Hyporhynchus armatus, Jens. iv. 66.
Provortex balticus, Schultze. iv. 67.
Plagiostoma sulphureum, v. Graff. iv. 68.
P. vittatum, Frey and Leuck. iv. 68.
Vorticeros auriculatum, O. F. Müll. iv. 69.
Allostoma pallidum, Van Ben. iv. 69.
Cylindrastoma quadrioculatum, Leuck. iv. 70.
C. inerme, Hallez. iv. 21.
Monotus lineatus, O. F. Müll. iv. 70.
M. fuscus, Oe. iv. 71.

III. TRICLADIDA.

Planaria littoralis, Van Ben. 10th A. R.

NEMERTEA.

[See Report by W. I. BEAUMONT, in 'Fauna,' vol. iv.]

Malacobdella grossa, O. F. M. i. 145.
Cephalothrix bioculata, Oersted. iv. 217, 452.

Carinella linearis, Mont. i. 145, 332.
C. Aragoi, Joubin. iv. 451.
? *C. annulata*, Mont. iv. 217.
Lineus marinus, Mont. i. 332.

CHÆTOGNATHA.

GEPHYREA.

HIRUDINEA.

CHÆTOPODA.

[See Reports by Prof. R. J. H GIBSON in 'Fauna,' vol. i. p. 144, and by
Mr. JAMES HORNELL in vol. iii. p. 126.]

Section B—SEDENTARIA.

Ophelia limacina, Rathke.   iii. 150.
Ammotrypane aulogaster, H. Rathke.
   iii. 150.
Chætopterus rariopedatus, Ren. iii.158.
Spio seticornis, Fabr.   i. 156, iii. 157.
Nerine vulgaris, J.   i. 156, iii. 158.
   N. cirratulus, Delle C.   iii. 157.
Leucodora ciliata, J.   iii. 158.
Magelona papillicornis, F. Müll.   10th
   A. R.
Arenicola marina, Linn.   iii. 151.
   A. ecaudata, Johnst.   10th A. R.
Capitella capitata, Fab.   iii. 151.
Nicomache lumbricalis. Fab.   iii. 154.
Axiothea catenata, Malmgren.   iii. 155.
Owenia filiformis, Delle C.   iii. 155.
Scoloplos armiger, Müll.   iii. 155.
Cirratulus cirratus, O. F. Müll.   i. 156,
   333, iii. 156.
   C. tentaculatus, Montagu.   iii. 156.
Chætozone setosa, Malmgren.   iii. 157.
Sabellaria alveolata, Linn.   i. 58, 156,
   iii. 163.
   S. spinulosa, R. Leuckart.   iii. 163.
Pectinaria belgica, Pall.   i. 8, 157, 332,
   349, iii. 163.

Pectinaria auricoma, O. F. Müll.   i.
   157, iii. 162.
Ampharete Grubei, Malmgren. iii.161.
Trophonia plumosa, Müll.   iii. 159.
Flabelligera affinis, Mgrn.   10th A. R.
Terebella nebulosa, Mont.   i. 158, 333,
   iii. 160.
Amphitrite figulus, Dalzell.   iii. 160.
Lanice conchilega, Pall.   iii. 160.
Thelepus cincinnatus, Fab.   i. 158, iii.
   160.
Nicolea venustula, Mont.   iii. 161.
Sabella pavonia, Sav.   iii. 164.
   Do., var. bicoronata, Hornell.   iii.164.
Dasychone Herdmani, Hornell.   iii. 165.
Amphicora fabricia, Müll.   iii. 166.
Serpula vermicularis, Ellis.   i. 3, 159.
   S. reversa, Mont.   iii. 167.
   S. triquetra, Linn.   i. 159.
Spirorbis borealis, Mörch.   i. 159, 333,
   iii. 167.
   S. lucidus, Mörch.   i. 160, iii. 167.
Filigrana implexa, Berkeley.   i. 12,
   160, 333, iii. 167.
Tomopteris onisciformis, Eschscholtz.
   i. 160, 333, iii. 150.

BRACHIOPODA.

Terebratula caput-serpentis, Linn. 7th | Crania anomala, Müller. iii. 62; 6th
   A. R., p. 28. | A.R., p. 25 ; 7th, pp. 15, 29 ; 8th, p. 15.

POLYZOA.

[See Mr. LOMAS' Reports in 'Fauna,' vols. i. and ii., and the various lists and additions
   made by Miss L. R. THORNELY in the Annual Reports since.]

CHEILOSTOMATA.

Aetea anguina, Linn.   i. 164; ii. 94.
   A. recta. Hincks.   i. 164; ii. 94; 9th
   A. R., pp. 20, 34.
   A. truncata, Lands. i. 164 ; ii. 94.
Eucratea chilata, Linn.   i. 164 ; ii. 94.
   Do., var. repens.   i. 164 ; ii. 94.
   Do., var. gracilis.   i. 165 ; ii. 94.
   Do., var. elongata, Lomas. i.165; ii.94.
Gemellaria loricata, Linn.   i. 165; ii.
   94.
Cellularia Peachii, Busk. i.166; ii. 94.
Scrupocellaria scruposa, Linn.   i. 166,
   ii. 94.
   S. scrupea, Busk.   i. 166, ii. 94.
   S. reptans, Linn.   i. 166, ii. 94.
Bicellaria ciliata, Johnst.   i. 167, ii.
   94.
Bugula avicularia, Linn.   i. 168, ii.
   94.
   B. turbinata, Alder.   i. 167, ii. 94,
   3rd A. R., p. 23.
   B. flabellata, J. V. Thomp.   i. 167,
   ii. 94.
   B. plumosa, Pallas.   i. 168, ii. 94.
   B. purpurotincta, Norm.   i. 168, ii.
   94.
Beania mirabilis, Johnst.   i. 168, ii. 94.

Cellaria fistulosa, Linn.   i. 169, ii. 92.
   C. sinuosa, Hass.   ii. 88, iii. 29.
Flustra foliacea, Linn.   i. 170, ii. 94.
   F. carbasea, Ell. & Sol.   i. 170.
   F. papyracea, Ell. & Sol. i. 170,ii. 94.
   F. securifrons, Pall.   i. 170, ii. 94.
Membranipora lacroixii, Aud.   i. 170,
   ii. 94.
   M. monostachys, Busk.   i. 171, ii. 94.
   M. catenularia, Jameson. i. 171,ii. 94.
   M. pilosa, Linn.   i. 171, ii. 94.
   Do., var. dentata.   4th A. R., p. 25.
   M. membranacea, Linn.   i. 171,ii. 95.
   M. hexagona, Busk.   i. 171.
   M. lineata, Linn.   i. 172, ii. 95.
   M. craticula, Ald.   i. 172, ii. 95.
   M. spinifera, Johnst.   10th A. R.
   M. discreta, Hincks. 9th A.R.,pp.11,
   34.
   M. Dumerilii, Aud.   i. 172, ii. 95.
   M. solidula, Ald. and Hincks.   9th
   A. R., pp. 11, 34.
   M. aurita, Hincks.   i. 172, ii. 95.
   M. imbellis, Hincks. 7th A. R.,p.18.
   M. Flemingii, Busk.   i. 172, ii. 94.
   M. Rosselii, Aud.   i. 172, ii. 95.
   M. nodulosa, Hincks.   9th A. R., pp.
   11, 34.

Bowerbankia pustulosa, E. & Sol. i. 187, ii. 97.
Farella repens, Farre, var. *elongata*, i. 188, ii. 97.
Buskia nitens, Ald. i. 188, ii. 97.
Cylindrœcium giganteum, Busk. 10th A. R.
C. dilatatum, Hincks. i. 188, ii. 97.
Anguinella palmata, Van Ben. i. 188, ii. 97.
Triticella Boeckii. Sars. 8th A. R., pp. 6, 15 ; 9th, p. 10.

Valkeria uva, Linn. i. 189, ii. 97.
V. uva, var. *cuscuta*. i. 189, ii. 97.
V. tremula, Hincks. i. 189, ii. 97.
Mimosella gracilis, Hincks. i. 189, ii. 97.
ENTOPROCTA.
Pedicellina cernua, Pall. i. 190, ii. 97.
Do., var. *glabra*. 4th A. R., p. 25.
Barentsia nodosa, Lomas. i. 190, ii. 90, 97.
Loxosoma phascolosomatum, Vogt. 10th A. R.

LIST OF THE CRUSTACEA.

MALACOSTRACA.

[For *Podophthalmata* and *Cumacea* see Mr. A. O. WALKER's Revision (Rev.), in ' Fauna,' vol. iii. p. 50; for the other groups of *Malacostraca* see Mr. Walker's other reports and papers in the ' Fauna.']

BRACHYURA.
Cancer pagurus, Linn. Rev.
Xantho tuberculatus, Couch. Rev.
Pilumnus hirtellus (Linn.) Rev.
Pirimela denticulata (Mont.) Rev., Addenda.
Carcinus mœnas, Penn. Rev.
Portunus puber (Linn.) Rev.
P. depurator (Linn.) Rev.
P. corrugatus (Penn.), off Calf of Man, 23 fath. 10th A. R.
P. arcuatus, Leach. Rev.
P. holsatus, Fabr. Rev.
P. pusillus, Leach. Rev.
Portumnus latipes (Penn.) ii 180.
Corystes cassivelaunus (Penn.) Rev.
Atelecyclus septem-dentatus (Mont.) Rev.
Thia residua (Herbst.) Rev.
Pinnotheres pisum (Linn.) Rev.
P. veterum, Bosc. Rev., Addenda.
Macropodia rostrata (Linn.) Rev.
M. longirostris (Fabr.) Rev.
Inachus dorsettensis (Penn.) Rev.
I. dorynchus, Leach. Rev., Addenda.
Hyas araneus (Linn.) Rev.
H. coarctatus, Leach. Rev.
Pisa biaculeata, Leach. 8th A. R., p. 25.
Eurynome aspera (Penn.) Rev.
Ebalia tuberosa (Penn.) Rev.
E. tumefacta (Mont.) Rev.
ANOMALA.
Eupagurus bernhardus (Linn.) Rev.
E. prideaux (Leach.) Rev.
E. cuanensis (Thomp.) Rev.
E. pubescens (Kröyer.) Rev.
Anapagurus lævis (Thomp.) Rev.
Porcellana platycheles (Penn.) Rev.
P. longicornis (Linn.) Rev.
Galathea squamifera, Leach. Rev.
G. nexa, Embleton. Rev.

Galathea dispersa, Bate. Rev.
G. intermedia, Lillj. Rev.
Munida rugosa, Fabr. ii. 70.
MACRURA.
Calocaris Macandreæ, Bell. 7th A. R., p. 18 ; Rev.
Palinurus vulgaris, Latr. 4th A. R., p 29.
Nephrops norvegicus (Linn.) Rev.
Astacus gammarus (Linn.) [Common Lobster]. Rev.
Crangon vulgaris (Linn.) Rev.
C. Allmanni, Kin. Rev.
C. trispinosus, Hailstone. Rev.
C. nanus, Kröyer. Rev.
C. sculptus, Bell. Rev.
C. fasciatus, Risso. Rev.
Pontophilus spinosus, Leach. 9th A. R., p. 13.
Nika edulis, Risso. County Down coast, 12 miles S.S.W. of Chicken Rock, in whiting's stomach. 10th A. R.
Caridion Gordoni, Sp. Bate. Rev.
Hippolyte varians, Leach. 7th A. R., p. 35.
Spirontocaris spinus (Sow.) Rev.
S. Cranchii (Leach.) Rev.
S. pusiola (Kr.) i. 222, ii. 170.
Pandalus Montagui, Leach. Rev.
P. brevirostris, Rathke. Rev.
Leander serratus (Penn.) Rev.
L. squilla (Linn.) Rev.
Palæmonetes varians (Leach.) Rev.
SCHIZOPODA.
Nyctiphanes norvegica (M. Sars.) Rev.
Macromysis flexuosa (Müll) Rev.
M. neglecta (Sars) iii. 245.
M. inermis (Rathke.) Rev.
Schistomysis spiritus (Norm.) Rev.

Schistomysis ornata (Sars.); and var.
Kervillei (Sars.) iii. 245. Rev.
Hemimysis Lamornæ (Couch.) ii. 178.
Rev.
Neomysis vulgaris (Thomp.) ii. 178.
Rev.
Leptomysis lingoura (Sars.) Rev.
Mysidopsis gibbosa (Sars.) 7th A. R.,
p. 25.
Erythrops elegans (Sars.) 7th A. R.,
p. 24.
Cynthilia (Siriella) norvegica. (Sars.)
iii., p. 244. Rev.
C. armata (M.-Edw.).
Gastrosaccus spinifer (Goes.).
G. sanctus (Van Ben.) 6th A. R., p.
38 ; 7th A. R., p. 25.
Haplostylus Normanni (Sars.) 7th A. R.,
p. 25.

PHYLLOCARIDA.

Nebalia bipes, M.-Edw. 10th A. R.
CUMACEA.

Cuma scorpioides (Mont.) iii. 246.
C. pulchella, Sars. 8th A. R., p. 25.
Iphinoë trispinosa (Goodsir.) 8th A. R.,
p. 25.
I. tenella, Sars. 9th A. R., p. 14.
Cumopsis Goodsiri (Van Ben.) Rev.
Eudorella truncatula, Sp. Bate. Rev.
E. nana. Sars. 8th A. R., p. 25 ; Brit.
Ass. Rep., 1895, p. 459.
Campylaspis macrophthalma, Sars.
8th A. R., p. 25.
Pseudocuma longicornis, Sp. Bate. Rev.
Petalosarsia decliris, Sars. 8th A. R.,
p 25.
Lamprops fasciata, Sars. iii. 247.
Pl. 16. Rev.
Hemilamprops assimilis, Sars. 9th
A. R., p. 14.
Diastylis Rathkei (Kr.) Rev.
D. spinosa (Norm.) ii 178, iii. 247.
D. biplicata (Sars.) 7th A. R., p.25.
D. rugosoides, Walker. 9th A. R.,
p. 14 ; B.A. Rep., 1895, p. 459.
Nannastacus unguiculatus, Bate. B. A.
Rep., 1894, p. 326.
ISOPODA.

Paratanais Batei, Sars. 7th A.R., p.25.
Leptognathia laticaudata, Sars. 7th
A. R., p. 25.
Anthura gracilis (Mont.) 7th A. R.,
p. 25.
Gnathia maxillaris (Mont.) Port Erin.
10th A. R.
Cirolana borealis, Lillj. 9th A. R.,
p. 14.
Conilera cylindracea (Mont.) iii. 241.
Eurydice achatus (Slabber.) i. 218.
Sphæroma serratum (Fabr.) i. 219.
S. rugicauda, Leach. R. Colwyn
Bay.
Cymodoce emarginata, (Leach.) 8th
A. R., p. 25. iii. 241, 248.

Dynamene rubra (Mont.) ♀ ii. 72.
Næsa bidentata, Leach. ♂ ii. 72.
Dynamene Montagui, Leach. ♂jr. ii. 72.
Limnoria lignorum (Rathke.) i. 219.
Idotea marina, Linn. i. 219.
I. linearis, Linn. i. 219.
Astacilla longicornis (Sow.) iii. 248.
A. gracilis (Goodsir.) 7th A. R., p.25.
Janira maculosa, Leach. i. 219.
Jæra albifrons (Mont.) i. 219.
Munna Fabricii, Kr. ii. 71.
Ligia oceanica (Linn.) i. 220, ii. 72.
BOPYRIDÆ.

Pleurocrypta nexa, Steb. 7th A. R.,
p. 43.
P. intermedia, G. & B. 7th A. R.,
p. 43.
P. galateæ, Hesse. 7th A. R., p. 43.
AMPHIPODA.

[See Mr. Walker's ' Revision ' (Rev.),
in ' Fauna,' iv.]

Hyperia galba (Montagu.) Rev.
Hyperoche tauriformis (Bate.) Rev.
Parathemista oblivia (Kr) Rev.
Talitrus locusta (Pall.) Rev.
Orchestia littorea (Mont.) ii. 171 ; 7th
A. R., p. 37. Rev.
Hyale Nilssonii (Rathke.) Rev.
Lysianax longicornis (Lucas.) ii. 73
(*L. ceratinus,* Walker.) Rev.
Socarnes erythrophthalmus, Robertson.
Rev.
Perrierella Audouiniana (Bate.) ii. 76.
Rev.
Callisoma crenata (Bate.) Rev.
Hippomedon denticulatus (Bate.) ii. 76.
Rev.
Orchomenella nana, Kr. ii. 172 (*T.
ciliata,* Sars.). Rev.
Nannonyx Goësii (Boeck.) Rev.
N. spinimanus, Walker. Rev.
Tryphosa Sarsi (Bonn.) Rev.
T. Hörringii, Boeck. Rev.
Tryphosites longipes, Bate. Rev.
Hoplonyx similis, Sars. Rev.
Lepidepecreum carinatum, Bate. Rev.
Euonyx chelatus, Norm. Rev.
Bathyporeia norvegica, Sars. Rev.
B. pelagica, Bate. Rev.
Haustorius arenarius (Slabber.) Rev.
Urothoë brevicornis, Bate. Rev.
U. elegans, Bate. Rev.
U. marinus, Bate. Rev.
Phoxocephalus Fultoni, T. Scott. Rev.
Paraphoxus oculatus, Sars. Rev.
Harpinia neglecta, Sars. Rev.
H. crenulata, Boeck. Rev.
H. lævis, Sars. Rev.
Ampelisca typica, Bate. Rev.
A. tenuicornis, Lillj. Rev.
A. brevicornis (Costa.) Rev.
A. spinipes, Boeck. Rev.
A. macrocephala, Lillj. Rev.

Haploops tubicola, Lillj. Rev.
Amphilochus manudens, Bate. Rev.
A. melanops, Walker. 7th A. R., p. 27.
    Rev. Pl. XVIII.
Amphilochoides pusillus, Sars. Rev.
Gitana Sarsii, Boeck. Rev.
Cyproidia brevirostris, T. & A. Scott
    Rev.
Stenothoë marina (Bate.) Rev.
S. monoculoides (Mont.) Rev.
Metopa Alderi, Bate. Rev.
M. borealis, Sars. Rev.
M. pusilla, Sars. Rev.
M. rubro-vittata, Sars. Rev.
M. Bruzelii (Goës.) Rev.
Cressa dubia (Bate.) Rev.
Leucothoë spinicarpa (Abildg.) Rev.
L. Lilljeborgii, Boeck. Rev.
Monoculodes carinatus, Bate. Rev.
Perioculodes longimanus (Bate.) Rev.
Pontocrates arenarius (Bate.) ii. p. 172.
Synchelidium haplocheles (Grube.) Rev.
Paramphithoë bicuspis (Kr.) ii. 173. Rev.
P. assimilis, Sars. Rev.
Stenopleustes nodifer, Sars. Rev.
Epimeria cornigera (Fabr.) Rev.
Iphimedia obesa, Rathke. Rev.
I. minuta, Sars. Rev.
Laphystius sturionis, Kr. Rev.
Syrrhoë fimbriatus, Stebb. & Rob. Rev.
Eusirus longipes, Boeck. Rev.
Apherusa bispinosa (Bate.) Rev.
A. Jurinii (M.-Edw.) ii. 79 (*Calliopius
    norvegicus*). Rev.
Calliopius leviusculus (Kr.) ii. 79. Rev.
Paratylus Swammerdamii (M.-Edw.)
    Rev.
P. falcatus (Metzger.) iii. 249. Rev.
P. uncinatus (Sars.) iii. 249. Rev.
P. vedlomensis (Bate.) Rev.
Dexamine spinosa (Mont.) Rev.
D. thea, Boeck. Rev.
Tritæta gibbosa (Bate.) iii. 249. Pl. 16.
    Rev.
Guernea coalita, Norm. Rev.
Melphidippella macera (Norm.) Rev.
Amathilla homari (Fabr.) ii. 175. Rev.
Gammarus marinus, Leach. Rev.
    G. locusta (Linn.) Rev.
    G. campylops, Leach. Brackish pond,
        Colwyn Bay. 10th A. R.
    G. pulex (De Geer.) Rev.
Melita palmata (Mont.) Rev.

Melita obtusata (Mont.) Rev.
Mæra othonis, M.-Edw. Rev.
M. semi-serrata, Bate. Rev.
M. Batei, Norm. Rev.
Megaluropus agilis, Norm. Rev.
Cheirocratus Sundevalli (Rathke.) ii.
    175. Rev.
C. assimilis (Lillj.) Rev.
Lilljeborgia pallida, Bate. Rev.
L. Kinahani (Bate.) Rev.
Aora gracilis, Bate. Rev.
Autonoë longipes (Lillj.) Rev.
Leptocheirus pilosus, Zaddach. Rev.
L. hirsutimanus (Bate.) Rev.
Gammaropsis maculata (Johnst.) Rev.
    G. nana, Sars. Rev.
Megamphopus cornutus, Norm. 6th
    A. R., p. 37. Rev.
Microprotopus maculatus, Norm. Rev.
Photis longicaudatus (Bate.) Rev.
    P. pollex, Walker. 9th A. R., p. 15.
    Rev.
Podoceropsis excavata (Bate.) Rev.
Amphithoë rubricata (Mont.) Rev.
Pleonexes gammaroides, Bate. Rev.
Ischyrocerus minutus, Lillj. ii. 82,
    iii. 250. Pl. 16 (*Podocerus isopus*,
    Walker). Rev.
Podocerus falcatus (Mont.) Rev.
    P. pusillus, Sars. Rev.
    P. Herdmani, Walker. 6th A. R., p.
        37. Rev.
    P. variegatus, Leach. Rev.
    P. ocius, Bate. Rev.
    P. cumbrensis, Stebb. and Rob. Rev.
Janassa capillata (Rathke.) ii. 81. Rev.
Erichthonius abditus (Temp.) Rev.
    E. difformis, M.-Edw. Rev.
Siphonœcetes Colletti, Boeck. Rev.
Corophium grossipes (Linn.) Rev.
    C. crassicorne, Bruzelius. Rev.
    C. bonellii, M.-Edw. ii. 84 (*C. crassi-
        corne*). Rev.
Unciola crenatipalmata (Bate.) Rev.
    U. planipes, Norm. Rev.
Colomastix pusilla, Grube. Rev.
Chelura terebrans, Phil. Rev.
Dulichia porrecta, Bate. Rev.
Phtisica marina, Slabber. Rev.
Protella phasma (Mont.) Rev.
Pariambus typicus (Kr.) Rev.
Caprella linearis (Linn.) Rev.
    C. acanthifera, Leach. Rev.

ENTOMOSTRACA.

OSTRACODA.
    [Identified by Professor G. S. BRADY
        (see 8th Ann. Rep., p. 20), Mr. A.
        SCOTT and Dr. CHASTER.]
Pontocypris trigonella, Sars. 8th A. R.,
    p. 20.
P. mytiloides, Norm. 8th A. R., p. 20.

Pontocypris serrulata, Sars. 8th A. R.,
    p. 20.
? *Argillœcia cylindrica*, Sars. 10th
    A. R.
Bairdia inflata, Norm. 8th A. R., p. 20.
B. acanthigera, Brady. 10th A. R.
Cythere Jonesii, Baird. 8th A. R., p. 20.

Cythere tuberculata, Sars. 8th A. R., p. 20.
C. tenera, Brady. 8th A. R , p. 20.
C. finmarchica, Sars. 8th A. R., p. 20.
C. confusa, B. & N. 8th A. R., p. 20.
C. albomaculata, Baird. 10th A. R.
C. globulifera, Brady. 10th A. R.
C. concinna, Jones. 8th A. R., p. 20.
C. dunelmensis, Norm. 8th A. R., p. 20.
C. antiquata, Baird. 8th A. R , p. 20.
C. emaciata, Brady. 8th A. R., p. 20.
C. convexa, Baird. 8th A. R., p. 20.
C. villosa, Sars. 8th A. R., p. 20.
C. lutea, O. F. M. 10th A. R.
C. Robertsoni, Brady. 10th A. R.
Eucythere argus, Sars. 8th A. R., p. 20.
E. declivis, Norm. 10th A. R.
Krithe bartonensis, Jones. 8th A. R., p. 20.
Loxoconcha impressa, Baird. 8th A. R., p. 20.
L. guttata, Norm. 8th A. R.. p. 20.
L. tamarindus, Jones. 8th A. R., p. 20.
L. pusilla, Brady. 8th A. R., p. 20.
L. multifora, Norm. 8th A. R.. p. 20.
Cytherura cornuta, Brady. 8th A. R., p. 20.
C. angulata, Brady. 8th A. R., p. 20.
C. cellulosa, Norm. 8th A. R., p. 20.
C. striata, Sars. 8th A. R., p. 20.
C. sella, Sars. 8th A. R., p. 20.
C. nigrescens, Baird. 10th A. R.
C acuticostata, Sars. 10th A. R.
Pseudocythere caudata, Sars. 8th A. R., p. 20
Cytheropteron latissimum, Norm. 8th A. R., p. 20.
C. pyramidale, Brady. 8th A. R., p. 20.
C. alatum, Sars. 8th A. R., p. 20.
C. punctatum, Brady. 10th A. R.
Sclerochilus contortus, Norm. 8th A. R., p. 20.
Paradoxostoma Normani, Brady. 8th A. R., p. 20.
P. ensiforme, Brady. 8th A. R., p. 20.
P. variabile, Baird. 8th A. R., p. 20.
P. hibernica, Brady. 8th A. R., p. 20.
P. flexuosum, Brady. 10th A. R.
Philomedes interpuncta, Baird. 8th A R., p. 20.
Cytheridea papillosa, Bosquet. 8th A. R., p. 21.
C. punctillata, Brady. 8th A. R., p. 21.
C. elongata, Brady. 10th A. R.
C. torosa, Jones. 10th A. R.
Cytherideis subulata, Brady. 10th A. R.
Bythocythere acuta, Norm. 8th A. R., p. 21.
B. constricta, Sars. 8th A. R., p. 21.
B. turgida, Sars. 8th A. R., p. 21.
B. simplex, Norm. 10th A. R.
Machærina tenuissima, Norm. 8th A. R., p. 21.

CLADOCERA
Evadne Nordmanni, Loven. i. p. 325.
Podon intermedium, Lillj. 4th A. R., p. 25.
COPEPODA.
[See Mr. I. C. THOMPSON's Reports, especially the 'Revision' in 'Fauna,' iv. p. 81.]
Calanus finmarchicus, Gunn. iv. 87.
Metridia armata, Boeck. iv. 87.
Pseudocalanus elongatus, Baird. iv. 87.
P. armatus, Boeck. iv. 87.
Paracalanus parvus, Claus. iv. 87.
Acartia Clausii, Giesbrecht. iv. 88.
A. discaudata, Giesb. iv. 88.
Temora longicornis, Müll. iv. 88.
Eurytemora affinis, Poppe. iv. 88.
E. Clausii, Boeck. iv. 88.
Scolecithrix hibernica, A. Scott. 10th A. R.
Isias clavipes, Boeck. iv. 88.
Centropages hamatus, Lillj. iv. 89.
C. typicus, Kr. (Missed reporting.)
Parapontella brevicornis, Lubb. iv. 89.
Labidocera Wollastoni, Lubb. iv. 89.
L. acutum, Dana. iv. 90.
Anomalocera Patersoni, Temp. iv. 90.
Euchæta marina, Prest. iv. 90.
Pseudocyclopia stephoides, Thomp. iv. 314.
Misophria pallida, Boeck. iv. 91.
Pseudocyclops crassiremis, Brady. 10th A. R.
P. obtusatus, Brady & Rob. 10th A. R.
Cervinia Bradyi, Norm. iv. 91.
Herdmania stylifera, Thomp. iv. 92.
Oithona spinifrons, Boeck. iv. 93.
Cyclopina littoralis, Brady. iv. 93.
C. gracilis, Claus. iv. 94.
Giardella callianassæ, Canu. iv. 95.
Hersiliodes puffini, Thomp. iv. 95.
Thorellia brunnea, Boeck. iv. 95.
Cyclops Ewarti, Brady. iv. 318.
C. magnoclavus, Cragin. iv. 317.
C. marinus, Thomp. iv. 94.
Notodelphys Allmani, Thorell. iv. 95.
Doropygus pulex, Thor. iv. 95.
D. poricauda, Brady. iv. 96.
D. gibber, Thorell. iv. 96.
Botachus cylinaratus, Thorell. iv. 96.
Ascidicola rosea, Thorell. iv. 96.
Notopterophorus papilio, Hesse. iv. 96.
Lamippi proteus, Clap. 10th A. R.
L. Forbesi, T. Scott. 10th A. R.
Longipedia coronata, Claus. iv. 97.
L. minor, T. & A. Scott. 8th A. R. p. 19.
Canuella perplexa, T. & A. Scott. iv. 318.
Sunaristes paguri, Hesse. 9th A. R. p. 11.
Ectinosoma atlanticum, B. & R. iv. 98.
E. curticorne, Boeck. iv. 98.
E. erythrops, Brady. iv. 98.
E. melaniceps, Brady. iv. 98.
E. spinipes, Brady. iv. 98.

Ketinosoma Normani, T. & A. Scott. 8th
A. R. p. 19.
K. elongata,T. & A. Scott. 8th A.R. p.19.
E. gracile, T. & A. Scott. 8th A. R. p. 20.
E. pygmæum, T. & A. Scott. 8th
A. R. p. 20.
E. Herdmani, T. & A. Scott. 8th A. R.
p. 20.
Bradya typica, Boeck. iv. 102.
B. minor, T. & A. Scott. 8th A. R. p. 20.
Tachidius brevicornis, Müll. iv. 98.
T. littoralis, Pop. iv. 99.
Euterpe acutifrons, Dana. iv. 99.
Robertsonia tenuis, Br. & Rob. iv. 99.
Amymone longimana, Claus. iv. 99.
A. sphærica, Claus. iv. 99.
Stenhelia hispida, Brady. iv. 99.
S. ima, Brady. iv. 100.
S. denticulata, Thomp. iv. 100.
S. hirsuta, Thomp. iv. 100.
S. Herdmani, A. Scott. 10th A. R.
S. similis, A. Scott. 10th A. R.
S. reflexa, T. Scott. 9th A. R. p. 11.
Ameira longipes, Boeck. iv. 101.
A. attenuata, Thomp. iv. 101.
A. longicaudata, T. Scott. 8th A. R.
p. 20.
A. exigua, T. Scott. 8th A. R. p. 20.
A. gracile, A. Scott. 9th A. R. p. 35.
A. reflexa, T. Scott. 9th A. R. p. 35.
A. longiremis, T. Scott. 8th A. R. p. 20.
Jonesiella fusiformis, Br. & Rob. iv. 102.
J. hycenæ, Thomp. iv. 102.
Delavalia palustris, Brady. iv. 103.
D. reflexa, Br. & Rob. iv. 103.
Canthocamptus palustris, Brady. 10th
A. R.
Mesochra Lilljeborgii, Boeck. iv. 103.
S. Macintoshi, T. & A. Scott. 9th A. R.
p. 35.
Paramesochra dubia, T. Scott. iv. 103.
Tetragoniceps Bradyi, T. Scott. iv. 103.
T. consimilis, T. Scott. 9th A. R. p. 35.
T. trispinosus, A. Scott. 10th A. R.
Diosaccus tenuicornis, Claus. iv. 103.
D. propinquus, T. & A. Scott. 8th
A. R. p. 20.
Laophonte serrata, Claus. iv. 104.
L. spinosa, Thomp. iv. 104.
L. thoracica, Boeck. iv. 105.
L. horrida, Norm. iv. 105.
L. similis, Claus. iv. 105.
L. curticauda, Boeck. iv. 105.
L. lamellifera, Claus. iv. 106.
L. hispida, Br & Rob. iv. 106.
L. propinqua, T. & A. Scott. 9th
A. R. p. 11.
L. intermedia, T. Scott. 9th A. R. p. 11.
L. inopinata, T. Scott. 8th A. R. p. 20.
Pseudolaophonte aculeata, A. Scott.
9th A. R. p. 35.
Laophontodes bicornis, A. Scott. 10th
A. R.
Normanella dubia, Br. & Rob. iv. 106.

Normanella attenuata, A. Scott. 9th
A. R. p. 35.
Cletodes limicola, Brady. iv. 106.
C. longicaudata, Br. & Rob. iv. 106.
C. linearis, Claus. iv. 106.
C. monensis, Thomp. iv. 106.
C. similis, T. Scott. 9th A. R. p. 11.
Enhydrosoma curvatum, Br. & Rob.
iv. 107.
Nannopus palustris, Brady. 9th A. R.
p. 11.
Platychelipus littoralis, Brady. iv. 107.
Dactylopus tisboides, Claus. iv. 107.
D. stromii, Baird. iv. 107.
Dactylopus tenuiremis, B. & R. iv. 108.
D. flavus, Claus. iv. 108.
D. brevicornis, Claus. iv. 108.
D. minutus, Claus. iv. 108.
D. rostratus, T. Scott. 8th A. R. p. 20.
Thalestris helgolandica, Claus. iv. 108.
T. rufoincta, Norm. iv. 108.
T. harpactoides, Claus. iv. 109.
T. Clausii, Norm. iv. 109.
T. rufo-violescens, Claus. iv. 109.
T. serrulata, Brady. iv. 109.
T. hibernica, Br. & Rob. iv. 109.
T. longimana, Claus. iv. 109.
T. peltata, Boeck. iv. 109.
T. forficuloides, T. & A. Scott. 10th
A. R.
Pseudothalestris pygmæa, T. & A. Scott.
8th A. R. p. 20.
P. major, T. & A. Scott. 10th A. R.
Westwoodia nobilis, Baird. iv. 110.
Harpacticus chelifer, Müll. iv. 110.
H. fulvus, Fischer. iv. 110.
H. flexus, Br. & Rob. iv. 110.
Zaus spinatus, Goods. iv. 110.
Z. Goodsiri, Brady. iv. 110.
Canuella tubulata, Dal. iv. 319.
Alteutha depressa, Baird. iv. 110.
A. interrupta, Goods. iv. 111.
A. crenulata, Brady. iv. 111.
Porcellidium viride, Phil. iv. 111.
P. tenuicauda, Claus. iv. 111.
Idya furcata, Baird. iv. 111.
I. elongata, A. Scott. 10th A. R.
I. gracilis, T. Scott. 9th A. R. p. 35.
Scutellidium tisboides, Claus. iv. 111.
S. fasciatum, Boeck. iv. 111.
Cylindropsyllus lævis, Brady. iv. 112.
Monstrilla Danæ, Claparède. iv. 112.
M. anglica, Lubb. iv. 112.
M. rigida, Thomp. iv. 112.
M. longicornis, Thomp. iv. 112.
Modiolicola insignis, Auriv. 9th A. R.
p. 11.
Lichomolgus albens, Thorell. iv. 113.
L. agilis, Leydig. 8th A. R. p. 20.
L. fucicolus, Brady. iv. 113.
L. furcillatus, Thorell. iv. 113.
L. maximus, Thomp. iv. 114.
Pseudanthessius Sauvagei, Canu. 8th
A. R. p. 20.

Pseudanthessius liber, Br. & Rob. iv. 113.
P. Thorellii, Br. & Rob. iv. 113.
Hermanella rostrata, Canu. (Recorded as *Lichomolgus agilis*, T. & A.S. iv. 33.)
Sabelliphilus Sarsii, Clap. iv. 116.
Cyclopicera nigripes, Br. & Rob. iv. 116.
C. lata, Brady. iv. 116.
Dyspontius striatus, Thorell. iv. 118.
Artotrogus Boeckii, Brady. iv. 117.
A. magniceps, Brady. iv. 117.
A. Normani, Br. & Rob. iv. 117.
A. orbicularis. Boeck. iv. 117.
Parartotrogus Richardi, T. & A. Scott. 10th A. R.
Acontiophorus scutatus, Br. & Rob. iv. 117.
A. elongatus, T. & A. Scott. iv. 320.
Collocheres gracilicauda, Brady. iv. 116.
C. elegans, A. Scott. 10th A. R.
Dermatomyzon gibberum, T. & A. Scott. 9th A. R. p. 11.
Ascomyzon Thompsoni, A. Scott. 9th A. R. p. 35.
Chondracanthus merluccii, Holt. 10th A. R.
Lernentoma lophii, Johnst. iv. 117.

Caligus rapax, M.-Edw. iv. 117.
C. curtus, Leach. iv. 118.
Lepeopotheirus Stromii, Baird. iv. 118.
L. Nordmannii, M.-Edw. iv. 118.
L. hippoglossi, Kr. iv. 118.
L. obscurus, Baird. iv. 118.
L. pectoralis, Müller. iv. 320.
Lernœa branchialis, Linn. iv. 118.
Anchorella appendiculata, Kr., iv. 321.
A. uncinata, Müll. iv. 119.
Lerneonema spratta, Sow. 10th A. R.
Lerneopoda galei, Kroyer. 10th A. R.

CIRRIPEDIA.

[See Mr. MARRAT'S list in 'Fauna,' i. 209; and records in the Ann. Reports since.]
Balanus porcatus, Costa. i. 209.
B. Hameri. Ascan. i. 209.
B. balanoides, Linn. i. 210.
B. perforatus, Brug. i. 210.
B. orenatus, Brug. i. 210.
Chthamalus stellatus, Poli. i. 210.
Verruca Strömia, O. F. M. i. 210.
Lepas anatifera, Linn. i. 210.
Scalpellum vulgare, Leach. 9th A. R. p. 17, &c.
Sacculina carcini, Thomp. i. 211.

LIST OF THE PYCNOGONIDA.

[See Reports by Mr. HALHED in 'Fauna,' i. 227; and also a note in 9th Ann. Report, p. 15.]

Nymphon gracile, Leach. i. 228. 9th A. R., p. 15.
N. rubrum, Hodge. 10th A. R.
N. gallicum, Hoek. 9th A. R., p. 15.
Ammothea echinata (Hodge.) i. 229.
A. lœvis (Hodge.) i. 229 (as *A. hispida*).
Chœtonymphon hirtum, (Kr.) 9th A. R., p. 15.

Pallene brevirostris (Johnst.) i. 230.
P. producta, Sars. 9th A. R., p. 15.
Phoxichilidium femoratum (Rathke.) i. 230.
Anoplodactylus petiolatus (Kr.) 9th A. R., p. 15.
Phoxichilus spinosus (Mont.) i. 230.
Pycnogonum litorale (Ström.) i. 231.

[NOTE.—A few of the marine insects and mites have been identified, but the lists are so far from complete that it would be useless to print them.]

LIST OF THE MOLLUSCA.

[See Reports by Mr. R. D. DARBISHIRE in 'Fauna,' i. 232 ; and by Mr. F. ARCHER in iii. 59, with additions by Mr. A. LEICESTER and Dr. CHASTER.]

LAMELLIBRANCHIATA.

Anomia ephippium, L. i. 234, 248, 320, 337 ; iii. 62.
Do., var. *squamula*, L. iii. 62.
Do., var. *aculeata*, Müll. iii. 62,
Do., var. *cylindrica*, Gm. iii. 62.
A. patelliformis, L. i. 5, 6, 235, 241, 248 ; iii. 62.
Ostrea edulis, L. i. 235, 248, 337 ; iii. 62.

Pecten pusio, L. i. 5, 235, 241, 248, 319, 337.
P. varius, L. i. 5, 235, 248, 337 ; iii. 32, 62.
P. opercularis, L. i. 235, 248, 337 ; iii. 215, 62.
P. tigrinus, Müll. i. 235, 248 ; iii. 62.
P. tigrinus, var. *costata*, Jeff. i. 13, 235, 337.

Pecten Testæ, Biv. 7th A. R., pp. 15, 28.
P. striatus, Müll. iii. 62; 6th A. R., p. 25.
P. similis, Lask. i. 248, 319, 337; iii. 63.
P. maximus, L. i. 241, 248, 319, 337; ii. 14.
Lima elliptica, Jeff. i. 13, 235, 248, 337.
(?) *L. subauriculata*, Mont. i. 248.
L. loscombii, G. B. Sow. i. 7, 13, 235, 248, 319, 337; iii. 63.
L. hians, Gm. i. 248.
Mytilus edulis, L. i. 31, 235, 241, 248, 337.
M. modiolus, L. i. 241, 249, 337; iii. 63.
? M. barbatus, L. i. 6, 235, 249; iii. 63.
M. adriaticus, Lmk. i. 249.
M. phaseolinus, Phil. Isle of Man, South. 10th A. R.
Modiolaria marmorata, Forb. i. 13, 31, 235, 249, 320, 321, 337; ii. 120, 121, 127.
M. discors, L. i. 249, 337; iii. 63.
Nucula sulcata, Brown, 7th A. R., pp. 16, 28.
N. nucleus, L. i. 235, 249, 337.
Do., var. *radiata*, F. & H. iii. 63.
N. nitida, G. B. Sow. i. 249, iii. 63.
Leda minuta, Müll. i. 249. iii. 63.
Do., var. *brevirostris*, Jeff. 10th A. R.
Pectunculus glycimeris, L. i. 13, 236, 249, 319, 323, 337.
Arca lactea, L. i. 249.
A. tetragona, Poli. i. 249, 319, 337.
Lepton squamosum, Mont. i. 249, iii. 64.
L. nitidum, Turt. iii. 64.
? L. sulcatulum, Jeff. 6th A. R., p. 26.
L. Clarkiæ, Cl. 7th A. R., p. 28.
Montacuta substriata, Mont. i. 249, iii. 64.
M. bidentata, Mont. i. 250, iii. 64.
M. ferruginosa, Mont. i. 250.
Lasæa rubra, Mont. i. 250.
Kellia suborbicularis, Mont. i. 250.
Loripes lacteus, L. i. 250.
Lucina spinifera, Mont. iii. 64.
L. borealis, L. i. 250. iii. 64.
Axinus flexuosus, Mont. i. 250.
Diplodonta rotundata, Mont. i. 250.
Cyamium minutum, Fabr. i. 250, iii. 64.
Cardium echinatum, L. i. 236, 241, 250, iii. 61, 64.
C. fasciatum, Mont. i. 250.
C. nodosum, Turt. iii. 64.
C. edule, L. i. 31, 241, 251.
C. minimum, Phil. 7th A.R., p. 28; 8th A. R., p. 27.
C. norvegicum, Speng. i. 6, 236, 251, 337; iii. 64.
Isocardia cor, L. 7th A. R., p. 28; 8th A. R., p. 30.
Cyprina islandica, L. i. 242, 251.

Astarte sulcata, Da C. i. 13, 236, 251, 337; iii. 65.
A. sulcata, var. *scotica*, M. & R. iii. 65.
A. triangularis, Mont. i. 251.
Circe minima, Mont. i. 251.
Venus exoleta, L. i. 236, 242, 251, 337.
V. lincta, Pult. i. 242, 251.
V. chione, L. iii. 65.
V. fasciata, Da C. i. 18, 236, 242, 251, 337; iii. 65.
V. casina, L. i. 13, 236, 251, 337.
V. ovata, Penn. i. 31, 236, 251; iii. 65.
V. gallina, L. i. 236, 251, 337.
Tapes virgineus, L. i. 236, 242, 251, 337; iii. 65.
T. pullastra, Mont. i. 242, 251.
Do., var. *perforans*, Mont. i. 8, 236.
Tapes decussatus, L. i. 242, 251.
Lucinopsis undata, Penn. i. 242, 251; iii. 65.
Tellina crassa, Penn. i. 242, 251.
T. balthica, L. i. 31, 236, 251, 337.
T. tenuis, Da C. i. 11, 251.
T. fabula, Gron. i. 251, iii. 65.
T. squalida, Pult. iii. 65.
T. donacina, L. i. 6, 236, 252; iii. 65.
T. pusilla, Phil. i. 252.
Psammobia tellinella, Lmk. i. 237, 252, 337; iii. 66.
P. ferroënsis, Chem. i. 237, 252.
P. vespertina, Chem. 8th A. R., p. 27.
Donax vittatus, Da C. i. 252; iii. 61.
Mactra solida, L. i. 5, 237, 239, 252.
M. solida, var. *truncata*, Mont. Puffin Island. 10th A. R.
M. solida, var. *elliptica*, Bro. i. 237, 337.
M. subtruncata, Da C. i. 252.
Do., var. *striata*, Brown. 10th A. R.
Do., var. *inæqualis*, Jeff. 10th A. R.
M. stultorum, L. i. 237, 242, 252.
Do., var. *cinerea*, Mont. i. 5.
Lutraria elliptica, Lmk. i. 237, 252.
Scrobicularia prismatica, Mont. i. 6, 237, 252; iii. 66.
S. nitida, Müll. 8th A. R., p. 27.
S. alba, Wood. i. 6, 237, 252.
S. tenuis, Mont. i. 253.
S. piperata, Gm. i. 253.
Solecurtus candidus, Ren. i. 253.
S. antiquatus, Pult. i. 253.
Ceratisolen legumen, L. i. 242, 253.
Solen pellucidus, Penn. i. 242, 253; iii. 66.
S. ensis, L. i. 242, 253.
S. siliqua, L. i. 253.
S. vagina, L. i. 242, 253.
Pandora inæquivalvis, L. iii. 66.
Lyonsia norvegica, Chem. i. 253; iii. 66.
Thracia prætenuis, Pult. i. 13, 237, 253, 337.
T. papyracea, Poli. i. 213, 253.
T. convexa, W. Wood. i. 242, 253.

Thracia distorta, Mont. i. 253.
Corbula gibba, Olivi. i. 7, 237, 253;
iii. 66.
Mya arenaria, L. i. 238, 253.
M. truncata, L. i. 31, 248, 253.
M. Binghami, Turt. i. 6. 238, 253;
iii. 66.
Panopea plicata, Mont. 10th A. R.
Saxicava rugosa, L. i. 6, 238, 243, 253,
320, 337; ii. 120; iii. 12, 149.
Do., var. *arctica*. L. Isle of Man,
South. 10th A. R.

Pholas candida, L. i. 4, 5, 31, 243.
254.
P. crispata, L. i. 8, 31, 238, 243, 254,
322, 337.
Pholadidea papyracea, Turt. i.
254.
Teredo navalis, L. iii. 67.
T. megotara, Han., and var. *mionota*,
Jeff. Southport. 10th A. R.
T. norvegica, sp., var. *divaricata*,
Desh. 10th A. R.

SCAPHOPODA.

Dentalium entale, L. i. 6, 13, 238, 254,
338; iii. 67.
D. tarentinum, Lmk. i. 254;
iii. 67.

Siphonodentalium lofotense, Sars. 10th
A. R.

POLYPLACOPHORA.

Chiton fascicularis, L. i. 244, 255;
iii. 67.
C. discrepans, Bro. iii. 29.
C. Hanleyi, Bean. 7th A. R., p. 42.
C. cancellatus, G. B. Sow. i. 238, 255,
338.
C. cinereus, L. i. 18, 238, 255, 338.

Chiton albus, L. i. 238, 255, 338.
C. marginatus, Penn. 8th A. R.,
p. 27.
C. ruber, Lowe. i. 255.
C. lævis, Mont. i. 238, 255, 338.
C. marmoreus, Fabr. i. 255.

GASTROPODA.

Patella vulgata, L. i. 239, 255, 338.
Do., var. *athletica*, Bean. i. 321, 338.
Helcion pellucidum, L. i. 239, 244,
255, 320.
Do., var. *lævis*, Penn. i. 320, 338.
Tectura testudinalis, Müll. i. 244,
255; iii. 67.
T. virginea, Müll. i. 255; iii. 67.
Propilidium ancyloides, Forb. 7th
A. R., p. 28; 8th, p. 27.
Puncturella noachina, L. iii. 67.
Emarginula fissura, L. i. 239, 255,
338; iii. 67.
Do., var. *elata*, Jeff. 10th A. R.
E. rosea, Bell. i. 255.
Fissurella græca, L. i. 13, 239, 255,
319, 338; iii. 67.
Capulus hungaricus, L. i. 255.
? *Cyclostrema cutlerianum*, Cl. 6th
A. R., pp. 26, 39.
C. nitens, Phil. 6th A. R., pp. 26, 39.
C. serpuloides, Mont. iii. 68.
Trochus helicinus, Fabr. 7th A. R., p. 28.
T. magus, L. i. 239, 256, 323, 338.
T. tumidus, Mont. i. 239, 256, 338; iii.
68.
T. cinerarius, L. i. 18, 239, 256, 338.
T. umbilicatus, Mont. i. 256.
T. Montaouti, W. Wood. i. 256; iii. 68.
T. striatus, L. i. 256.
T. millegranus, Phil. i. 256; iii. 68.
T. granulatus, Born. i. 257; iii. 68.
T. zizyphinus, L. i. 5, 13, 239, 257,
319, 322, 338; iii. 68.

Trochus zizyphinus, var. *humilior*, Jeff.
10th A. R.
Do., var. *Lyonsii*, Leach. iii. 68.
Do., var. *lævigata*, J. Sow. iii. 68.
Phasianella pullus, L. i. 13, 239, 257,
319, 338; iii. 35, 36.
Lacuna crassior, Mont. i. 31, 257; iii.
68.
L. divaricata, Fabr. i. 31, 257, 338;
iii. 68, 69.
L. puteolus, Turt. iii. 69.
L. pallidula, Da C. i. 257; iii. 69.
Littorina obtusata, L. i. 257, 338.
L. rudis, Maton. i. 257.
L. littorea, L. i. 31, 257, 338.
? *Rissoa striatula*, Mont. Waterloo.
10th A. R.
R. cancellata, Da C. iii. 69.
R. calathus, F. and H. iii. 69.
R. reticulata, Mont. i. 257.
R. puncturu, Mont. i. 257; iii. 69.
R. abyssicola, Forb. 7th A. R. pp.
16, 28.
R. zetlandica, Mont. Isle of Man,
South. 10th A. R.
R. costata, Ad. i. 257; iii. 69.
R. parva, Da C. i. 257; iii. 69.
Do., var. *interrupta*, Ad. iii. 36, 69.
R. inconspicua, Ald. 8th A. R., p. 27.
R. violacea, Desm. 7th A. R., p. 28.
R. striata, Ad. i. 258; iii. 69.
Do., var. *arctica*, Lov. Puffin Island.
10th A. R.
Do., var. *distorta*, Mar. 10th A. R.

Pleurotoma rufa, Mont. i. 263.
P. *turricula*, Mont. i. 5, 240, 263, 338 ; iii. 73.
Cypraea Europaea, Mont. i. 13, 240, 263, 317, 338.
Cylichna umbilicata, Mont. 7th A. R., p. 28.
C. *cylindracea*, Penn. i. 244, 264.
Utriculus truncatulus, Brug. iii. 73.
Do., var. *pellucida*, Bro. Puffin Island. 10th A. R.
U. *obtusus*, Mont. i. 31, 264; iii. 73.
U. *hyalinus*, Turt. 6th A. R., p. 39 ; 7th A. R., p. 28.
U. *mamillatus*, Phil. 10th A. R.
Actaeon tornatilis, L. i. 244, 264.
Bulla hydatis, L. 6th A. R., p. 35.

Bulla utriculus, Broc. 6th A. R., p. 39
7th A. R., p. 28.
Scaphander lignarius, L. i. 244, 264 ; iii. 73.
Philine scabra, Müll. 7th A. R., p. 28.
P. *catena*, Mont. Isle of Man, South. 10th A. R.
P. *angulata*, Jeff. 7th A. R., 28 8th A R., p. 27.
P. *punctata*, Cl. iii. 74.
P. *nitida*, Jeff. iii. 74.
P. *aperta*, L. i. 12, 31, 240, 265, 317; iii. 28, 74.
Aplysia punctata, Cuv. i. 13, 240, 265. 323, 339 ; iii. 137.
Pleurobranchus membranaceus, Mont. i. 13, 240, 265, 322, 339; iii. 74.
P. *plumula*, Mont. i. 13; iii. 74.

NUDIBRANCHIATA.

[See Reports by Professor HERDMAN and Mr. CLUBB in ' Fauna,' i. 268, ii. 98, and iii. 131.]

Archidoris tuberculata, Cuv. i. 268.
A. *Johnstoni*, Ald. & Han. i. 268.
A. *flammea*, Ald. & Han. i. 268.
Doris, sp. (?). 9th A. R., p. 11.
Lamellidoris bilamellata, Linn. i. 268.
L. *depressa*, Ald. & Han. i. 269.
L. *proxima*, Ald. & Han. i. 269.
L. *aspera*, Ald. & Han. 9th A. R., p. 11.
Ægirus punctilucens, D'Orb. 9th A. R., p. 11.
Acanthodoris pilosa, O. F. M. i. 269.
A. *quadrangulata*, Ald. & Han. i. 269.
Goniodoris nodosa, Mont. i. 269.
G. *castanea*, Ald. & Han. i. 270.
Triopa claviger, O. F. M. i. 270.
Polycera Lessoni, D'Orb. i. 270.
Do., var. *ocellata*, Ald. & Han. i. 270.
P. *quadrilineata*, O. F. M. i. 270.
Ancula cristata, Alder. i. 270; iii. 134.
Tritonia Hombergi, Cuv. i. 270.
T. *plebeia*, Johnst. i. 271.
Dendronotus arborescens, O. F. M. i. 271; ii. 101.
Lomanotus genei, Ver. 9th A. R., p 11.
Doto coronata, Gm. i. 272.
D. *fragilis*, Forbes. i. 272.
Janus cristatus, D. Ch. i. 272.
J. *hyalinus*, Ald. & Han. i. 272.
Eolidia papillosa, Linn. i. 273.
Eolidiella glauca, Ald. & Han. i. 273.
Facelina coronata, Forb. i. 273.
F. *Drummondi*, Thomp. i. 273.
Coryphella lineata, Lov. i. 274.

Coryphella gracilis, Ald. & Han. i. 274.
C. *Landsburgi*, Ald. & Han. i. 274.
C. *rufibranchialis*, Johnst. i. 274; iii. 140.
Favorinus albus, Ald. & Han. 9th A. R., p. 11.
Carolina angulata, Ald. & Han. 7th A. R., p. 45.
Cratena concinna, Ald. & Han. i. 274.
C. *olivacea*, Ald. & Han. i. 274.
C. *amoena*, Ald. & Han. i. 274.
C. *aurantiaca*, Ald. & Han. i. 275.
C. *arenicola*, Forb. i. 275.
C. *viridis*, Forb. i. 275.
Cuthona nana, Ald. & Han. i. 275.
C. *aurantiaca*, Ald. & Han. 9th A. R., p. 11.
Galvina picta, Ald. & Han. i. 275.
G. *tricolor*, Forbes. i. 275.
G. *Farrani*, Ald. & Han. 9th A. R., p. 11.
Tergipes despecta, Johnst. i. 276.
T. *exigua*, Ald. & Han. i. 276.
Embletonia pallida, Ald. & Han. i. 276.
E. *pulchra*, Ald. & Han. 9th A. R., p. 11.
Fiona marina, Forsk. ii. 108.
Elysia viridis, Mont. 9th A. R., p. 11.
Runcina Hancocki, Forb. 9th A. R., p. 11.
Actaeonia corrugata, Ald. & Han. 9th A. R., p. 11.
Limapontia nigra, John. 9th A. R., p. 11.

PULMONIBRANCHIATA.

Melampus bidentatus, Mont. iii. 74.
Do., var. *alba*, Turt. Isle of Man, South. 10th A. R.

Melampus myosotis, Drap. 7th A. R., p. 28.
Otina otis, Turt. i. 265; iii. 74.

PTEROPODA.

Spirialis retroversus, Flem. 7th A. R., p. 15.

CEPHALOPODA.

[See Mr. HOYLE's list in ' Fauna,' i. 278, and additions in A. R. since.]

Sepiola atlantica, Lamk. i. 6, 11, 24, 246, 266, 279. 7th A. R., 28.
S. scandica, Stnp. 7th A. R., p. 28.
Rossia macrosoma, D. Ch. i. 245, 266.
Loligo media, Linn. i. 5, 7, 245, 266, 270.

Loligo Forbesi, Stnp. i. 245, 265. 7th A. R., 28.
Sepia officinalis, Linn. i. 29, 245, 266.
Eledone cirrosa, Lamk. i. 6, 24, 245, 266, 278; iii. 35.

LIST OF THE TUNICATA.

[See Professor HERDMAN's Report upon the Tunicata in the ' Fauna,' vol. i., and Second Report upon the Tunicata in the ' Fauna,' vol. ii., and various passing references and short lists in the Annual Reports.]

LARVACEA.

Oikopleura flabellum, J. Müll. i. 281; ii. 114.

Fritillaria, sp. Port Erin. 10th A. R.

ASCIDIACEA.

Polycyclus Savignyi, Hrdm. i. 283, ii. 114.
Botryllus morio, Giard (?). i. 284, 6th A. R., p. 35.
B. smaragdus, M.-Edw. i. 285, ii. 115.
B. violaceus, M.-Edw. i. 286, 6th A. R., p. 35.
B. Schlosseri, Pall. i. 287, ii. 115.
B. gemmeus, Sav. i. 287.
B. pruinosus, Giard (?). i. 287.
B. aurolineatus, Giard (?). 6th A. R., p. 35.
Botrylloides rubrum, M.-Edw. i. 287; ii. 115.
B. albicans, M.-Edw. i. 287; ii. 116.
B. Leachii, Sav. (?). i. 288; ii. 115.
B. sp. (?). i. 288.
Sarcobotrylloides, sp. (?). ii. 116.
Distoma rubrum, Sav.(?). i.288; ii.116.
D. vitreum, Ald. (?). i. 289.
D. sp. (?). i. 289.
Aplidium fallax, John. (?). i. 290.
Parascidia Forbesii, Ald. i. 290.
Morchellium argus, M.-Edw. i. 290; ii. 117.
Morchellioides Alderi, Hrdm. i. 291.
Amaroucium proliferum, M.-Edw. i. 293; ii. 117.
Amaroucium, sp. (?). i. 293.
Glossophorum sabulosum, Giard. 7th A. R., p. 17.
Leptoclinum durum, M.-Edw. i. 293; ii. 118.
L. maculatum, M.-Edw. i. 293; ii.117.
L. candidum, Sav. (?). i. 294; ii. 117.

Leptoclinum asperum, M.-Edw. i. 294.
Diplosoma punctatum, Forb. i. 294.
D.gelatinosum, M.-Edw. i. 295; ii.118.
D. crystallinum, Giard. i. 295.
Astellium spongiforme, Giard. 7th A. R., p. 17.
Clavelina lepadiformis, O. F. M. i. 296; ii. 118.
Perophora Listeri, Wieg. i. 297; ii.119.
Ciona intestinalis, Linn. i. 297, 362; ii. 119.
Ascidiella virginea, O. F. M. i. 298; ii. 124.
A. scabra, O. F. M. i. 299; ii. 125.
A. elliptica, A. & H. i. 299.
A. aspersa, O. F. M. i. 300; ii. 125.
A. venosa, O. F. M. ii. 122.
Ascidia mentula, O. F. M. i. 298; ii. 121.
A. plebeia, Ald. i. 300; ii. 121.
A. depressa, Ald. & H. i. 301; 6th A. R., p. 35.
A. prunum, O. F. M. i. 301.
Corella parallelogramma, O. F. M. i. 301; ii. 126.
Forbesella tessellata, Forbes. 3rd A.R., p. 37.
Styelopsis grossularia, V. Ben. i. 302; ii. 126.
Polycarpa rustica, Linn. (?). i. 303; ii. 127.
P. comata, Ald. i. 303; 8th A. R., p. 11.
P. pomaria, Sav. i. 304; ii. 127.
P. glomerata, Ald. A. R.
P. monensis, Hrdm. i. 305.
Cynthia echinata, Linn. ii. 127.

Cynthia morus, Forb. 7th A. R., *Molgula citrina*, A. & H. ii. 128; 6th
   p. 19.   A. R., p. 35.
Molqula occulta, Kupf. i. 307; ii. *M. Hancocki*, Hrdm. ii. 130.
   128.   *Eugyra glutinans*, Möll. i. 309; ii. 128.

CEPHALOCHORDA.

Branchiostoma lanceolatum, Pall. 10th A. R.

LIST OF THE FISHES.

[See lists by Mr. P. M. C. KERMODE in 'Zoologist,' 1893, and by Prof. HERDMAN
in 'Transactions' Liverpool Biological Society for 1893.]

Labrax lupus, Cuv.
Serranus cabrilla, C. and V.
Mullus barbatus, var. *surmuletus*, Linn.
Cantharus lineatus, Mont
Pagellus centrodontus, C. and V.
Sebastes norvegicus, Ascan.
Cottus scorpius, Linn.
   C. bubalis, Euph.
Trigla hirundo, Linn.
   T. cuculus, Linn.
   T. lineata, Gm.
   T. gurnardus, Linn.
Agonus cataphractus, Bl.
Lophius piscatorius, Linn.
Trachinus draco, Linn.
   T. vipera, C. and V.
Scomber scomber, Linn.
S. Colias, Gm.
Orcynus germo, Lac.
Thynnus pelamys, Linn.
Lampris luna, Gm.
Caranx trachurus, Lac.
Zeus faber, Cuv.
Xiphias gladius, Linn.
Sciæna aquila, Risso.
Gobius niger, Linn.
   G. Ruthensparri, Euph.
   G. minutus, Gm.
   G. paganellus, Gm.
   G. pictus, Malm.
   G. quadrimaculatus, C. and V.
   G. Parnelli, Day.
Aphia pellucida, Nard.
Callionymus lyra, Linn.
Cyclopterus lumpus, Linn.
Liparis Montagui, Don.
   L. vulgaris, Flem.
Lepadogaster Gouanii, Lac.
   L. bimaculatus, Don.
Carelophus Ascanii, Coll
Blennius pholis, Linn.
   B. ocellaris, Linn.
   B. galerita, Linn.
   B. gattoruginæ, Bl.
Centronotus gunnellus, Bl.
Zoarces viviparus, Linn.
Gasterosteus aculeatus, Linn.
   G. spinachia, Linn.
   G. pungitius, Linn.
Mugil chelo, Cuv.

Labrus maculatus, Bl.
   L. mixtus, Fries and Eks.
Centrolabrus exoletus, Linn.
Crenilabrus melops, Cuv.
Ctenolabrus rupestris, Linn.
Gadus morrhua, Linn.
   G merlangus, Linn.
   G. virens, Linn.
   G. æglefinus, Linn.
   G. luscus, Linn.
   G. minutus, Linn.
   G. pollachius, Linn.
Merluccius vulgaris, Cuv.
Molva vulgaris, Flem.
Loto vulgaris, Cuv.
Phycis blennoides, Bl.
Motella tricirrata, Nils.
   M. cimbria, Linn.
   M. mustela, Linn.
Raniceps raninus, Linn.
Ammodytes lanceolatus, Les.
   A. tobianus, Lion.
Rhombus maximus, Cuv.
   R. lævis, Rond.
Hippoglossus vulgaris, Flem.
Hippoglossoides limandoides, Bloch.
Zeugopterus punctatus, Bl.
   Z. unimaculatus, Risso.
   Z. norvegicus, Günth.
Arnoglossus megastoma, Don.
   A. laterna, Walb.
Pleuronectes platessa, Linn.
   P. limanda, Linn.
   P. flesus, Linn.
Pleuronectes microcephalus, Don.
   P. cynoglossus, Linn.
Solea vulgaris, Quen.
   S. lutea, Risso.
   S. aurantiaca, Günth.
   S. lascaris, Risso.
   S. variegata, Don.
Maurolicus Pennantii, Walb.
Argentina sphyræna, Linn.
Salmo salar, Linn.
   S. trutta, Linn.
   S. fario, Linn.
Osmerus eperlanus, Linn.
Belone vulgaris, Flem.
Engraulis encrasicholus, Linn.
Clupea harengus, Linn.

Clupea sprattus, Linn.
  C. finta, Cuv.
Anguilla vulgaris, Turt.
Conger vulgaris, Cuv.
Siphonostoma typhle, Linn.
Syngnathus acus, Linn.
Nerophis æquoreus, Linn.
  N. ophidion, Linn.
  N. lumbriciformis, Willugh.
Orthagoriscus mola, Linn.
Carcharias glaucus, Cuv.
Acipenser sturio, Linn.
Galeus vulgaris, Flem.
Mustelus vulgaris, Müll.
Lamna cornubica, Gm.
Alopias vulpes, Gm.
Selache maxima, Gunner.

Scyllium canicula, Cuv.
  S. catulus, Cuv.
Pristiurus melanostomus, Raf.
Acanthias vulgaris, Risso.
Rhina squatina, Linn.
Torpedo nobiliana, Bonap.
Raia batis, Linn.
  R. oxyrhynchus, Linn.
  R. alba, Lacép.
  R. clavata, Linn.
  R. maculata, Mont.
  R. circularis, Couch.
  R. macrorhynchus, Raf.
  R. radiata, Don.
Trygon pastinaca, Linn.
Petromyzon marinus, Linn.
  P. fluviatilis, Linn.

LIST OF THE MARINE MAMMALIA.

[See Report on Seals and Whales, by Mr. MOORE, in ' Fauna,' ii. p. 134.]

PINNIPEDIA.
Phoca grœnlandica, Fabr.   ii. 136.
  P. vitulina, Linn.   10th A. R.
Halichœrus grypus, Fabr.   ii. 136.
Cystophora cristata, Erxl.   ii. 137.
CETACEA.
Megaptera longimana, Rud.   ii. 139.
Hyperoodon rostratus, Chem.   ii. 140.

Balænoptera musculus, Linn.   10th
  A. R.
Phocæna communis, F. Cuv.   ii. 142.
Orca gladiator, Lac.   ii. 143.
Lagenorhynchus albirostris, Gray.   ii.
  144.
Delphinus delphis, Linn.   ii. 147.
Tursiops tursio, Fabr.   ii. 148.